Advances in Marine Engineering: Geological Environment and Hazards—3rd Edition

Advances in Marine Engineering: Geological Environment and Hazards—3rd Edition

Guest Editors

Xingsen Guo
Xiaolei Liu
Tingkai Nian
Thorsten Stoesser

MDPI

Basel • Beijing • Wuhan • Barcelona • Belgrade • Novi Sad • Cluj • Manchester

Guest Editors
Xingsen Guo
Department of Civil, Environmental and Geomatic Engineering
University College London (UCL)
London
UK

Xiaolei Liu
College of Environmental Science and Engineering
Ocean University of China
Qingdao
China

Tingkai Nian
State Key Laboratory of Coastal and Offshore Engineering
Dalian University of Technology
Dalian
China

Thorsten Stoesser
Department of Civil, Environmental and Geomatic Engineering
University College London
London
UK

Editorial Office
MDPI AG
Grosspeteranlage 5
4052 Basel, Switzerland

This is a reprint of the Special Issue, published open access by the journal *Journal of Marine Science and Engineering* (ISSN 2077-1312), freely accessible at: https://www.mdpi.com/journal/jmse/special_issues/WQ2HA5G358.

For citation purposes, cite each article independently as indicated on the article page online and as indicated below:

Lastname, A.A.; Lastname, B.B. Article Title. *Journal Name* **Year**, *Volume Number*, Page Range.

ISBN 978-3-7258-7362-3 (Hbk)
ISBN 978-3-7258-7363-0 (PDF)
https://doi.org/10.3390/books978-3-7258-7363-0

Contents

Preface

This Reprint presents a curated selection of published articles from the Special Issue focused on hydrodynamics, geological environments, and hazards in estuarine, coastal, offshore, and deep-sea settings. The scope covers the coupled processes of flows, waves, sediments, and seabed soils, and how these interactions govern seabed evolution, stability, and engineering performance. The aim of this Reprint is to consolidate recent scientific and engineering progress that supports safer and more resilient planning, design, construction, and operation of marine infrastructure, including offshore energy and other seabed developments. The motivation for compiling this Reprint is the growing need to translate rapid research advances into an accessible reference that highlights both fundamental mechanisms and practical implications for hazard assessment and mitigation. This Reprint will be addressed to researchers, engineers, and decision makers working in coastal and offshore geoscience, hydraulics, and geotechnical engineering, as well as early career scholars seeking an up-to-date overview of methods and current challenges.

Xingsen Guo, Xiaolei Liu, Tingkai Nian, and Thorsten Stoesser
Guest Editors

Journal of
Marine Science
and Engineering

MDPI

Editorial

Advances in Marine Engineering: Geological Environment and Hazards—3rd Edition

Xingsen Guo [1,2,3,4,5,6], Xiaolei Liu [1,2,*], Tingkai Nian [7,*] and Thorsten Stoesser [3]

1 Shandong Provincial Key Laboratory of Marine Engineering Geology and the Environment, Ocean University of China, Qingdao 266100, China; xingsen.guo@ouc.edu.cn
2 Sanya Oceanographic Institution, Ocean University of China, Sanya 572024, China
3 Department of Civil, Environmental and Geomatic Engineering, University College London, London WC1E 6BT, UK; t.stoesser@ucl.ac.uk
4 Department of Engineering, University of Cambridge, Cambridge CB2 1PZ, UK
5 State Key Laboratory of Intelligent Construction and Healthy Operation and Maintenance of Deep Underground Engineering, Sichuan University, Chengdu 610065, China
6 State Key Laboratory of Ocean Engineering, Shanghai Jiao Tong University, Shanghai 200240, China
7 State Key Laboratory of Coastal and Offshore Engineering, Dalian University of Technology, Dalian 116024, China
* Correspondence: xiaolei@ouc.edu.cn (X.L.); tknian@dlut.edu.cn (T.N.)

1. Introduction

Following the success of the first two Special Issues [1–4] on "*Advances in Marine Engineering: Geological Environment and Hazards*," this third installment continues to present cutting-edge research in the field. This Special Issue (SI) was organized by Prof. Dr. Xingsen Guo from Ocean University of China, University College London, and the University of Cambridge; Prof. Dr. Xiaolei Liu from Ocean University of China; Prof. Dr. Tingkai Nian from Dalian University of Technology; and Prof. Dr. Thorsten Stoesser from University College London. The primary aim of this SI is to gather together high-quality papers highlighting the latest advancements in marine geological environments and hazards, marine engineering geology, marine hydrodynamics, marine fluid mechanics, and marine geotechnical engineering.

As marine engineering continues to advance toward deeper waters and more challenging environments, the complexity of geohazards and the uncertainty of seafloor conditions pose increasing risks to offshore infrastructure. The contributions in this Special Issue collectively address these critical challenges by spanning the full spectrum from seabed characterization and environmental process mechanics to engineering response and risk mitigation. Through multi-source geophysical data integration, controlled physical experiments, cutting-edge numerical simulations, and data-driven prediction frameworks, the included studies provide new insights into submarine gravity flows and fluid migration risks, hydrodynamic transformation under climate-driven sea level rise, contaminant mobilization under storm-induced seabed liquefaction, and the deformation and stability of marine soils and foundations under cyclic or impact loading. In parallel, advanced computational and machine learning approaches enhance predictive capability for novel ocean engineering systems—from deep sea mining vehicles to bio-inspired hydrodynamic structures.

Altogether, this Special Issue comprises 1 review paper [5] and 11 research papers [6–16], contributed by 63 scholars from leading universities, research institutions, and industry worldwide. Collectively, these contributions present state-of-the-art concepts, innovative methodologies, and high-quality data that advance both the knowledge and practice of marine geological environments and hazards. The following sections provide a concise synthesis of the main findings and significant contributions of each paper included in this SI.

2. Papers Details

Zhang et al. [5] presented a comprehensive review of the interactions between submarine gravity flows and seafloor pipelines, synthesizing recent advances in understanding how flow characteristics, pipe geometry, seabed conditions, and environmental factors jointly influence hydrodynamic loading and pipeline response. The authors highlight key controlling parameters—including Reynolds number, pipeline span ratio, impact angle, surface roughness, and temperature—and discuss the additional complexity introduced by multi-pipe configurations and flow-induced vibration. They identify major limitations in existing research, such as the scarcity of physical experiments, oversimplified seabed boundary conditions, and uncertainties associated with non-Newtonian rheological models commonly used for debris flows. Zhang et al. further outline practical mitigation strategies and emphasize future research priorities, particularly the development of fully coupled 3D multi-physics models and improved assessments of fatigue and stability under extreme gravity flow events. This review provides valuable guidance for the risk assessment and design optimization of offshore pipeline systems subject to submarine gravity flow hazards.

Tao et al. [6] employed a multi-source data fusion strategy integrating 3D seismic data, shipborne multibeam bathymetry, and high-resolution AUV surveys to construct a refined seabed terrain inversion model for the Qiongdongnan Basin. Their analysis successfully identified complex geomorphologic features—including escarpments, fissures, faults, and multi-phase mass transport deposits—and revealed marked improvements in inversion accuracy over single-data approaches. By combining seismic attribute extraction with geostatistical evaluation, the authors further established a vertical risk zonation framework for shallow gas, demonstrating that the most significant enrichment occurs at depths between 200 and 300 m under the joint influence of fault-guided migration and sedimentary compaction structures. These results provide valuable technical support for deep water engineering design, hazard mitigation, and safe offshore energy development in the South China Sea.

Yu et al. [7] developed a two-dimensional MIKE 21 hydrodynamic model to evaluate how projected sea level rise alters tidal wave propagation throughout the Yangtze River Estuary under varying river discharge conditions. Their simulations demonstrate that sea level rise consistently elevates average water levels while modifying tidal amplitudes in a spatially heterogeneous manner, with the most pronounced changes occurring near Xuliujing, where tidal dynamics transition between estuarine and riverine regimes. The study reveals that rising sea levels enhance tidal current velocities in most estuarine sections and reduce tidal wave attenuation along upstream reaches, causing the tidal limit to shift further inland. These findings highlight the increasing sensitivity of the estuary's hydrodynamics to future sea level scenarios and provide important guidance for flood risk assessment, navigation safety, and estuarine management in the Yangtze River basin.

Lu et al. [8] performed controlled wave flume experiments to quantify how wave-induced resuspension and sediment liquefaction mobilize petroleum-derived PAHs from silty seabeds in the Yellow River subaqueous delta. Their results show that increasing wave intensity—from static conditions to 5 cm and then 12 cm waves—substantially enhances suspended sediment concentration and triggers excess porewater pressure leading to sediment liquefaction. This liquefaction state dramatically increases PAH release, with dissolved PAH concentrations rising up to fivefold compared with static diffusion, and facilitates deeper vertical migration of low-molecular-weight PAHs within the sediment column. The study highlights the critical role of hydrodynamic disturbances in controlling PAH mobility and underscores the elevated pollution risks during storm events, providing essential insights for environmental assessment and coastal management in deltaic environments.

Liu et al. [9] investigated the mechanical behavior of marine soft soil under cyclic loading by integrating dynamic triaxial testing with discrete element simulations to capture both macroscopic deformation and micromechanical evolution. Their results show that increasing water content, confining pressure, and dynamic stress ratios significantly amplify strain accumulation, hysteresis loop expansion, and energy dissipation, while high moisture levels (e.g., 60%) trigger pronounced shear band development and structural degradation. The DEM analysis elucidates particle rearrangement, lubrication effects, and porosity evolution, offering mechanistic insights consistent with laboratory observations. The combined macro–micro framework provides a more comprehensive understanding of soil softening and instability under repeated loading, delivering valuable guidance for the design, assessment, and reinforcement of marine soft soil foundations for coastal urban infrastructure.

Ge et al. [10] proposed an explainable ensemble learning framework for predicting the compression index of clays, a critical parameter in geotechnical design for clay-rich coastal environments. Using a global dataset of 1080 samples, the authors evaluated multiple ensemble models—including random forest, gradient boosting decision trees, extreme gradient boosting, and a stacking ensemble—and demonstrated that the stacking model achieved the highest predictive accuracy. To handle missing data, they employed a K-nearest neighbors imputation approach, which effectively preserved the underlying data distribution and enhanced model robustness across varied soil conditions. Furthermore, SHAP-based interpretability analysis identified the initial void ratio and natural water content as the most influential factors governing compression index predictions. This work highlights the strong capability of ensemble learning in modeling clay compressibility and provides useful insights for geotechnical and coastal infrastructure design, particularly in regions with heterogeneous soil properties and incomplete datasets.

Wei et al. [11] conducted centrifuge model tests to investigate the uplift behavior of single-cylinder suction caisson foundations in high-sensitivity marine clay, focusing on performance degradation under vertical cyclic loading. By preparing sensitive soil through a freeze–thaw procedure, they demonstrated that cyclic uplift capacity decreases rapidly with increasing soil sensitivity and load cycles, accompanied by accelerated cumulative displacement until stabilization. The results further reveal that the secant stiffness of the load–displacement hysteresis loops follows a logarithmic relationship with both cyclic loading number and soil sensitivity. This study provides valuable insights into the uplift capacity and stiffness evolution of suction caisson foundations in soft, highly sensitive seabeds, offering practical guidance for the design and long-term safety of offshore wind energy infrastructure in clay-rich regions.

Liu et al. [12] performed wave flume experiments supported by image analysis to examine the formation and evolution of interfacial waves within liquefied submarine sediments and their role in driving sliding zone development. Their results show that interfacial waves emerging after seabed liquefaction govern the progressive downslope movement of sediment and dynamically control both the position and thickness of the sliding zone, which corresponds closely to the liquefaction depth. The interfacial wave height increases with greater liquefaction depth—reaching up to 17.5% of the surface wave height—and yields a maximum sliding zone width ratio of 0.25. These waves effectively transfer surface-wave energy into deeper soil layers, extending the lower boundary of the sliding zone. The study provides important theoretical insights for predicting and mitigating seabed sliding during extreme storm events.

https://doi.org/10.3390/jmse14030303

Dai et al. [13] employed high-fidelity large-eddy simulations to investigate the flow field around a cuboid artificial reef under different inflow angles, aiming to reveal how reef orientation affects hydrodynamics and turbulence characteristics. Their simulations, validated against flume experiments, demonstrate that an inflow angle of 45° produces the most favorable flow environment, yielding the longest recirculation bubble, enhanced upwelling, and the largest wake volume compared with 0° and 90° configurations. The study also identifies intensified shear-layer interactions and the generation of torn horseshoe vortices and hairpin vortices as key mechanisms promoting turbulent mixing. These findings provide important guidance for optimizing artificial reef design and deployment strategies to improve ecological performance and marine habitat enhancement.

Gong et al. [14] used high-resolution large-eddy simulations to analyze flow dynamics and momentum exchange mechanisms in an isolated semi-circular lateral cavity influenced by longshore currents. Their numerical model, validated against flume measurements, revealed that increasing the Reynolds number intensifies shear-layer instabilities, promotes upstream migration of the primary vortex core, and enhances secondary vortex development inside the cavity. The study further showed that Kelvin–Helmholtz vortices dominate periodic mass and momentum exchange across the cavity mouth, transitioning from viscosity-damped turbulence at lower Reynolds numbers to fully developed shedding downstream. These findings advance the understanding of cavity-induced hydrodynamics and provide useful implications for optimizing nature-based coastal engineering strategies and sediment retention designs in curved shoreline systems.

Qu et al. [15] performed a comparative seismic and well-based analysis of deep water sedimentary architectures on the central West African passive margin, contrasting the Lower Congo Basin and the Niger Delta by integrating 4039 km of seismic lines, RMS amplitude slices, well logs from 18 key wells, and volumetric sediment budget metrics. Their work demonstrates a striking dichotomy: since the Eocene, the Lower Congo Basin developed a retrogradational, narrow-shelf system dominated by erosive channel–levee complexes and terminal lobes, whereas the Niger Delta evolved a progradational, broad-shelf system characterized by fan lobes and delta-fed turbidites. Quantitative analysis identifies diapir-induced seafloor relief and basement uplift as first-order controls that redirect sediment pathways and amplify facies heterogeneity, while the shelf-break configuration and oceanic relief play subordinate roles. By explicitly linking salt/mud tectonics, shelf slope geometry, and sediment supply, the study refines source-to-sink models and offers practical guidance for predicting reservoir distribution and reducing exploration risk in diapir-affected passive margins.

Zeng et al. [16] introduced a hybrid computational framework that combines the Coupled Eulerian–Lagrangian (CEL) finite element method with a random forest (RF) machine learning algorithm to predict the landing penetration of a full-scale deep sea mining vehicle (DSMV) on ultrasoft seabed sediments. Their high-fidelity CEL simulations incorporated strain-softening and rate-dependent soil behavior, showing that penetration depth increases with soil sensitivity, descent velocity, and vehicle weight, but decreases with higher undrained shear strength and larger track contact area. To address the substantial computational cost of FEM analyses, the authors trained an RF model using the simulation dataset, achieving strong predictive performance ($R^2 > 0.94$) for normalized penetration depth. This integrated FEM–ML approach offers an efficient and reliable tool for optimizing track geometry and landing strategies, ultimately improving the operational safety and stability of DSMVs in deep sea mining applications.

Author Contributions: Conceptualization, X.G., X.L., T.N. and T.S.; investigation, X.G.; resources, X.G.; writing—original draft preparation, X.G.; writing—review and editing, X.G., X.L., T.N. and T.S. All authors have read and agreed to the published version of the manuscript.

https://doi.org/10.3390/jmse14030303

Funding: Financial support for this work is provided by the National Key Research and Development Program of China (2024YFC2815400), the Young Taishan Scholars Program of Shandong Province (TSQN202507107), the Shandong Provincial Natural Science Foundation (ZR2025MS647), the European Commission (HORIZON MSCA-2024-PF-01, 101200637), the Opening fund of the State Key Laboratory of Intelligent Construction and Healthy Operation and Maintenance of Deep Underground Engineering (SDGZ2529), and the Opening fund of State Key Laboratory of Ocean Engineering (GKZD010090).

Institutional Review Board Statement: Not applicable.

Informed Consent Statement: Not applicable.

Data Availability Statement: No new data were created or analyzed in this study. Data sharing is not applicable to this article.

Acknowledgments: We express our sincere gratitude to all the authors, the Academic Editors, and the reviewers for their invaluable contributions to this work. Additionally, we extend our thanks to the Editorial Board for their support, with special appreciation to the Contact Editor for their assistance in managing the daily procedures.

Conflicts of Interest: The authors declare no conflicts of interest.

References

1. Guo, X.; Liu, X.; Stoesser, T. Advances in Marine Engineering: Geological Environment and Hazards. *J. Mar. Sci. Eng.* **2023**, *11*, 475. [CrossRef]
2. Guo, X.; Liu, X.; Stoesser, T. Advances in Marine Engineering: Geological Environment and Hazards II. *J. Mar. Sci. Eng.* **2024**, *12*, 1253. [CrossRef]
3. Liu, X.; Stoesser, T.; Guo, X. *Advances in Marine Engineering: Geological Environment and Hazards*; MDPI Books: Basel, Switzerland, 2023.
4. Liu, X.; Guo, X.; Stoesser, T. *Advances in Marine Engineering: Geological Environment and Hazards II*; MDPI Books: Basel, Switzerland, 2024.
5. Zhang, C.; Tang, T.; Zhang, F.; Ren, C.; Zhang, H.; Wu, G. A State-of-the-Art Review of the Hydrodynamics of Offshore Pipelines Under Submarine Gravity Flows and Their Interactions. *J. Mar. Sci. Eng.* **2025**, *13*, 1654. [CrossRef]
6. Tao, H.; Li, Y.; Jiang, Q.; Huang, B.; Zuo, H.; Liu, X. Topographic Inversion and Shallow Gas Risk Analysis in the Canyon Area of Southeastern Qiongdong Basin Based on Multi-Source Data Fusion. *J. Mar. Sci. Eng.* **2025**, *13*, 1897. [CrossRef]
7. Yu, K.; Ding, W.; Ni, Y.; Chen, W.; Jiang, M.; Dai, Z. Influence of Sea Level Rise on Tidal Waves of the Yangtze River Estuary. *J. Mar. Sci. Eng.* **2025**, *13*, 1846. [CrossRef]
8. Lu, F.; Song, Q.; Liu, W. Mobilization of PAHs by Wave-Induced Resuspension and Liquefaction in Silty Sediment. *J. Mar. Sci. Eng.* **2025**, *13*, 1661. [CrossRef]
9. Liu, X.; Li, X.; Zhang, H.; Wang, Y.; Zhang, Q.; Wei, H.; Guo, X. An Experimental Investigation of the Effect of Interfacial Waves on the Evolution of Sliding Zones in a Liquefied Seabed. *J. Mar. Sci. Eng.* **2024**, *12*, 1355. [CrossRef]
10. Ge, Q.; Xia, Y.; Shu, J.; Li, J.; Sun, H. Explainable Ensemble Learning Approaches for Predicting the Compression Index of Clays. *J. Mar. Sci. Eng.* **2024**, *12*, 1701. [CrossRef]
11. Wei, M.; Ye, Y.; Zhao, W.; Wang, Z.; Ge, F.; Nian, T. Centrifugal Test Study on the Vertical Uplift Capacity of Single-Cylinder Foundation in High-Sensitivity Marine Soil. *J. Mar. Sci. Eng.* **2024**, *12*, 2152. [CrossRef]
12. Liu, Y.; Zhang, H.; Sun, Y.; Wu, K.; Xiao, W. Mechanical Behavior of Marine Soft Soil with Different Water Contents Under Cyclic Loading. *J. Mar. Sci. Eng.* **2024**, *12*, 2307. [CrossRef]
13. Dai, J.; Xu, Q.; Gong, Y.; Lu, Y.; Liu, X.; Mao, J. An LES Investigation of Flow Field Around the Cuboid Artificial Reef at Different Angles of Attack. *J. Mar. Sci. Eng.* **2025**, *13*, 463. [CrossRef]
14. Gong, Y.; Xu, Y.; Mao, J.; Dai, J.; He, L.; Zhang, H.; Xu, Q. Large-Eddy Simulation of Flows Past an Isolated Lateral Semi-Circular Cavity. *J. Mar. Sci. Eng.* **2025**, *13*, 859. [CrossRef]
15. Qu, F.; Gao, X.; Gong, L.; Yin, J. Contrastive Analysis of Deep-Water Sedimentary Architectures in Central West African Passive Margin Basins During Late-Stage Continental Drift. *J. Mar. Sci. Eng.* **2025**, *13*, 1533. [CrossRef]
16. Zeng, Y.; Xiu, Z.; Liu, L.; Xie, Q.; Sun, Y.; Yang, J.; Guo, X. Integration of the CEL and ML Methods for Landing Safety Prediction and Optimization of Full-Scale Track Design in a Deep-Sea Mining Vehicle. *J. Mar. Sci. Eng.* **2025**, *13*, 1584.

https://doi.org/10.3390/jmse14030303

Short Biography of Authors

Xingsen Guo is a Professor and Doctoral Supervisor at the Ocean University of China, a Marie Skłodowska-Curie Fellow at the University of Cambridge, and a Research Fellow at the University College London. His research primarily focuses on marine engineering geology and related environmental challenges. He earned his PhD from the Dalian University of Technology (DUT) in 2021. Prof. Guo has published over 90 peer-reviewed articles in leading journals such as Coastal Engineering and Engineering Geology, including multiple ESI hot and highly cited papers. He serves on the editorial boards of several journals, including Marine Georesources & Geotechnology. His accolades include the International Consortium on Geo-disaster Reduction (ICGdR) Outstanding Young Scientist Award, the BGA Poster Award, the JMSE Travel Award, and multiple Excellent Doctoral Dissertation Awards from ICGdR, the Chinese Society for Rock Mechanics & Engineering, Liaoning Province, and DUT. He has also been recognized among the World's Top 2% Scientists.

Xiaolei Liu is a Professor and Doctoral Supervisor at the Ocean University of China (OUC). He serves as the Director of the Shandong Provincial Key Laboratory of Marine Engineering Geology and the Environment and the Chairman of the Committee on Marine Geo-disaster and Geo-environment (TC-4) within the International Consortium on Geo-disaster Reduction (ICGdR). His research focuses on marine engineering geology and geohazards, specifically addressing the phase transition mechanisms of submarine sediments, the evolution of fluidized sediment movements, and advanced probing techniques for the seabed interface layer. Prof. Liu's work as a Principal Investigator is distinguished by a strong record of innovation and scholarship, including leadership of over 20 national and industry-sponsored projects, more than 130 publications in leading international journals, over 70 invention patents, and several prestigious awards.

Tingkai Nian is a Professor and the leader of the Marine Geohazard Research Group in the State Key Laboratory of Coastal and Offshore Engineering at Dalian University of Technology and a Senior Research Scholar at University College London. Prof. Nian's research interest is in developing advanced numerical, experimental and AI tools and their application to solve marine soil mechanics and geohazards problems. He has led more than 10 national-level research projects and co-authored over 150 peer-reviewed journal papers (SCI/EI-indexed). Prof. Nian has held over 40 invention patents and software copyrights, and obtained the Technology Invention Award from the International Consortium on Geo-disaster Reduction (ICGdR), First Prize for Scientific and Technological Progress in Liaoning Province, and appointment as a Leading Talent in Science and Technology Innovation. He also serves as Associate Editor-in-Chief of the journal of Geoenvironmental Disasters, Deputy Director of the Coastal and Offshore Engineering Disaster Environmental Protection Special Committee of the Chinese Seismological Society.

Thorsten Stoesser is the Leader of the Fluid Mechanics Research Group in the Department of Civil, Environmental and Geomatic Engineering at the University College London. His research interest is in developing advanced Computational Fluid Dynamics (CFD) tools and their application to solve environmental fluid mechanics problems. Thorsten has published over 100 peer-reviewed journal papers on developing, testing and applying advanced CFD methods to predict the hydrodynamics and transport processes in rivers, estuaries and coastal waters, fluid-structure interaction of marine turbines, and the nearfield dynamics of jets and plumes. For his research Prof. Stoesser received twice the American Society of Civil Engineers (ASCE) Karl Emil Hilgard Hydraulic Prize (2012 and 2016), in 2015 he won the International Association of Hydro-environmental Research (IAHR) Harold Shoemaker Award and in 2016 he won the Institution of Civil Engineers' George Stephenson medal. He has received over £5M in funding from industry, government institutions and research councils including the DFG (Germany), NSF (USA) and EPSRC (UK).

https://doi.org/10.3390/jmse14030303

Journal of
Marine Science and Engineering

MDPI

Article

Topographic Inversion and Shallow Gas Risk Analysis in the Canyon Area of Southeastern Qiongdong Basin Based on Multi-Source Data Fusion

Hua Tao [1], Yufei Li [1], Qilin Jiang [2], Bigui Huang [3], Hanqiong Zuo [3] and Xiaolei Liu [2,4,*]

1 China Oilfield Services Limited, Tianjin 300451, China; taohua@cnooc.com.cn (H.T.); liyf62@cnooc.com.cn (Y.L.)
2 Shandong Provincial Key Laboratory of Marine Environment and Geological Engineering, Ocean University of China, Qingdao 266100, China; jiangqilin97@163.com
3 China National Offshore Oil Corporation Research Institute Co., Ltd., Beijing 100028, China; huangbg3@cnooc.com.cn (B.H.); zuohq2@cnooc.com.cn (H.Z.)
4 Sanya Oceanographic Institution, Ocean University of China, Sanya 572024, China
* Correspondence: xiaolei@ouc.edu.cn

Abstract: The submarine topography in the canyon area of the Qiongdongnan Basin is complex, with severe risks of shallow gas hazards threatening marine engineering safety. To accurately characterize seabed morphology and assess shallow gas risks, this study employed multi-source data fusion technology, integrating 3D seismic data, shipborne multibeam bathymetry data, and high-precision AUV topographic data from key areas to construct a refined seabed terrain inversion model. For the first time, the spatial distribution characteristics of complex geomorphological features such as scarps, mounds, fissures, faults, and mass transport deposits (MTDs) were systematically delineated. Based on attribute analysis of 3D seismic data and geostatistical methods, the enrichment intensity of shallow gas was quantified, its distribution patterns were systematically identified, and risk level evaluations were conducted. The results indicate: (1) multi-source data fusion significantly improved the resolution and accuracy of terrain inversion, revealing intricate geomorphological details in deep-water regions; and (2) seismic attribute analysis effectively delineated shallow gas enrichment zones, clarifying their spatial distribution patterns and risk levels. This study provides critical technical support for deep-water drilling platform site selection, submarine pipeline route optimization, and engineering geohazard prevention, offering significant practical implications for ensuring the safety of deep-water energy development in the South China Sea.

Keywords: Qiongdongnan Basin; 3D seismic; terrain inversion; shallow gas; multi-source data fusion

1. Introduction

The Qiongdongnan Basin, with its unique geological structural background and hydrocarbon accumulation conditions, has become a key target area for oil and gas exploration and development in the western South China Sea [1–3]. As a critical structural unit in the western South China Sea, the Qiongdongnan Basin has undergone multiple phases of tectonic movement superposition, forming a complex geological framework dominated by fault systems with significant variations in sedimentary strata [4–6]. In recent years, as exploration and development accelerate into deepwater and ultra-deepwater areas, accurately characterizing seabed topography and precisely identifying shallow gas enrichment zones have become prerequisites for ensuring marine engineering safety and efficient development [7]. Therefore, achieving breakthroughs in terrain inversion accuracy through multi-source data fusion and accurately delineating shallow gas distribution patterns are key scientific challenges that urgently need to be addressed in the field of marine engineering geology.

Although significant progress has been made in the study of hydrocarbon accumulation mechanisms and deep structures in the Qiongdongnan Basin [8–11], research on fine-scale seabed topographic characterization and systematic risk assessment of shallow gas hazards remains insufficient. Previous studies have identified spatial correlations between shallow gas enrichment zones, fault trends, and paleochannels based on seismic reflection characteristics, but limitations in vertical resolution from traditional bathymetric data hinder the understanding of microscale topographic variations controlling gas migration pathways [12,13]. In recent years, shipborne multibeam bathymetry has enabled high-precision water depth data acquisition, yet beam angle effects in complex terrains create significant blind spots for steep slopes and fracture zones [14–16]. Against this backdrop, innovative applications of high-resolution topographic inversion techniques offer new solutions to overcome these bottlenecks [17–19]. Autonomous underwater vehicle (AUV)-mounted multibeam sonar systems can obtain high-resolution topographic data, which—when combined with sequence boundary extraction from 3D seismic data—enables multiscale terrain modeling from regional to localized key areas [20]. However, current research predominantly focuses on single-technique applications [21,22], lacking a systematic framework for collaborative inversion mechanisms integrating multi-source data (seismic, shipborne multibeam, AUV), thereby impeding comprehensive evaluation capabilities for complex seabed geological environments and hazard risks.

Facing these challenges, this study focuses on the canyon area of the Qiongdongnan Basin, aiming to address the critical issues of accurately characterizing complex submarine topography and understanding shallow gas hazard risks. By synergistically integrating 3D seismic data, shipborne multibeam bathymetric grids, and high-resolution AUV topographic point cloud data, the study significantly enhances the precision and completeness of seafloor terrain modeling, effectively overcoming the limitations of traditional single-data-source approaches. Building on this foundation, the research conducts an in-depth analysis of 3D seismic attributes to systematically identify the distribution patterns of shallow gas in the study area. The findings achieve refined representations of complex submarine geomorphological features (such as scarps, fissures, and MTDs) and reveal the controlling influence of high-resolution topographic elements on shallow gas distribution. This provides a solid scientific basis for the safe design of deepwater engineering projects and geohazard prevention in the Qiongdongnan Basin, while also offering research perspectives and technical pathways for a deeper understanding of complex submarine geological processes and their engineering implications in global continental margin basins.

2. Overview of the Study Area

The Qiongdongnan Basin is located at the western end of the northern continental margin rift zone of the South China Sea, adjacent to the Pearl River Mouth Basin [23,24]. It generally extends in a NE-SW direction, covering a total area of approximately 56,000 square kilometers (Figure 1). This region formed under the extensional rifting associated with the Eocene expansion of the South China Sea, undergoing three evolutionary stages: the Paleocene-Eocene rifting phase, the Oligocene-Miocene depression phase, and the Pliocene-Quaternary regional subsidence phase [25,26]. The basement primarily consists of pre-Paleogene metamorphic and igneous rocks, with the Neocene Lingshui Formation, Sanya Formation, and Meishan Formation serving as the main hydrocarbon reservoirs [27]. It is one of the most exploration-promising Cenozoic sedimentary basins along the passive continental margin of the northern South China Sea.

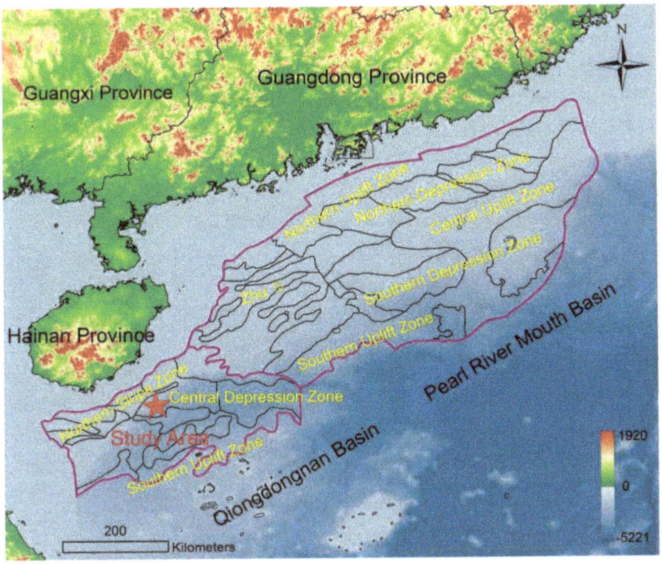

Figure 1. Geographic Location Map of Qiongdongnan Basin.

The study area is located in the central depression zone of the Qiongdongnan Basin, with water depths ranging from 158 to 2014 m. The seabed topography in this region descends stepwise from northwest to southeast [28]. According to the classification of the northern South China Sea shelf-slope, its geomorphological units can be divided into the shelf area, slope area, and deep-water area (Figure 2). The seabed in both the shelf and deep-water areas is generally flat, with an overall slope of less than 10°. The slope area has water depths of 250–1500 m, with a maximum slope reaching 70°, and features extensive submarine canyon development. The western slope area is a collapse-type slope, with canyons at depths of 250–1000 m and cross-sections exhibiting a "U" shape. The eastern slope area is a channel-type slope, where canyons are distributed at depths of 250–1500 m, with "V"-shaped cross-sections and dense canyon channel development.

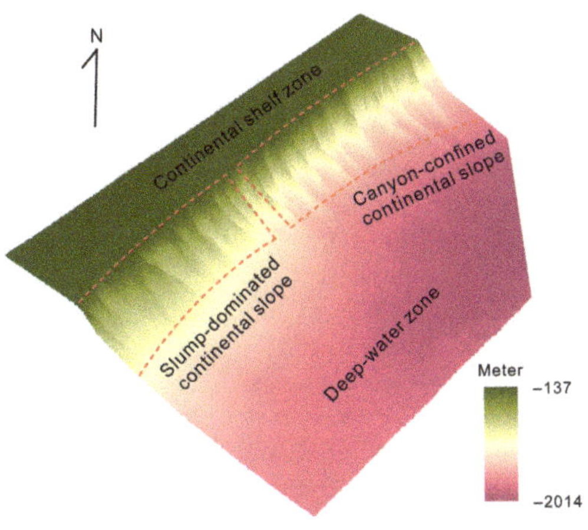

Figure 2. Division of Geomorphological Units in the Study Area.

3. Data and Methods

3.1. Data Sources

Multi-source data were obtained through 3D seismic exploration, shipborne multibeam surveys, and autonomous underwater vehicle (AUV) investigations, with specific technical parameters listed in Table 1. The region-wide 3D seismic exploration data formed the research foundation, with the seismic source and streamer deployed at a depth of 10 m. By precisely recording the first-arrival travel times of seabed reflection interfaces, these data provided core support for regional topographic inversion and shallow stratigraphic structure analysis. The shipborne integrated survey data included full-coverage seabed topographic data with 10 m spatial resolution collected by the EM712 multibeam system. Simultaneously acquired TOPAS PS18 parametric sub-bottom profiler data revealed high-resolution shallow subsurface stratigraphic structures, while the multichannel sparker system provided mid-to-deep stratigraphic profiles. These three datasets collectively characterized seabed topographic details, shallow-to-deep stratigraphic contacts, and potential geological anomalies such as gas chimneys. For initially identified high-risk hazard target areas, autonomous underwater vehicles (AUVs) equipped with SeaBat T50-S multibeam systems and EdgeTech 2205 side-scan sonar/sub-bottom profiler combination systems were deployed. The former acquired ultra-high-precision microtopographic models with 1 m grid resolution, while the latter enabled high-resolution observation of microtopographic boundary identification and shallow gas occurrence states through side-scan image texture analysis and shallow profiling.

Table 1. Classification and Technical Parameters of Multi-source Geophysical Data.

Data Type	Acquisition System /Method	Spatial Resolution /Detection Depth
3D seismic data	Source and streamer (deployment depth 10 m)	Vertical resolution: <5 m; Horizontal resolution: 25 m
Vessel-mounted multibeam bathymetric data	EM712 multibeam system	Terrain resolution: 10 m grid
	TOPAS PS18 parametric sub-bottom profiler	Vertical resolution: 0.15 m; Detection depth: <100 m
	Multi-channel sparker system	Vertical resolution: 1 m; Detection depth: >300 m
Key region AUV bathymetric data	SeaBat T50-S multibeam system	Terrain resolution: 1–2 m grid
	EdgeTech2205 Side Scan Sonar—Sub-bottom Profile	Side scan resolution: 0.1 m; Sub-bottom vertical resolution: 0.1 m

3.2. Analysis Methods

3.2.1. Seabed Topography Inversion

Seabed topography inversion is implemented based on global 3D seismic data. First, an automatic tracking algorithm is employed to precisely extract the one-way travel time (T) of the seabed reflection first arrival wave [29], and systematic delays caused by source and streamer depth are eliminated through specialized correction techniques [30] to ensure the accuracy of travel time data. Subsequently, time-depth conversion is performed using the seventh-order polynomial empirical formula (Formula (1)) proposed by Advocate & Hood (1993) [31]. Spatial interpolation is applied to the discrete water depth points obtained from inversion to construct a digital elevation model (DEM) with a 30 m grid spacing. By combining amplitude anomaly tracking technology and 3D seismic time slice analysis, key structural geomorphic features such as faults and scarps are systematically identified on the DEM. To rigorously validate inversion accuracy, shipborne multibeam measured topographic data and AUV topographic data from key areas are introduced for cross-comparison and error correction, significantly enhancing the accuracy of the topographic model.

$$D = 0.1105 - 5066.9193T + 468.6693T^2 - 554.7107T^3 + 340.7019T^4 - 116.9910T^5 + 20.7280T^6 - 1.4658T^7 \quad (1)$$

In the equation, D represents water depth, in feet; T denotes one-way travel time, in seconds.

3.2.2. Multi-Source Data Fusion and Interpretation

For the shipborne EM712 multibeam data, sound velocity profile correction and tidal compensation processing were performed, eliminating the influence of marine environmental factors and generating a high-precision seabed topographic DEM with a 10 m grid. Professional software was employed to conduct three-dimensional visualization fusion and comparative analysis of seismic inversion DEM, multibeam DEM, and AUV topographic data. Through the integrated interpretation system, the spatial distribution characteristics of complex geomorphic units such as scarps, protrusions, fissures, faults, and mass transport deposit systems (MTDs) were identified and delineated. TOPAS PS18 sub-bottom profiler and multi-channel sparker data were analyzed to accurately determine stratigraphic contact relationships and identify hazard indicators such as gas chimneys and shallow gas reservoirs. For AUV data, SeaBat T50-S multibeam data were motion-compensated to generate a 1 m grid micro-topographic model. Combined with the texture features of EdgeTech 2205 side-scan sonar images, the spatial distribution information of seabed pockmark boundaries and fissure networks was precisely extracted. The instantaneous amplitude extraction technique of sub-bottom profiler data was used to detect acoustic transparent zones (ATZ, refer to strata or intervals in seismic acoustic profiles characterized by a lack of internal reflections, appearing blank or nearly blank), enabling precise localization of shallow gas occurrence positions and extents.

3.2.3. Identification of Shallow Gas Distribution and Risk Level Classification

Using 3D seismic attribute analysis as the core technical approach, parameters such as amplitude, phase, and frequency are extracted. By enhancing the discontinuity between seismic traces, the boundaries of gas reservoirs, fault distributions, and special sedimentary bodies are effectively delineated. Combined with 3D seismic time-slice technology, the structural morphology and spatial distribution of amplitude anomalies are visually displayed. Integrating seismic attribute anomalies, ATZ zones identified by shallow profiling, pockmark micro-seepage landforms(Describe small, crater-like depressions on the seafloor formed by the seepage of gases or fluids from underlying sediments.) detected by AUV, and fault networks revealed by multi-source data, a comprehensive identification system for shallow gas spatial distribution is constructed. Geostatistical methods are introduced to quantify the enrichment intensity and spatial aggregation characteristics of shallow gas. Based on key factors such as shallow gas enrichment intensity, gas-controlling fault density, and sedimentary environment stability, a multi-indicator overlay risk assessment model for shallow gas hazards is established, enabling quantitative zoning of risk levels and providing a scientific basis for engineering disaster prevention.

4. Terrain Inversion Analysis

Based on the comprehensive interpretation of 3D seismic data, shipborne multibeam bathymetric data, and AUV topographic data from key areas, the inversion analysis of seabed topography in the study area reveals complex geological structural features and potential hazard risks. Through synergistic analysis of multi-source data, the distribution characteristics of steep slopes, protrusions, fractures, faults, and mass transport deposit (MTD) systems were identified (Table 2). The spatial distribution patterns of these topographic elements are visually presented in the topographic inversion spatial distribution map (as shown in Figure 3).

Table 2. Distribution Characteristics Table of Various Terrain Features.

Topographic Features	Distribution Range	Key Characteristics
Escarpments	Water depth 200–250 m, eastern part of the study area	NE-SW strike, length 8.5–9.0 km, width 400–900 m
Mounds	Water depth: 200–250 m	Nearly circular, scattered hill-like structures with diameters ranging 100–500 m and relative heights 2–10 m, another type exhibits linear distribution, 1.5–2.5 km long, 250–300 m wide, and 10–15 high.
Fissures	Near the continental slope fold belt	Length: 2–6 km, Width: 200–300 m, Span: 20–40 m, Maximum gradient: 26°
Faults	At the continental shelf edge, layer-controlled faults develop in deepwater areas.	The trend is near east–west, with a tendency toward the east. The top is buried 200 m below the seafloor and outcrops on the seafloor. The stratigraphic control fault lies 500–700 m below the seafloor.
Mass Transport Deposits (MTDs)	Deep Water Zone	Weak amplitude, semi-transparent, and diffuse reflection characteristics, with high amplitude displayed on the corresponding seismic attribute slice.

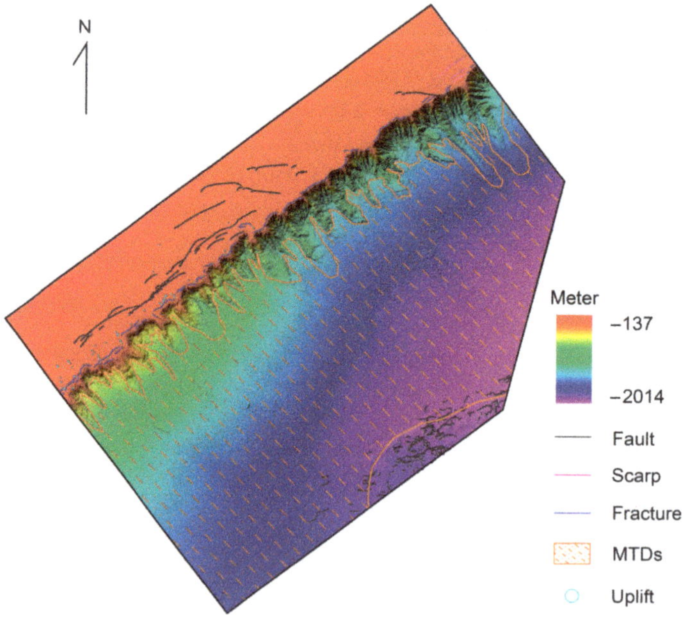

Figure 3. Topographic inversion results of the study area.

4.1. *Escarpments*

Submarine escarpments refer to steep slope terrains parallel or similar in height to the slope orientation on the seafloor [32,33]. Within the study area, escarpments primarily develop at water depths of 200~250 m on the outer continental shelf, spatially concentrated in the eastern part of the study area, exhibiting an overall NE-SW trend. Morphological analysis indicates that these escarpments extend 8.5~9.0 km in length, commonly displaying step-like features with widths ranging from 400~900 m. They exhibit significant topographic relief, with maximum water depth spans reaching 15~40 m, corresponding to a maximum slope of approximately 8°. Seismic data reveal no obvious fault displacement in the underlying strata of the escarpments. However, potential sliding surface structures have been identified in some adjacent areas (Figure 4), suggesting potential landslide geological risks in this region.

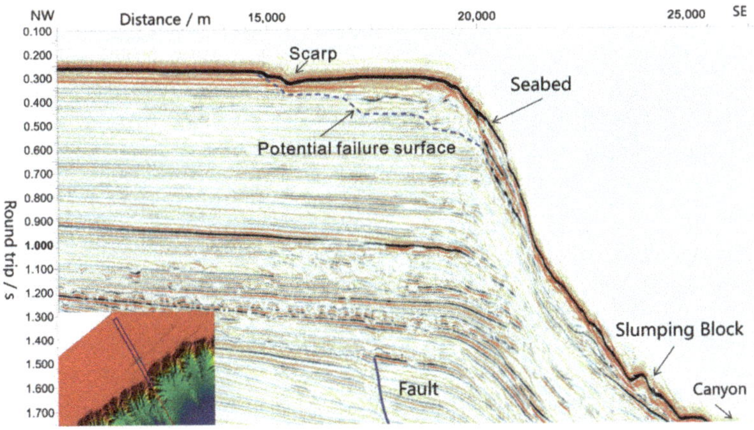

Figure 4. Cross-sectional view of submarine escarpments from 3D seismic data.

4.2. Mounds

Mounds in the study area are primarily distributed in the western region, gradually disappearing toward the east. These mounds are concentrated in the upper edge zone of the shelf break at water depths of 200–250 m. Morphologically, they exhibit two main types of features: one consists of approximately circular, sporadically distributed mound-like bodies with diameters of 100–500 m and relative heights of 2–10 m; the other appears as linear distributions, extending in a direction roughly parallel to the shelf break, with lengths of 1.5–2.5 km, widths of 250–300 m, and heights of 10–15 m. AUV sub-bottom profiling data have identified shallowly buried hard strata beneath these mounds and confirmed that some mound surfaces are exposed areas of harder seabed. These mound zones serve as exit points for deep fluid conduit systems on the seabed, connecting to buried high-pressure plastic fluid sources.

The formation mechanism of seabed mounds is closely related to deep fluid migration, often developing near the slope breaks of progradational and slump-type continental slopes [34,35]. Three-dimensional seismic data reveal deep fluid migration channels beneath the mounds, providing direct evidence for their fluid origin (Figure 5). Additionally, the distribution areas of the mounds show a high spatial correlation with ancient buried landslide bodies near the shelf edge slope break, indicating that their formation is also controlled by historical landslide activities. The region's high sedimentation rate causes the buried ancient landslide bodies to undergo differential settlement under regional tectonic activity and differential compaction(During sedimentation, the phenomenon where different degrees of compaction occur due to variations in sediment thickness or material composition), particularly around the landslide edges, forming weak zones in the overlying strata.

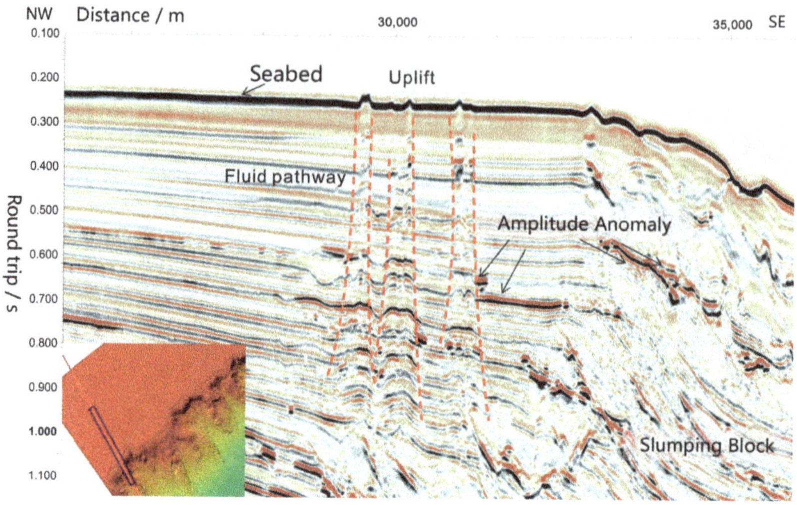

Figure 5. Three-dimensional seismic profile of seabed mounds.

4.3. Fissures

In the vicinity of the shelf break zone, numerous curved submarine fissures have developed, characterized by distinct submarine scarps [36,37]. These fissures measure 2–6 km in length, 200–300 m in width, and span 20–40 m in water depth, with local maximum slopes reaching 26°. Based on analysis of stratigraphic profile data (Figure 6), such fissures are primarily distributed near the head of ancient landslide bodies and genetically belong to tensile fissures at landslide heads, formed by sediment under differential compaction. The presence of tensile fissures serves as an important indicator of potential slope instability; moreover, under continuous gravitational forces and external dynamic factors such as seismic loading, these fissures exhibit a tendency for further expansion.

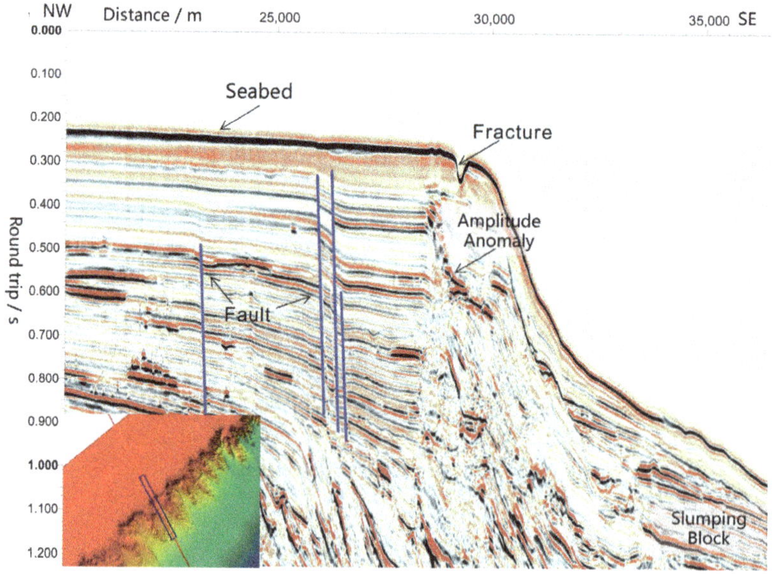

Figure 6. Submarine fissure profile from 3D seismic data.

4.4. Faults

A significant number of normal faults have been identified along the continental shelf margin within the study area. Among them, one set strikes NE–SW, consistent with the orientation of the shelf break, and dips SE; another set, distributed in the eastern part of the study area, is arranged in an en echelon pattern with an approximate EW strike and overall eastward dip. Based on the top depth of the faults, they can be classified into four categories: seafloor-exposed, buried less than 100 m, buried between 100–200 m, and buried greater than 200 m. The faults in the study area belong to the Central Subbasin Fault Zone (CSFZ), a typical active fault zone in the northern South China Sea [38]. This fault zone is exposed on the seabed in the western part of the study area, forming discontinuous linear scarps with lengths of 3.5–5.5 km, widths of 200–300 m, maximum water depth spans of 10–25 m, and slopes reaching up to 9°, gradually transitioning to a concealed state eastward (Figure 7). Activity analysis indicates that some of the seabed-exposed faults remain active today. Such active faults in marine environments can trigger various natural hazards, including earthquakes, tsunamis, and volcanic activity [39–41]. Particularly in areas with steep seabed slopes, such as deep-water troughs and continental slope margins, direct displacement of active faults or triggered earthquakes can easily induce submarine landslides. Active faults pose significant threats to submarine pipeline routes, as surface ruptures, associated ground deformation, and differential settlement caused by them can severely damage engineering facilities crossing or adjacent to the faults [42,43].

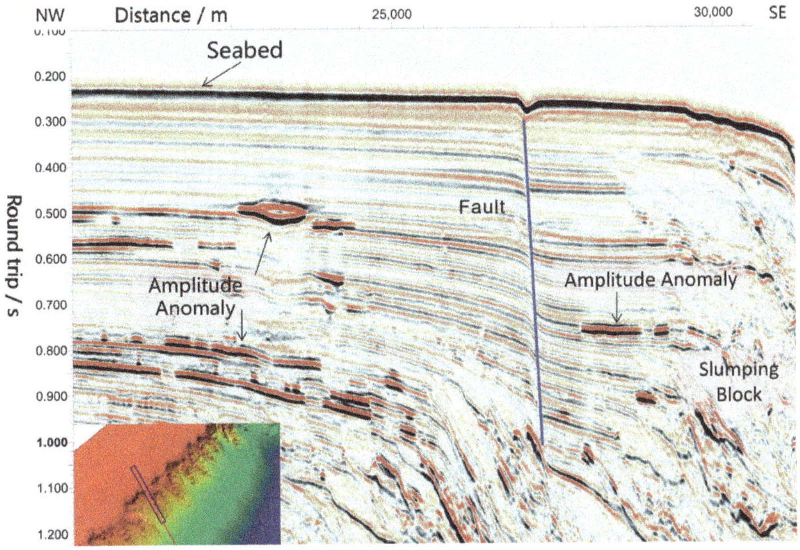

Figure 7. Shallow fault profile from 3D seismic data.

In addition to the faults in the shelf area, a series of stratabound faults have developed in the deep-water region at depths of 1800–2000 m, often associated with the migration of plastic fluids. These faults primarily occur within the depth interval of 500–700 m below the seabed, with some extending up to 200 m below the seabed. They are characterized by relatively small scales, dense development along bedding planes, and overall control by a single stratigraphic unit (Figure 8). Their formation mechanism is typically related to gravitational instability within the strata, leading to lateral creep and extension under the compaction of overlying sediments [44].

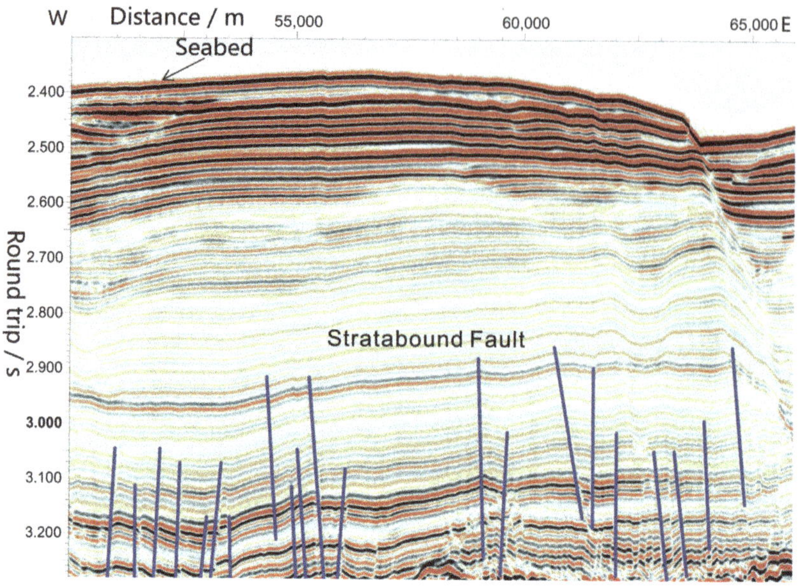

Figure 8. Stratabound fault profile from 3D seismic data.

4.5. *Mass Transport Deposits (MTDs)*

Mass Transport Deposits (MTDs) are generally regarded as stratigraphic units formed through redeposition processes after initial sedimentation [45,46]. Given that large-scale MTDs primarily develop in deep-water environments, their formation processes are difficult to observe directly. Research on their formation mechanisms and developmental history often relies on post-event investigations, integrating acoustic and geological detection techniques, along with inferences from seabed morphology and sediment deformation theories. Overall, MTDs commonly exhibit weak amplitude, semi-transparent, and chaotic reflection characteristics on seismic profiles, while displaying strong amplitude responses on related seismic attribute slices. During their migration toward deep-water areas in the study region, the tensile forces acting on MTDs gradually weaken, while their leading edges continuously undergo thrust compression, forming wedge-shaped or hummocky toe structures. Compressional ridges develop in the upper strata and taper laterally. The edges of these hummocks show strong-amplitude, continuous reflections with clear boundaries, while their interiors predominantly feature weak-amplitude parallel reflections, chaotic reflections, or blank reflections. As shown in Figure 9, multiple phases of MTDs developed sequentially from bottom to top in the study area during historical periods. These appear as semi-transparent and chaotic reflections on seismic profiles, typically with well-defined basal shear surfaces. Thrust compressional structures are identifiable in the lower sections, while the tops often exhibit undulating reflections associated with compressional structures. Due to differences in developmental timing, later-stage MTDs frequently erode and modify earlier-formed units. AUV data reveal thin-layer MTDs on the seabed surface, as illustrated in Figure 10, with thicknesses of 2–5 m. Continuous sedimentary strata underlie MTDs in localized areas, while in other regions, these continuous strata are eroded by MTDs. Owing to variations in developmental timing, later-stage MTDs often erode and reshape earlier-formed units.

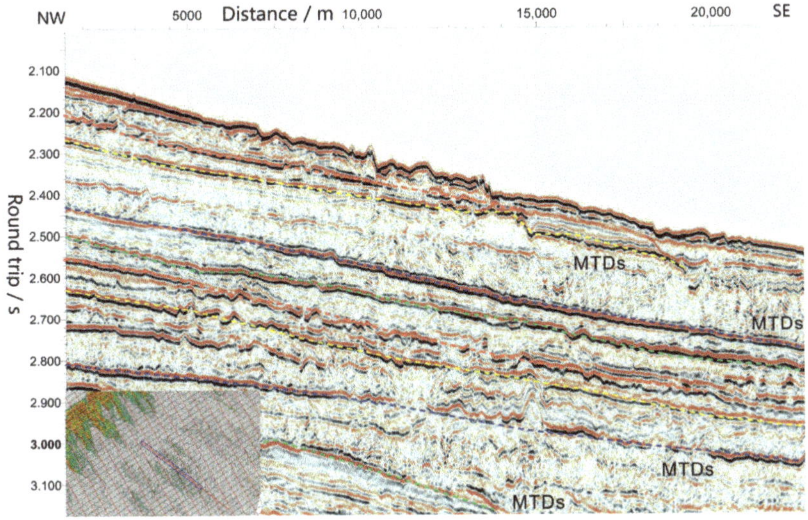

Figure 9. Multi-phase MTDs on 3D seismic profiles.

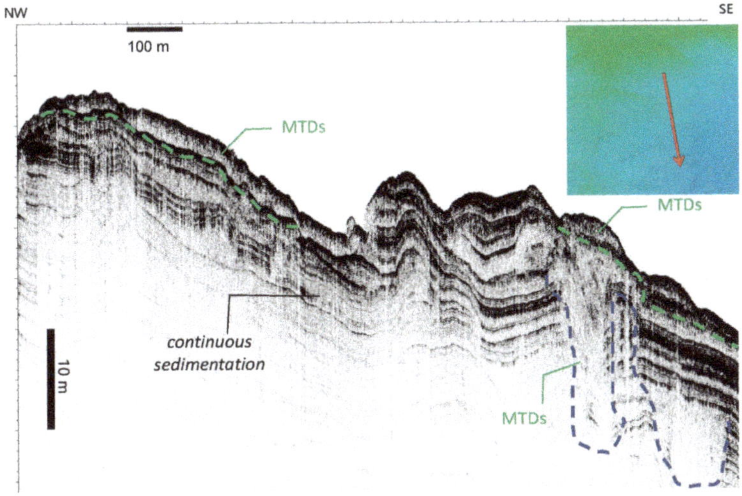

Figure 10. MTDs on AUV sub-bottom profiler data.

5. Shallow Gas Risk Analysis

Based on 3D seismic attribute analysis, identification of acoustic transparent zones (ATZ) from shallow stratigraphic profiles (TOPAS PS18 & EdgeTech 2205), detection of pockmark microtopography via AUV side-scan sonar, and multi-source data fusion to construct fault networks, this study systematically characterizes the spatial enrichment patterns and risk levels of shallow gas within 0–500 m depth in the canyon area of the Qiongdongnan Basin. By employing geostatistical kriging interpolation to quantify enrichment intensity and integrating factors such as gas-controlling structure density, sedimentary environment stability, and seepage landform scale, a shallow gas risk assessment system was established for the study area. Five depth intervals below the seabed were delineated (Figure 11), revealing their enrichment patterns and spatial risk differentiation.As shown in Table 3, the key characteristics of different depth intervals are clearly compared.

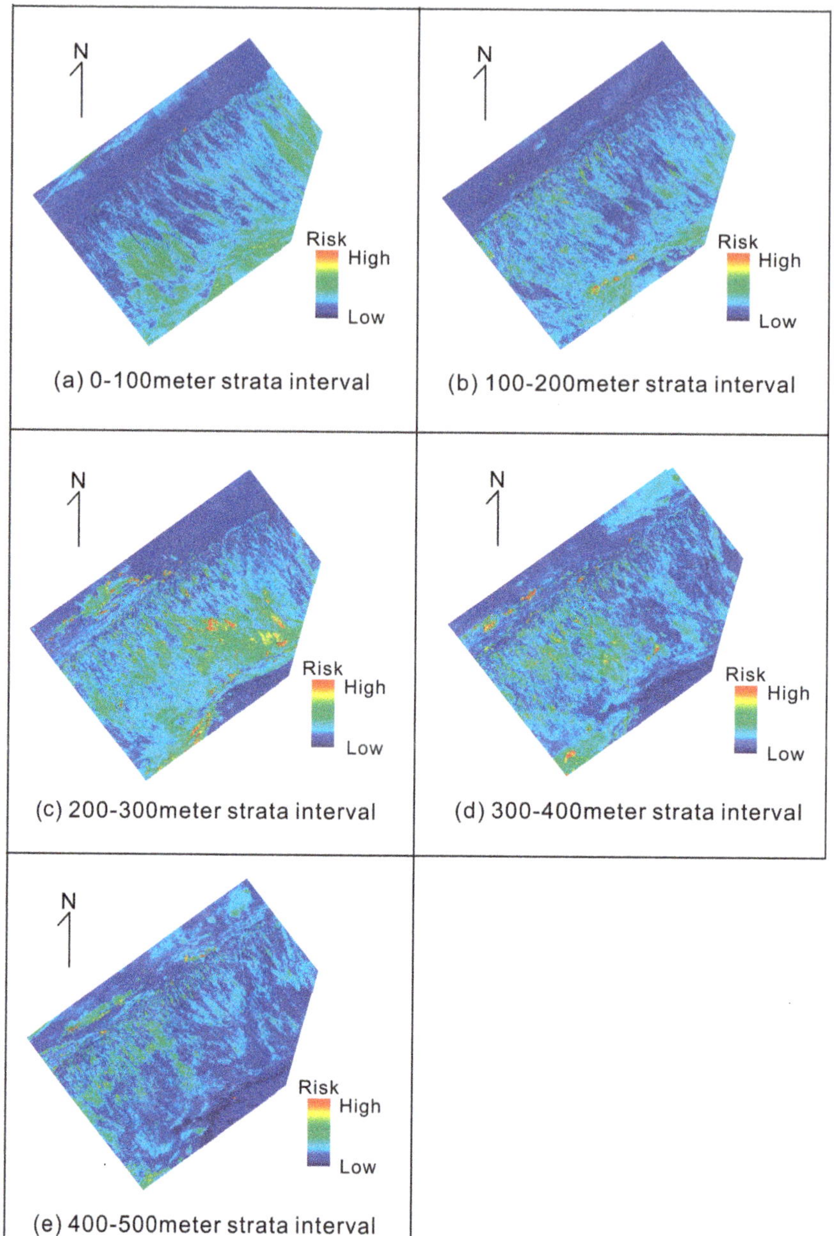

Figure 11. Shallow Gas Risk Zonation Map.

Table 3. Summary of Vertical Zonation Characteristics for Shallow Gas Risk in the Study Area.

Depth Below Sea Level/m	Spatial Distribution	Controlling Factors	Primary Gas Type	Risk Level
0–100	Discrete spots	Microtopography, weak sealing	Biogenic gas	Low-risk dominant
100–200	Local enrichment	Channels, MTDs migration	Deep gas	Fragmented low-risk
200–300	Continuous belt	Faults, MTDs traps	Pyrolytic gas	High-risk peak
300–400	Dispersed patches	Gas chimneys, fault networks	Pyrolytic gas	Dispersed high-risk
400–500	Sporadic occurrence	Intense compaction, homogenization	Pyrolytic gas	Low-risk stable

5.1. Spatial Distribution Characteristics of Shallow Gas Risk by Depth Interval

5.1.1. 0–100 m Interval

The high-risk zones of shallow gas exhibit a "discrete spotty" distribution without forming continuous accumulation belts, sporadically concentrated in the central part of the slope break zone (Figure 11a). This distribution is primarily controlled by microtopographic features such as near-seabed steep scarps and fractures. At this depth, biogenic gas or early diagenetic gas mainly escapes vertically, constrained by the low compaction and weak sealing capacity of overlying strata, making large-scale accumulation difficult [47,48]. The risk level gradually transitions from medium-risk to low-risk zones along the main sediment supply direction, reflecting differential control by sediment supply rates and early depositional structures on initial gas sealing capacity. Low-risk zones dominate this interval with relatively good spatial continuity.

5.1.2. 100–200 m Interval

The extent of high-risk zones expands significantly, displaying "local enrichment" characteristics, primarily clustered in the southern deep-water canyon channel systems and mass transport deposit (MTDs) development areas (Figure 11b). Canyon channels provide pathways for lateral gas migration, while the complex lithologic structures, differential compaction, and sedimentary disturbances within MTDs promote deep gas release [49]. Low-risk areas still dominate in spatial proportion, though their continuity is fragmented—either dissected or surrounded by southern high-risk zones and erosional areas along landslide flanks.

5.1.3. 200–300 m Interval

As shown in Figure 11c, the high-risk zones reach their peak both in quantity and spatial continuity, forming a relatively continuous, nearly basin-wide high-risk belt in the central and southern regions. This belt's spatial distribution closely aligns with the upward extension paths of deep-source gas faults and the main distribution areas of mass transport deposits (MTDs). The deep source rocks and pyrolytic gas supply are abundant, with the deep fault system serving as the primary vertical migration pathway [50]. The MTDs and channel-filling bodies exhibit excellent pore-fracture systems and differential compaction trap conditions [51]. Consequently, the tectonic-sedimentary coupling effect dominates large-scale gas accumulation and sealing. Low-risk zones are severely compressed, surviving only at the basin margins and localized tectonically sedimentarily stable areas.

5.1.4. 300–400 m Interval

The distribution pattern of shallow gas high-risk zones transitions from the concentrated, continuous distribution in the 200–300 m interval to a highly dispersed and widespread state, exhibiting coexisting patchy, banded, and stippled patterns (Figure 11d). This phenomenon results from well-developed gas chimney structures in this interval, densely interwoven fault networks, and widespread isolated small traps formed by intense differential compaction, leading to dispersed gas migration pathways. Low-risk zones in this interval display significant fragmentation characteristics, appearing as "islands" scattered across high positions at the basin margins and localized structural highs.

5.1.5. 400–500 m Interval

The number of high-risk zones sharply decreases, showing only a "sporadic" distribution pattern (Figure 11e). This is attributed to significantly enhanced formation homogenization, where intense compaction drastically reduces porosity and deteriorates permeability, thereby increasing gas migration resistance and making enrichment and sealing considerably more difficult. Consequently, the deep environment generally returns to a relatively stable state with low disturbance. Low-risk zones resume a continuous distribution pattern along the basin margins, disturbed only by sporadic high-risk patches formed by leakage at the termini of deep-seated faults or isolated turbidite sand body traps.

5.2. Vertical Distribution Patterns of Shallow Gas Hazards

The shallow gas hazards in the study area exhibit vertical zonation, fundamentally controlled by the depth-dependent evolution of the "gas source supply migration pathway" system. The shallow hazard zone (0–200 m) is dominated by microtopography and early sedimentation, with high-risk areas displaying discrete patchy distributions. The gas sources are primarily biogenic gas and early diagenetic gas, with migration modes dominated by vertical escape and short-range lateral migration. Microtopographic features like steep slopes are the key controlling factors, while faults have weaker influence. The mid-deep core hazard zone (200–400 m) is strongly controlled by tectonic-sedimentary coupling, with the 200–300 m interval reaching peak hazard intensity and continuity. Gas sources are diversified, with significant contributions from deep thermogenic gas, where deep fault systems serve as main vertical migration pathways. Internal heterogeneity within MTDs and channel sand bodies provides superior storage spaces, while differential compaction and tectonic stress synergistically enhance local sealing, forming the core disaster interval. The deep hazard attenuation zone (>400 m) sees suppressed gas enrichment due to strong compaction and formation homogenization, where sharply reduced porosity and permeability lead to drastically increased migration resistance. Only fault termini or isolated lithologic lenses maintain sporadic high-risk spots, with the overall system reverting to a stable state with low gas reservoir disturbance.

6. Discussion

6.1. Resolution Gradient Comparison of Terrain Inversion Technologies

In marine geological surveys, the precise characterization of seabed microtopography is crucial, with its accuracy directly controlled by the vertical and horizontal resolution of detection data. Due to significant differences in acquisition platforms, acoustic frequencies, and operational distances among 3D seismic exploration, shipborne multibeam bathymetric systems, and AUV terrain detection, there exists a gradient effect in their capabilities to depict water depth resolution and terrain details, thereby affecting the identification accuracy of microtopographic features [52,53]. As shown in Figure 12 and Table 4, noticeable discrepancies exist in water depth and slope values at the same location when surveyed by different detection methods. Therefore, improving the identification accuracy of seabed microtopography urgently requires the integration of multi-source detection data to overcome the gradient effect.

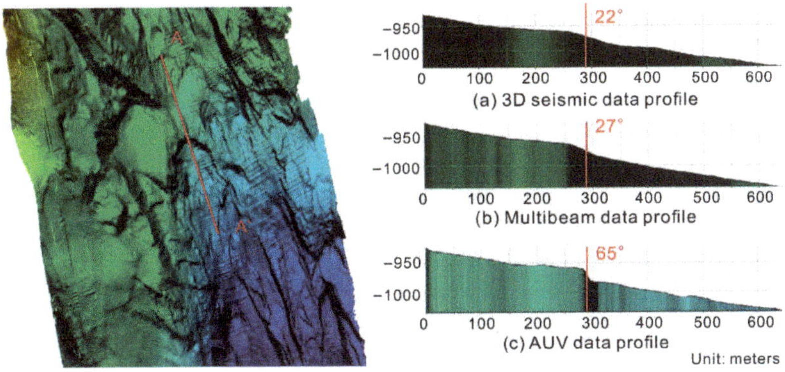

Figure 12. Comparison Diagram of AA' Section.

Table 4. Comparison of Multiple Data Results for AA Section.

Resolution Level Difference	Resolution/m	Water Depth		Slope	
		Value/m	Relative Error	Value/°	Relative Error
3D seismic data profile	30 × 30	971	+0.25%	22	−66.15%
Shipborne multibeam data profile	10 × 10	976.5	+0.82%	27	−58.46%
AUV data profile	1 × 1	968.6	0.00%	65	0.00%

Although 3D seismic data can penetrate subsurface strata to reveal structural frameworks and the boundaries of large-scale mass transport deposits (MTDs), their horizontal resolution is also constrained by trace spacing, severely limiting their ability to depict microtopography on the seabed surface. Shipborne multibeam systems operate at higher frequencies, significantly enhancing vertical resolution. These systems can effectively identify topographic features such as submarine landslide scarps but remain insensitive to surface textures. The sampling density and signal-to-noise ratio for microtopography are further compromised by the shipborne platform's speed limitations and altitude above the seabed, which are constrained by sea conditions. AUV-mounted near-bottom high-resolution multibeam/sidescan systems, by flying close to the seabed, enable detailed characterization of terrain and microtopography.

6.2. Shallow Gas Risk Discrimination Using Seismic Attributes

Three-dimensional digital seismic data can only identify shallow gas and reflection amplitudes that exceed the minimum detectable size threshold. Shallow gas layers or gas clusters below this threshold, as well as dispersed geological features, cannot be detected using 3D digital seismic data. To study the geological reflection anomalies associated with shallow gas, seismic workstation processing and interpretation software modules were employed. These modules extracted parameters such as the maximum negative amplitude, maximum positive amplitude, and root mean square (RMS) amplitude for each seismic trace, enabling amplitude analysis of the study area data. The amplitude extraction method used a background threshold approximately five times the average amplitude of the surrounding area. Amplitudes exceeding this threshold were evaluated as potential anomalies caused by shallow gas. The spatial extent and distribution of amplitude anomalies were mapped using the seismic workstation and manually verified.

In 3D seismic data volumes, the seafloor produces a high-impedance acoustic response, manifesting as a positive-polarity, positive-amplitude peak reflection interface. Potential shallow gas accumulation zones within the strata generate low-impedance acoustic responses, appearing as strong negative-polarity amplitude anomalies. The identification of shallow gas integrates one or more seismic wave attributes, such as strong amplitude, abrupt phase changes, polarity reversals, velocity reduction, acoustic diffraction, attenuation, and masking of underlying stratum reflections. However, not all amplitude anomalies are related to shallow gas. Gas hydrates, shallow water, conventional lithological variations, organic deposits, masking of stratum reflections, localized lithological changes, and tuning effects from adjacent target reflections can also produce strong amplitude anomalies. Therefore, only reflection anomalies exhibiting strong amplitude, low impedance, and at least several of the aforementioned characteristics are interpreted as potential shallow gas accumulation zones.

7. Conclusions

This study focuses on the key issues of precise characterization of complex submarine topography and risk assessment of shallow gas hazards in the canyon area of the Qiongdongnan Basin, conducting systematic research based on multi-source data fusion technology. The findings are summarized as follows:

(1) By systematically integrating 3D seismic data, shipborne multibeam bathymetric data, and AUV high-precision topographic data, this study constructed a multi-source data fusion inversion model, significantly improving the resolution and accuracy of the submarine topographic model. This approach overcomes the limitations of traditional single-data-source characterization of complex submarine topography, achieving the first systematic revelation of the spatial distribution characteristics of microtopographic features (such as steep slopes, protrusions, fissures, faults, and MTDs) in the canyon area of the Qiongdongnan Basin.

(2) The research findings clearly delineate the distribution patterns of gravity-flow-derived landforms such as mass transport deposits (MTDs), indicating active deep-water sedimentary dynamics and tectonic activity in this region. The precise identification of such geomorphic units holds significant scientific value for understanding the geological evolution, sediment transport mechanisms, and seabed stability assessment in the deep-water area of the northern South China Sea continental margin.

(3) By employing seismic attribute analysis and geostatistical methods, this study achieved spatial quantification and risk classification of shallow gas enrichment intensity. The research reveals that shallow gas distribution exhibits notable vertical zonation and spatial heterogeneity, with enrichment zones being jointly controlled by deep gas source supply and dominant migration pathways (such as faults and fracture systems), reflecting the dynamic geological processes of gas generation, migration, and accumulation.

(4) The research findings of this paper indicate that the risk pattern of shallow gas is not randomly distributed but is jointly constrained by deep hydrocarbon systems and shallow geological structures. Faults and fracture systems, serving as preferential migration pathways for fluids, along with the connectivity to deep gas sources, collectively determine the dispersion and accumulation behavior of shallow gas. This understanding holds theoretical significance for studying fluid escape mechanisms in deepwater regions.

(5) The research outcomes of this paper can support platform site selection, optimization of submarine pipeline routes, and geological hazard prevention and control, providing critical geological safety support for deepwater energy development in the South China Sea. This has significant practical implications for ensuring the safety of deepwater energy development in the region.

Author Contributions: Conceptualization, H.T. and Y.L.; methodology, B.H.; data analysis, H.T. and Q.J.; writing—original draft preparation, H.T. and H.Z.; writing—review and editing, X.L.; project administration, B.H. and X.L.; funding acquisition, X.L. All authors have read and agreed to the published version of the manuscript.

Funding: This work was supported by the National Key Research and Development Program of China (No. 2024YFF0506803), the Shandong Provincial Natural Science Foundation (ZR2025QA02), the Fundamental Research Funds for the Central Universities (202441003), and the Shandong Province National-Level Leading Talent Supporting Project (2022GJJLJRC-15).

Data Availability Statement: The data are available on request due to restrictions from the data provider.

Acknowledgments: The authors extend their sincere thanks to the editors and reviewers for their careful reading and fruitful suggestions.

Conflicts of Interest: Author Bigui Huang and Hanqiong Zuo are employed by the company CNOOC Research Institute Co. Ltd. The remaining authors declare that the research was conducted in the absence of any commercial or financial relationships that could be construed as a potential con-flict of interest.

References

1. Gong, C.; Wang, D.; Zhang, Q.; Yang, H.; Li, Z.; Zhao, Z.; Guo, S. Characteristics of coal-measure source rock and gas accumulation belts in marine-continental transitional facies fault basins: A case study of the Oligocene deposits in the Qiongdongnan Basin located in the northern region of the South China Sea. *Energy Explor. Exploit.* **2019**, *37*, 1752–1778. [CrossRef]
2. Zhang, W.; Liang, J.; Yang, X.; Su, P.; Wan, Z. The formation mechanism of mud diapirs and gas chimneys and their relationship with natural gas hydrates: Insights from the deep-water area of Qiongdongnan Basin, northern South China Sea. *Int. Geol. Rev.* **2020**, *62*, 789–810. [CrossRef]
3. Xu, Y.; Li, X.; Fan, C.; Tan, J.; Liu, K.; Liu, Y.; Wu, J. The axial channel provenance system and natural gas accumulation of the Upper Miocene Huangliu Formation in Qiongdongnan Basin, South China Sea. *Pet. Explor. Dev.* **2016**, *43*, 570–578. [CrossRef]
4. Liu, J.; Sun, Z.; Wang, Z.; Sun, Z.; Zhao, Z.; Wang, Z.; Zhang, J. Tectonic differences between eastern and western sub-basins of the Qiongdongnan Basin and their dynamics. *Mar. Geophys. Res.* **2015**, *36*, 61–79. [CrossRef]
5. He, B.; Zheng, M. Structural characteristics and formation dynamics: A review of the main sedimentary basins in the continent of China. *Acta Geogr. Sin.* **2016**, *90*, 1156–1194.
6. Zhao, Z.; Sun, Z.; Sun, L.; Wang, Z.; Sun, Z. Cenozoic tectonic subsidence in the Qiongdongnan basin, northern South China Sea. *Basin Res.* **2018**, *30*, 269–288. [CrossRef]
7. Liu, X.; Jeng, D.-S.; Moon, V.; Stoesser, T.; Zhang, L. Preface to the special issue on "Marine Engineering Geology in Developing Ocean Resources and Energy". *Eng. Geol.* **2025**, *354*, 108206. [CrossRef]
8. He, D.; Jia, C.; Zhao, W.; Xu, F.; Luo, X.; Liu, W.; Zheng, N. Research progress and key issues of ultra-deep oil and gas exploration in China. *Pet. Explor. Dev.* **2023**, *50*, 1333–1344. [CrossRef]
9. Wu, K.; Xie, X.; Pei, J.; Ren, J.; You, L.; Quan, Y. Deep architecture of hyperextended marginal basin and implications for hydrocarbon exploration: A case study of Qiongdongnan Basin. *Oil Gas Geol.* **2023**, *44*, 651–661.
10. Zhang, J.; Yan, D.; Pei, J.; Luo, W.; Liu, E.; Song, P.; Lin, X. Shallow gas accumulation mechanism in the Qiongdongnan Basin, South China Sea. *Mar. Pet. Geol.* **2025**, *177*, 107389. [CrossRef]
11. Yang, G.; Yin, H.; Gan, J.; Wang, W.; Zhu, J.; Jia, D.; Xu, W. Structural segmentation in the Qiongdongnan Basin, South China Sea: Insights from analogue models and implications for hydrocarbon exploration. *AAPG Bull.* **2024**, *108*, 907–941. [CrossRef]
12. Qiu, N.; Wang, Z.; Xie, H.; Sun, Z.; Wang, Z.; Sun, Z.; Zhou, D. Geophysical investigations of crust-scale structural model of the Qiongdongnan Basin, Northern South China Sea. *Mar. Geophys. Res.* **2013**, *34*, 259–279. [CrossRef]

13. Liu, K.; Cheng, P.; Tian, H.; Song, P.; Hu, W. Development model of shallow lithologic traps and natural gas accumulation mechanisms in marine deep-water areas: A case study in the Qiongdongnan basin, south China sea. *Mar. Pet. Geol.* **2023**, *151*, 106211. [CrossRef]
14. Chen, M.; Takahashi, S.; Takamasu, K. Multi-beam angle sensor for flatness measurement of mirror using circumferential scan technology. *Int. J. Precis. Eng. Manuf.* **2016**, *17*, 1093–1099. [CrossRef]
15. Li, H.; Li, C.; Gao, H.; Wu, S.; Fang, G. Study of moving targets tracking methods for a multi-beam tracking system in terahertz band. *IEEE Sens. J.* **2021**, *21*, 6520–6529. [CrossRef]
16. Misiuk, B.; Brown, C.J. Multiple imputation of multibeam angular response data for high resolution full coverage seabed mapping. *Mar. Geophys. Res.* **2022**, *43*, 7. [CrossRef]
17. Li, Z.; Peng, Z.; Zhang, Z.; Chu, Y.; Xu, C.; Yao, S.; Ma, J. Exploring modern bathymetry: A comprehensive review of data acquisition devices, model accuracy, and interpolation techniques for enhanced underwater mapping. *Front. Mar. Sci.* **2023**, *10*, 1178845. [CrossRef]
18. Fan, D.; Li, S.; Meng, S.; Lin, Y.; Xing, Z.; Zhang, C.; Qu, Z. Applying iterative method to solving high-order terms of seafloor topography. *Mar. Geod.* **2020**, *43*, 63–85. [CrossRef]
19. Ge, B.; Guo, J.; Kong, Q.; Zhu, C.; Huang, L.; Sun, H.; Liu, X. Seafloor topography inversion from multi-source marine gravity data using multi-channel convolutional neural network. *Eng. Appl. Artif. Intell.* **2025**, *139*, 109567. [CrossRef]
20. Orange, D.L.; Teas, P.A.; Decker, J.; Gharib, J. Use of multibeam bathymetry and backscatter to improve seabed geochemical surveys—Part 1: Historical review, technical description, and best practices. *Interpretation* **2023**, *11*, T215–T247. [CrossRef]
21. Liu, Y.; Li, S.; Zou, Z.; Sun, Y. Techniques and methods for seafloor topography mapping: Past, present, and future. *Intell. Mar. Technol. Syst.* **2025**, *3*, 1–15. [CrossRef]
22. Xu, G.; Hefner, B.T.; Jackson, D.R.; Ivakin, A.N.; Wendelboe, G. A Physics-Based Inversion of Multibeam Sonar Data for Seafloor Characterization. *IEEE J. Ocean. Eng.* **2024**, in press. [CrossRef]
23. Hu, B.; Wang, L.; Yan, W.; Liu, S.; Cai, D.; Zhang, G.; Sun, B. The tectonic evolution of the Qiongdongnan Basin in the northern margin of the South China Sea. *J. Asian Earth Sci.* **2013**, *77*, 163–182. [CrossRef]
24. Song, Y.; Zhao, C.Y.; Zhang, G.C.; Song, H.B.; Shan, J.N.; Chen, L. Tectonic-Thermal Modeling for Qiongdongnan Basin and Pearl River Mouth Basin in the Northern South China Sea. *Chin. J. Geophys.* **2011**, *54*, 942–955.
25. Meng, M.; Liang, J.; Kuang, Z.; Ren, J.; He, Y.; Deng, W.; Gong, Y. Distribution characteristics of quaternary channel systems and their controlling factors in the Qiongdongnan Basin, South China Sea. *Front. Earth Sci.* **2022**, *10*, 902517. [CrossRef]
26. Gong, C.; Wang, Y.; Zhu, W.; Li, W.; Xu, Q.; Zhang, J. The Central Submarine Canyon in the Qiongdongnan Basin, northwestern South China Sea: Architecture, sequence stratigraphy, and depositional processes. *Mar. Pet. Geol.* **2011**, *28*, 1690–1702. [CrossRef]
27. Liu, K.; Cheng, P.; Fan, C.W.; Song, P.; Huang, Q.T. Evolutions of sedimentary facies and palaeoenvironment and their controls on the development of source rocks in continental margin basins: A case study from the Qiongdongnan Basin, South China Sea. *Pet. Sci.* **2023**, *20*, 2648–2663. [CrossRef]
28. Xu, C.; You, Y. North slope transition zone of Songnan-Baodao sag in Qiongdongnan Basin and its control on medium and large gas fields, South China Sea. *Pet. Explor. Dev.* **2022**, *49*, 1229–1242. [CrossRef]
29. Liu, L.; Liao, Z.; Chen, C.; Chen, J.; Niu, J.; Jia, Y.; Liu, T. A seabed real-time sensing system for in-situ long-term multi-parameter observation applications. *Sensors* **2019**, *19*, 1255. [CrossRef]
30. Hill, D.; Combee, C.; Bacon, J. Over/under acquisition and data processing: The next quantum leap in seismic technology? *First Break* **2006**, *24*, 81–92. [CrossRef]
31. Advocate, D.M.; Hood, K.C. An empirical time-depth model for calculating water depth, northwest Gulf of Mexico. *Geo-Mar. Lett.* **1993**, *13*, 207–211. [CrossRef]
32. Dorokhov, D.V.; Lugovoy, N.N.; Dorokhova, E.V.; Budanov, L.M.; Dudkov, I.Y. Morphology and origin of the palaeo cliff area in the Sambia Peninsula nearshore (SE Baltic Sea). *Quat. Int.* **2022**, *630*, 17–33. [CrossRef]
33. Khan, H.A. Numerical Analysis of Erosive Sea Cliff-A Case Study. Ph.D. Thesis, Politecnico di Torino, Torino, Italy, 2025.
34. Ercilla, G.; Vázquez, J.T.; Alonso, B.; Bárcenas, P.; Casas, D.; d'Acremont, E.; Gorini, C. Seafloor morphology and processes in the Alboran Sea. In *Alboran Sea-Ecosystems and Marine Resources*; Springer International Publishing: Cham, Switzerland, 2021; pp. 157–205.
35. Grobbelaar, M. A Comparison Between Diamictites at the Witteberg-Dwyka Contact in Southern South Africa. Master's Thesis, Stellenbosch University, Stellenbosch, South Africa, 2015.
36. Nomikou, P.; Papanikolaou, D.; Dietrich, V.J. Geodynamics and volcanism in the Kos-Yali-Nisyros volcanic field. In *Nisyros Volcano: The Kos-Yali-Nisyros Volcanic Field*; Dietrich, V.J., Lagios, E., Eds.; Springer: Cham, Switzerland, 2018; pp. 13–55.
37. Walsh, P.T.; Atkinson, K.; Boulter, M.C.; Shakesby, R.A. The Oligocene and Miocene outliers of west Cornwall and their bearing on the geomorphological evolution of Oldland Britain. *Philos. Trans. R. Soc. Lond. Ser. A* **1987**, *322*, 211–245. [CrossRef]
38. Zeng, F.; Wang, D.; Li, Z.; Wang, W.; Dai, X.; Sun, Y.; Zhang, P. The discovery of an active fault in the Qiongdongnan Basin of the northern South China Sea. *Mar. Pet. Geol.* **2024**, *163*, 106777. [CrossRef]

39. Harbitz, C.B.; Løvholt, F.; Bungum, H. Submarine landslide tsunamis: How extreme and how likely? *Nat. Hazards* **2014**, *72*, 1341–1374. [CrossRef]
40. Haridhi, H.A.; Huang, B.S.; Wen, K.L.; Mirza, A.; Rizal, S.; Purnawan, S.; Phung, V.B. Tsunami scenario triggered by a submarine landslide offshore of northern Sumatra Island and its hazard assessment. *Nat. Hazards Earth Syst. Sci.* **2023**, *23*, 507–523. [CrossRef]
41. Gogou, M.; Mavroulis, S.; Evelpidou, N.; Lekkas, E. Earthquakes and Tsunamis: Natural hazards over the Aegean Archipelago. *Tectonophysics* **2022**, *825*, 229502.
42. Ko, Y.Y.; Tsai, C.C.; Hwang, J.H.; Hwang, Y.W.; Ge, L.; Chu, M.C. Failure of engineering structures and associated geotechnical problems during the 2022 ML 6.8 Chihshang earthquake, Taiwan. *Nat. Hazards* **2023**, *118*, 55–94. [CrossRef]
43. Yan, G.; Gao, B.; Shen, Y.; Zheng, Q.; Fan, K.; Huang, H. Shaking table test on seismic performances of newly designed joints for mountain tunnels crossing faults. *Adv. Struct. Eng.* **2020**, *23*, 248–262. [CrossRef]
44. Liu, S.; Yang, K.; Zhang, T.; Tang, C. Rib spalling 3D model for soft coal seam faces with large mining height in protective seam mining: Theoretical and numerical analyses. *Geofluids* **2020**, *2020*, 8828844. [CrossRef]
45. Racey, A.; Ridd, M.F. Petroleum geology of the Rakhine region, Myanmar. In *Geology of Myanmar*; The Geological Society of London, Ed.; The Geological Society of London: London, UK, 2015; Volume 45, pp. 93–108.
46. Cukur, D.; Kim, S.P.; Kong, G.S.; Bahk, J.J.; Horozal, S.; Um, I.K.; Kim, J.K. Geophysical evidence and inferred triggering factors of submarine landslides on the western continental margin of the Ulleung Basin, East Sea. *Geo-Mar. Lett.* **2016**, *36*, 425–444. [CrossRef]
47. Lin, C.M.; Li, Y.L.; Zhuo, H.C.; Shurr, G.W.; Ridgley, J.L.; Zhang, Z.P.; Xue, T. Features and sealing mechanism of shallow biogenic gas in incised valley fills (the Qiantang River, eastern China): A case study. *Mar. Pet. Geol.* **2010**, *27*, 909–922. [CrossRef]
48. Cartwright, J.; Santamarina, C. Seismic characteristics of fluid escape pipes in sedimentary basins: Implications for pipe genesis. *Mar. Pet. Geol.* **2015**, *65*, 126–140. [CrossRef]
49. Liu, L.; Chu, F.; Wu, N.; Zhang, L.; Li, X.; Li, H.; Wang, X. Gas sources, migration, and accumulation systems: The shallow subsurface and near-seafloor gas hydrate deposits. *Energies* **2022**, *15*, 6921. [CrossRef]
50. Ren, Z.; Cui, J.; Qi, K.; Yang, G.; Chen, Z.; Yang, P.; Wang, K. Control effects of temperature and thermal evolution history of deep and ultra-deep layers on hydrocarbon phase state and hydrocarbon generation history. *Nat. Gas Ind. B* **2020**, *7*, 453–461. [CrossRef]
51. Li, W.; Alves, T.M.; Wu, S.; Völker, D.; Zhao, F.; Mi, L.; Kopf, A. Recurrent slope failure and submarine channel incision as key factors controlling reservoir potential in the South China Sea (Qiongdongnan Basin, South Hainan Island). *Mar. Pet. Geol.* **2015**, *64*, 17–30. [CrossRef]
52. Correa, T.B.; Eberli, G.P.; Grasmueck, M.; Reed, J.K.; Correa, A.M. Genesis and morphology of cold-water coral ridges in a unidirectional current regime. *Mar. Geol.* **2012**, *326*, 14–27. [CrossRef]
53. Agarwala, N. Using robotics to achieve ocean sustainability during the exploration phase of deep seabed mining. *Mar. Technol. Soc. J.* **2023**, *57*, 130–150. [CrossRef]

Journal of
Marine Science and Engineering

MDPI

Article

Influence of Sea Level Rise on Tidal Waves of the Yangtze River Estuary

Kai Yu [1], Wei Ding [2], Yunlin Ni [3], Wei Chen [3], Mingtao Jiang [4,*] and Zili Dai [4]

1 Zhejiang Dayang Bihai Planning and Design Co., Ltd., Hangzhou 310030, China; 18007402195@163.com
2 Zhoushan Ecological and Environmental Emergency Management and Monitoring Center, Zhoushan 316000, China; weiding@live.com
3 College of Marine Engineering Equipment, Zhejiang Ocean University, Zhoushan 316022, China; oceannyl@zjou.edu.cn (Y.N.); chenwei_112233@163.com (W.C.)
4 School of Mechanics and Engineering Science, Shanghai University, Shanghai 200400, China; zilidai@shu.edu.cn
* Correspondence: mtjiang@shu.edu.cn

Abstract: In order to study the influence of sea level rise (SLR) on tidal wave in tidal reach, a hydrodynamic numerical model of the Yangtze River Estuary was established using MIKE 21. The verified model was used to simulate tidal wave changes from Wuhu to the estuary under SLR scenarios of 0.5 m, 1.0 m, and 2.0 m. The results show the following: (1) The impact of SLR on water level within the tidal reach can be divided into two segments: from the estuary to approximately 35 km upstream of Xuliujing and from there further upstream to the tidal limit. (2) The average water level increase is less than the magnitude of SLR. The attenuation depends on both river discharge and location within the reach; higher river discharge and increasing proximity to inland areas reduce the influence of SLR. (3) Water level fluctuation amplitude increases with SLR, showing minimal sensitivity to river discharge within the estuarine section but significant sensitivity to discharge fluctuations in the upstream section. Finally, tidal current velocity generally increases in response to SLR, with the spatial extent of the increase controlled by river discharge.

Keywords: MIKE21; tidal reach of the Yangtze River; sea level rise; tidal wave change

1. Introduction

The Yangtze River Estuary (YRE) is the largest estuary in China, and the discharge is only less than that of the Amazon River and the Congo River, ranking third in the world. Under the combined influence of runoff and tide, the YRE has developed a unique morphology characterized by three distributaries and four passages to the sea [1]. Tidal reach is affected by both upstream runoff and downstream tidal action. Within this tidal reach, the tidal current limit refers to the point furthest upstream along the river where the tide encounters resistance from both the riverbed and the opposing downstream flow of freshwater. At this location, the velocity generated by the incoming tide counterbalances the river's flow velocity, causing the net upstream movement of tidal water to cease. It marks the upper boundary of significant tidal hydrodynamic influence. Further upstream, the tidal limit signifies where the tidal wave's energy is ultimately depleted, causing the tidal range to diminish to zero. It represents the absolute farthest extent of periodic tidal water level fluctuations. Both the tidal limit and the tidal current limit are dynamic boundaries, shifting in response to upstream discharge volume, tidal intensity, and seasonal variations.

The tidal limit of the Yangtze River is generally near Datong (117°37′ E, 30°46′ N), Anhui province [2,3]. After analyzing the hydrological data of the lower reaches of the Yangtze River from 2007 to 2016, Shi et al. concluded that the tidal limit extends up to approximately 900 km upstream of the estuary during extreme dry seasons and retreats to 640–665 km upstream during flood seasons [4]. The tidal current limit is near Jiangyin [5,6]. Under the combined influence of tide and runoff, the average water level within this tidal reach (from the estuary to the tidal limit) gradually rises further upstream, while the change in water level fluctuation amplitude decreases [7].

Sea level change in a certain region is mainly driven by changes in sea water mass (land glacier, storage loss), specific volume (sea water temperature, salinity), and seabed elevation. Sea level rise poses multifaceted threats, including the flooding of coastal zones [8,9], saltwater intrusion [10,11], and even an increased risk of geological hazards [12,13]. At the global scale, Cazenave et al. showed that about 30% of climate-induced sea level change is attributed to ocean temperature changes, while the melting of land glaciers accounted for about 55% [14]. Horton et al. predicted that under a global temperature rise of 4.5 °C in 2100, SLR would be more than 1 m [15]. In the East China Sea, specific volume is affected by Yangtze River runoff and Kuroshio current [16], playing a dominant role in the regional sea level change, which presents obvious periodic variations [17]. Significant anthropogenic alterations in the lower Yangtze basin have dramatically altered sediment properties, leading to intensified riverbed scour and increasing estuary water depth [18]. The morphological shift transformed the estuarine tidal wave from a dissipative progressive wave into a standing wave system, enhancing susceptibility to tidal amplification [19]. Broader mechanistic research includes Chen's analysis of dominant sea level change modes using AVISO satellite altimeter data, alongside modeling studies investigating long-term trends and their impacts on the North Pacific Ocean [20]. Focusing on the impact of the Yangtze River on the Yangtze estuary, Kuang et al. [21] adopted Mike 21 to establish a hydrodynamic model of the East China Sea, evaluating seasonal and long-term changes under characteristic discharge scenarios (75th percentile, median, 25th percentile and multi-year monthly average discharge during 1950–2011). Simulations indicate that a 20,000 m^3/s increase in Yangtze River discharge typically induced a 5–10 mm sea level rise across most areas of Hangzhou Bay.

A recent study on the environmental impacts of SLR mainly integrate field measurements with numerical models. Wang et al. used MIKE 21 to compare the effect of SLR, land subsidence, and water depth changes on the coastal environment in the Shanghai coastal area [22]. Liu applied a three-dimensional EFDC hydrodynamic model in the Pearl River Estuary, demonstrating that SLR intensifies estuarine water column stratification [23,24]. Projected salinity intrusion lengths under 1.0 m SLR would increase by 21.37, 9.64, 9.75, and 4.82 km at the Humen, Jiaomen, Hongqili, and Hengmen outlets, respectively. Chen simulated the influence of a 100 cm SLR on the ebb and flood duration within the upper YRE using a two-dimensional tidal model under flood, dry, and average annual discharge conditions [25]. Yan used the SPRC (source–pathway–receptor–consequence) model combined with MIKE 21 to assess the potential combined impacts of SLR and storm surges [26].

Previous studies have extensively addressed SLR effects on estuarine saline intrusion and tidal currents, few studies have specifically investigated the impacts of SLR on tidal wave propagation characteristics. In this paper, a two-dimensional hydrodynamic model of the Yangtze Estuary is established using MIKE 21 to examine changes in tidal wave propagation characteristics in the tidal reach of the Yangtze River under different discharge conditions.

2. Study Area

This study investigates the tidal-influenced reach of the Yangtze River and its adjacent coastal waters, a region characterized by complex interactions between the freshwater runoff and marine tidal dynamic. The Yangtze River discharges substantial freshwater that interacts with vigorous Kuroshio Current-influenced tides from the East China Sea. Anthropogenic activities, such as upstream damming (e.g., the Three Gorges Dam), have significantly reduced sediment supply to the estuary. This reduction has triggered alterations in sediment dynamics, leading to widespread erosion and channel bed incision, thereby modifying tidal wave characteristics. The study area extends from Wuhu, Anhui Province (~440 km upstream from the estuary mouth) eastward to the Yangtze's discharge point into the East China Sea and encompasses the entire Yangtze River Estuary (YRE), Hangzhou Bay, Zhoushan Archipelago, and adjacent East China Sea waters. Figure 1 shows the detailed study area, key locations, and monitoring stations.

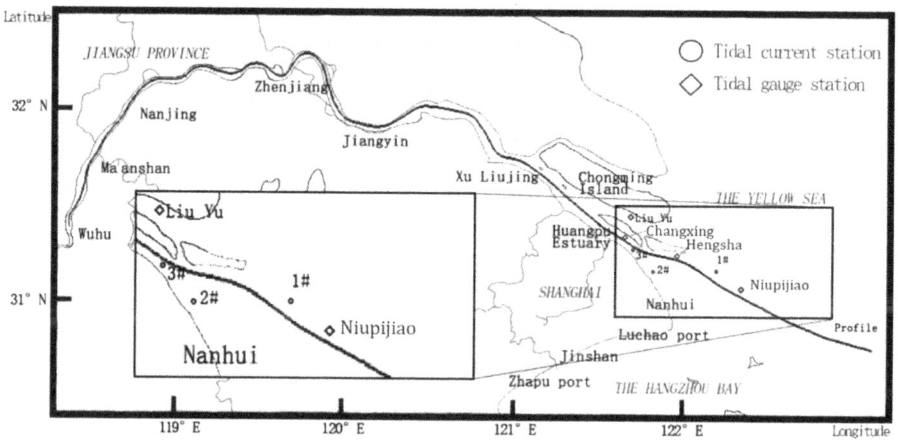

Figure 1. Study area: distribution of hydrodynamic monitoring stations and analysis profile.

3. Model Setup

3.1. Model Introduction

MIKE is a modeling system developed by the Danish Hydraulic Institute (DHI), which has strong post-processing capability [27]. It provides a comprehensive framework integrating specialized numerical modules, and is suitable for hydrodynamic environment simulation across diverse conditions.

3.2. Computational Domain

The model extends from Wuhu of the Yangtze River in the west to longitude 125° E in the east, to Lianyungang in the north, and to latitude 27.4° N in the south, including Hangzhou Bay, Zhoushan Islands, and other sea areas, covering the main tidal reaches of the lower reaches of the Yangtze River. To minimize boundary effects, the lateral width of the domain is about 653 km, and the longitudinal length is about 740 km, covering a computational area of approximately 273,135 km^2. Due to the long and tortuous coastline of the YRE and the numerous islands with complex terrain in the Zhoushan sea area, an unstructured triangular mesh was employed, which enables the model to simulate the shoreline situation more realistically. The g mesh was optimized with resolutions ranging from 500 to 24,000 m to balance computational efficiency and accuracy, with higher resolution applied in the estuary area. As shown in Figure 2, the model contains 18,114 nodes and 33,544 grids.

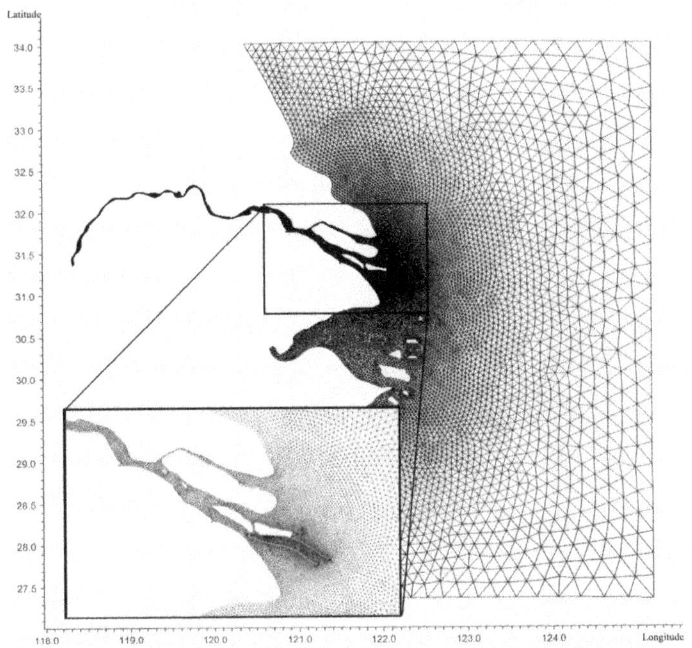

Figure 2. The computational domain and grids.

The accuracy of underwater terrain data is critical for model calculation. A wide-range underwater topography across the computation domain was obtained through the GIS digitization of charts provided by the Naval Aviation Protection Department. River depths were assigned based on navigational chart data, with all elevation data referencing the National 1985 Elevation Datum [28].

3.3. Boundary Conditions and Parameter Settings

The time step of the model was automatically adjusted in the range of 0.01~30 s. The Courant number limit was set to 0.8. Bed roughness was represented by the Manning coefficient, which varied with water depth within a range of 20 to 80. A discharge boundary condition was applied at the upstream river boundary. Tidal level forcing at the open boundaries was derived from the MIKE 21 Tide Prediction Tool, encompassing 10 tidal constituents: M_2, S_2, K_2, N_2, S_1, K_1, O_1, P_1, Q_1, and M_4. Land and island boundaries were treated as impermeable, with a free-slip condition applied for water particles tangent to the coastline.

The horizontal eddy viscosity coefficient was calculated using the Smagorinsky formulation, which accounts for subgrid-scale effects. The expression is $A = c_s^2 l^2 \sqrt{2S_{ij}S_{ij}}$, where Cs is a constant, and mixing length l is calculated as $l = C_s \cdot \sqrt{\Delta x \cdot \Delta y}$ (where Δx and Δy are grid cell dimensions), and S_{ij} is defined by $S_{ij} = \frac{1}{2}\left(\frac{\partial u_i}{\partial x_j} + \frac{\partial u_j}{\partial x_i}\right)(i, j = 1, 2)$. Additionally, the Coriolis force varies with latitude, calculated as $f = 2\omega sin\varphi$ (where ω is the angular velocity of Earth's rotation, and ϕ is the geographic latitude).

In this study, a modified model of the static water depth was adopted to simulate the influence of the rising sea level in the East China Sea on the tidal reach of the Yangtze River.

To minimize the impact of the initial static water depth increase on simulated river levels, a cold start was employed, with the initial water level set uniformly to −5 m.

3.4. Model Calibration and Verification

The hydrodynamic model was calibrated against field observations before validation to ensure accurate tidal dynamics within the study domain. Tidal level and current velocity/direction measurements were used to optimize key parameters, with a primary focus on the spatially distributed bed roughness coefficient. Initial Manning values were assigned based on sedimentological properties and established research values, with subsequent localized adjustments based on bathymetric and seabed sediment distribution variations. Additional parameters, including the horizontal eddy viscosity coefficient, were determined using established empirical formulations. The model employed a wet–dry moving boundary scheme. The critical depth [29] for dry, flood, and wet points was set to 0.005 m, 0.05 m, and 0.1 m, respectively.

Measured tidal level, velocity, and direction data from four tidal stations and three flow stations were used for model validation. Station locations are shown in Figure 1. Tidal levels at Liuyu and Hengsha (tidal stations) were validated using extended observational records spanning 14–24 August 2005. To broaden the validation period coverage and include the Yangtze Estuary–Yellow Sea confluence zone, tidal level data from Niupijiao and Changxing, along with velocity and direction data from stations #1–#3 during the spring tide period (5–6 May 2005), were also included.

Figure 3 shows the tidal level verification results. As shown in the figure, the calculated low tide levels at the Liuyu, Hengsha, Changxing, and Niupijiao stations are slightly higher than the measured values. The calculated high tide level at the Liuyu station is slightly lower than the measured value. Overall, the verification accuracy is better for short-term sequences than for long-term sequences. Figure 4 presents the current velocity and direction verification process for stations #1–#3. The results show that the calculated maximum flood current velocity at #2 is less than the measured value; the calculated maximum flood current velocities at stations #1 and #3 are also slightly less than the measured values. For the remaining parts, the measured current velocities and directions match the calculated results well. Due to limitations in onsite measurements, tidal and seasonal velocity verification for the upstream tidal reach is currently not feasible. This study employed typical flow scenarios for subsequent simulations. In general, the verification results indicate that this model can effectively simulate the fundamental characteristics of tidal current movement in the Yangtze Estuary.

The model performance was quantitatively assessed using the skill model proposed by Wilmott [30]. The skill model is formulated as below:

$$d = 1 - \frac{\sum_{i=1}^{N} (P_i - O_i)^2}{\sum_{i=1}^{N} \left[\left| P_i - \overline{O} \right| + \left| O_i - \overline{O} \right| \right]^2} \tag{1}$$

where P_i is the simulated value, O_i is the value of the in situ observed data, $\overline{O}$ is the average value of the observed data, and N is the number of i observations. A skill value of 1.0 indicates perfect model performance, excellent between 0.65 and 1, very good in the range of 0.5–0.65, good in the range of 0.2–0.5, and poor when less than 0.2 [31]. According to the calculated results Table 1, the validation generally supports the applicability of the hydrodynamic model in the YRE.

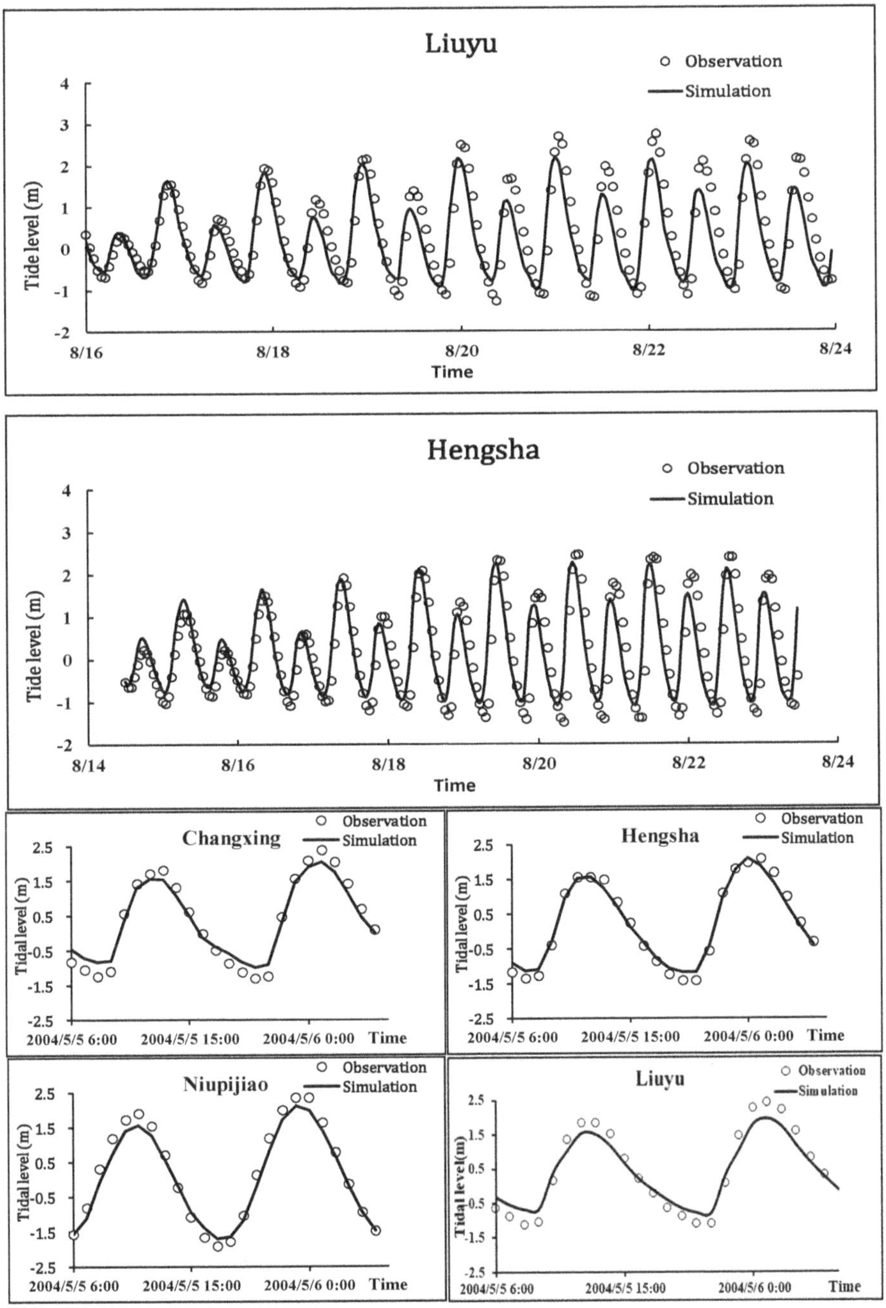

Figure 3. Tide level verification.

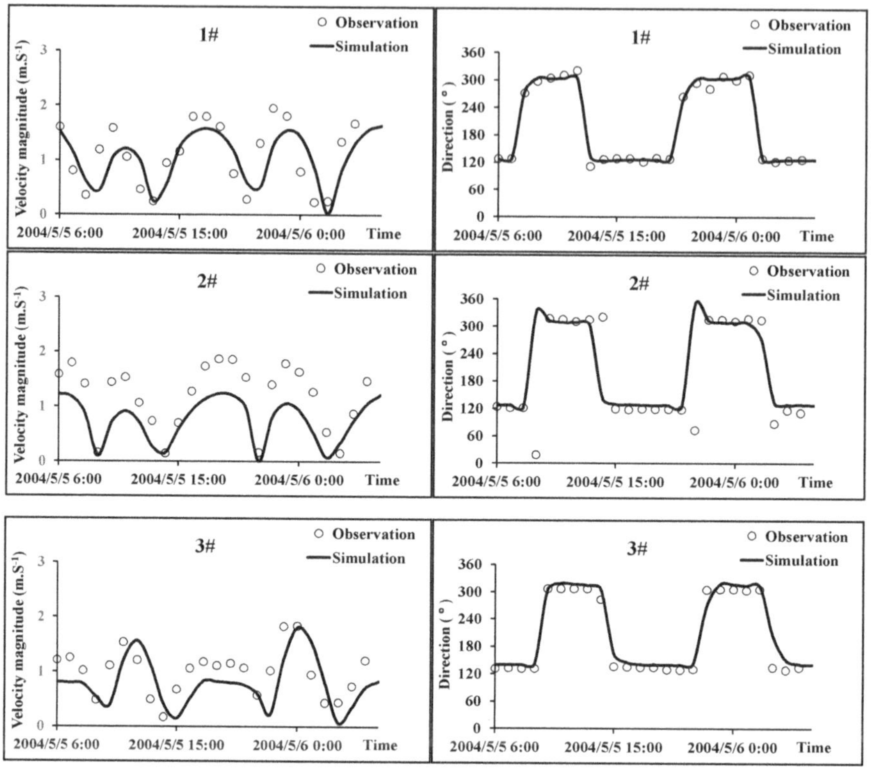

Figure 4. Current verification.

Table 1. The index of agreement.

Quantity	Stations	*d*	
Tidal level	Niupijiao Liuyu	0.9812 0.9768	Excellent Excellent
	Changxing Hengsha	0.9852 0.9751	Excellent Excellent
Flow velocity	1# 2# 3#	0.9449 0.9394 0.9457	Excellent Excellent Excellent
Flow direction	1# 2# 3#	0.9976 0.7676 0.9855	Excellent Excellent Excellent

4. Computational Conditions Analysis

Yangtze River discharge exhibits significant seasonal variation between flood and dry seasons. The discharge increases significantly from May to October, accounting for 71.7% of the annual discharge, and decreases from November to April of the following year, accounting for 28.3%.

Based on China Sea Level Bulletin 2019, the rate of SLR along the coastal area of China is 3.4 mm/a from 1980 to 2019, and it is expected that in the next 30 years, the coastal sea level of Zhejiang province will rise by 51–179 mm [32]. Cheng et al. believed that due to the superimposed effects of channel erosion in estuaries, including land reclamation projects in estuaries and water level rise caused by marine construction projects, the observed actual sea level rise would be larger than that of the predicted results [33]. The IPCC 2013 assessment projected approximately 1 m of cumulative sea level rise from 1700 to 2100 under the RCP 8.5 scenario [34]. Based on the historical relationship between global average land surface temperature and sea level change rate, Rahmstorf integrated the semi-empirical model of global average land surface temperature, concluding that the global sea level rise in 2100 could reach a maximum of 2.05 m [35]. Based on various predictions of SLR, this study simulated the effects of an SLR of 0.5 m, 1.0 m, and 2.0 m on tidal waves in the YRE.

Based on the observed discharge data of the Yangtze River, combined effects of SLR and river discharges on tidal waves in the YRE were investigated using the maximum flood peak discharge, minimum dry discharge, and annual average discharge in the observation period. The time of the three discharge conditions corresponds, respectively, to the monthly average flow from July to September in the year of 1954 (76,900 m^3/s), the monthly average flow from January to March in the year of 1963 (7800 m^3/s), and the multi-year average flow from 1950 to 2000 (28,560 m^3/s).

5. Results and Discussions

Based on the numerical results, water level data in May 2004 (the section position of which can be seen in Figure 1) were extracted. The average, maximum, and minimum water level values under different SLR scenarios at the section nodes are calculated. The maximum water level fluctuation amplitude, the ratio R of average water level to fluctuation amplitude, and the change in tidal current along the river are computed [36]. The harmonic analysis of tidal levels in Shanghai, Xuliujing, Jiangyin, Zhenjiang, Nanjing, and Wuhu was carried out to analyze the amplitude η of the M_2, S_2, K_1, O_1, and M_4 tidal constituents and calculate the tidal wave attenuation rate D [37]. The formula is as follows:

$$D = \frac{2(\eta_{i+1} - \eta_i)}{(\eta_{i+1} + \eta_i)\Delta x} \quad (2)$$

where η_i is the sum of the amplitude value of partial tide in section nodes, namely $\eta_i = \eta_{M_2} + \eta_{S_2} + \eta_{K_1} + \eta_{O_1} + \eta_{M_4}$. Δx is the distance between two locations.

To clarify spatial relationships between stations, Wuhu is set as the reference point (0 km point), with along-estuary coordinates increasing eastward (Table 2).

Table 2. Main station locations and distances from Wuhu.

	Wuhu	Nanjing	Zhenjiang	Jiangyin	Xuliujing	Nanhui
Distance from Wuhu (km)	0	87	183	278	356	488

5.1. Changes in Tidal Waves

5.1.1. Effects of SLR Under Multi-Year Average Discharge

Figure 5 shows the average water level and corresponding water level changes along the river for different SLR scenarios (relative to the 0 m SLR baseline) under multi-year average discharge. Average water level denotes the tidal cycle averaged water surface elevation, referencing the local MSL in tidal-dominated zones and the Wusong Datum upstream, capturing both tidal and flow influences. Under multi-year average discharge conditions, the average water level gradually rises upstream. A rapid increase in average water level at approximately 95 km can be observed, along with another section exhibiting a rapid increase near Jiangyin. Sea level rise elevates the average water level throughout the tidal reach. Furthermore, the increase in the water level diminishes progressively upstream. At Wuhu, the average water level rise is only about 50% of the sea level rise.

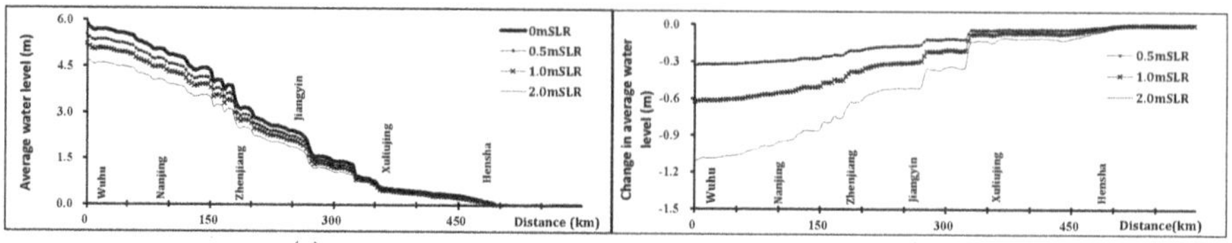

(a) (b)

Figure 5. Average water level and change in average water level for different SLR scenarios under multi-year average discharge (**a**) Average water level, (**b**) change in average water level.

Figure 6 shows the maximum water level fluctuation amplitude and its corresponding changes for different SLR scenarios under multi-year average discharge. The maximum water level fluctuation amplitude gradually decreases upstream within the YRE, with a significant point of decrease at a point 315 km downstream from Wuhu (near Jiangyin). Changes in these fluctuation amplitudes are proportional to the SLR magnitude. Within the reach between Xuliujing and Jiangyin, the amplitudes decrease sharply to a minimum and then increase sharply to a secondary peak. Subsequently, the fluctuation amplitudes decrease gradually upstream from Jiangyin.

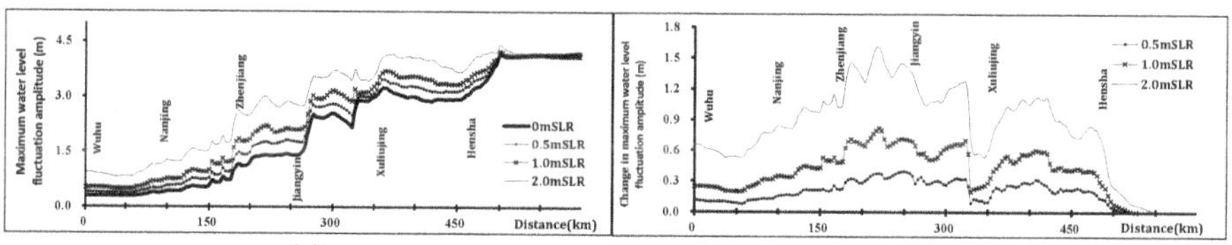

(a) (b)

Figure 6. Maximum water level and change in maximum water level for different SLR scenarios under multi-year average discharge. (**a**) Maximum water level fluctuation amplitude, (**b**) Change in maximum water level fluctuation amplitude.

5.1.2. Effects of SLR Under Maximum Flood Discharge

Figure 7 shows the average water level and its change under different SLR scenarios. Under maximum discharge, the spatial pattern of the average water level change along the estuary is similar to that under average annual discharge (Figure 5), but the changes are more significant. For 0.5 m and 1 m SLR, the amplitude of water level fluctuation tends to flatten in the upper reaches near Jiangyin. In contrast, under the 2 m SLR scenario, the amplitude continues to fluctuate and decrease along the reach from Jiangyin to Nanjing, stabilizing only in the upper reaches of Nanjing. Furthermore, the water level rise observed in Nanjing's upper reaches accounts for only about 20% of the 2 m SLR.

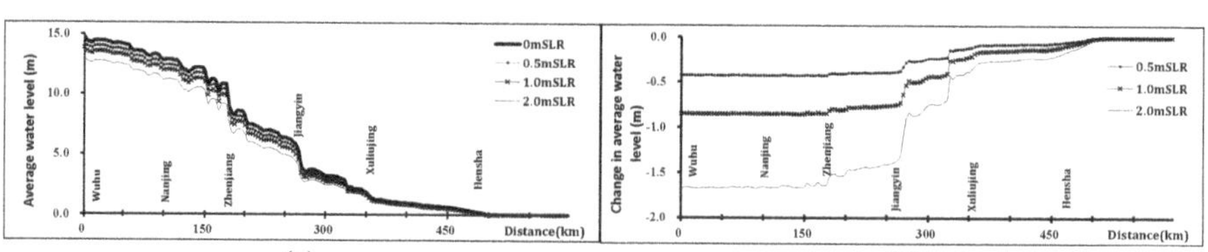

(a) (b)

Figure 7. Average water level and change in average water level for different SLR scenarios under peak flood discharge. (**a**) Average water level, (**b**) change in average water level.

Figure 8 shows the absolute maximum water level fluctuation amplitude and its change under different SLR scenarios. Under the maximum peak flood discharge, the absolute maximum water level fluctuation amplitude (Figure 8a) gradually decreases upstream along the Yangtze River (from Hensha to Wuhu). Upon entering the Yangtze Estuary region from downstream, the change in amplitude increases gradually. The fluctuation reaches its first peak near 400 km. The amplitude then peaks again near 315 km (Jiangyin). Further upstream, both the absolute maximum amplitude and its change decrease progressively (Figure 8b), approaching zero in the upper Nanjing reaches.

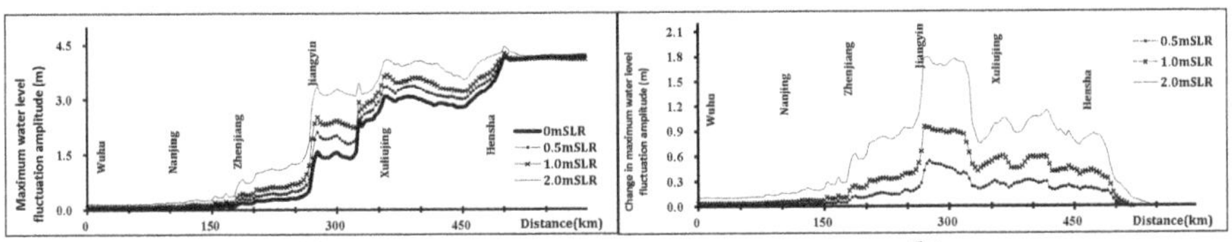

(a) (b)

Figure 8. Maximum water level and change in maximum water level for different SLR scenarios under peak flood discharge. (**a**) Maximum water level fluctuation amplitude, (**b**) change in maximum water level fluctuation amplitude.

5.1.3. Effects of SLR Under Minimum Dry Discharge

Figure 9 shows the average water level and its change under different SLR scenarios. In the case of minimum dry discharge, SLR induces water level rise along the tidal reach, which is largely equivalent to the magnitude of SLR. A sharp increase in water level occurs approximately 35 km upstream from Xuliujing. Beyond this point, the water level increase induced by SLR diminishes sharply. Under the influence of SLR, the water level change in the estuary decreases at first and then increases. Upstream from Xuliujing, the change in the water level decreases rapidly and stabilizes further upstream.

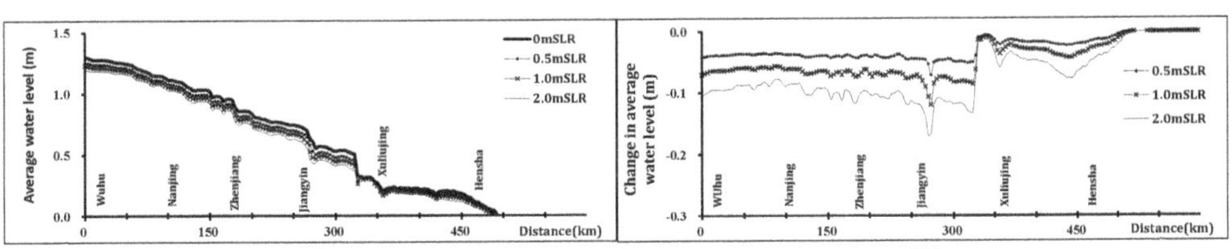

(a) (b)

Figure 9. Average water level and change in average water level for different SLR scenarios under minimum dry discharge. (**a**) Average water level, (**b**) change in average water level.

Figure 10 shows the maximum water level fluctuation amplitude and its change under different SLR scenarios. The maximum amplitude fluctuates and decreases upward along the river, with rapid (though not large) decreases observed near Jiangyin and approximately 35 km upstream of Xuliujing. The magnitude of the maximum amplitude change is proportional to the magnitude of SLR. At about 35 km upstream of Xuliujing, the amplitude change increases sharply and remains at this range until the second peak occurs between Jiangyin and Zhenjiang. Beyond this point, both the water level fluctuation amplitude and its change decrease progressively upstream.

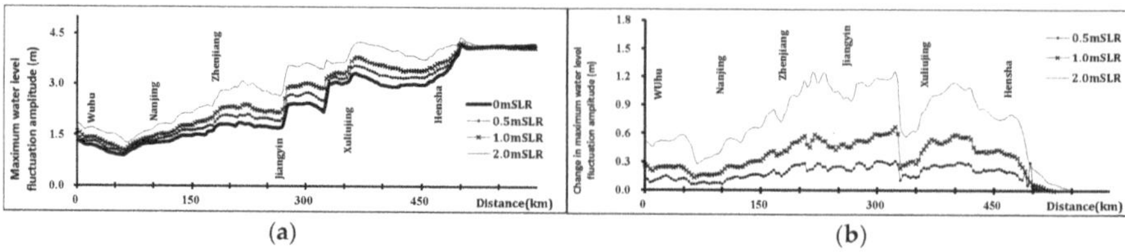

Figure 10. Maximum water level and change in maximum water level for different SLR scenarios under minimum dry discharge. (**a**) Maximum water level fluctuation amplitude, (**b**) change in maximum water level fluctuation amplitude.

5.1.4. Changes in Ratio of Average Water Level to Maximum Water Level Fluctuation Amplitude

Figure 11 shows the ratio (R) of the average water level to the maximum water level fluctuation amplitude. Under SLR scenarios, this ratio decreases in the tidal reach as sea level rises, but its specific value is significantly influenced by the river discharge.

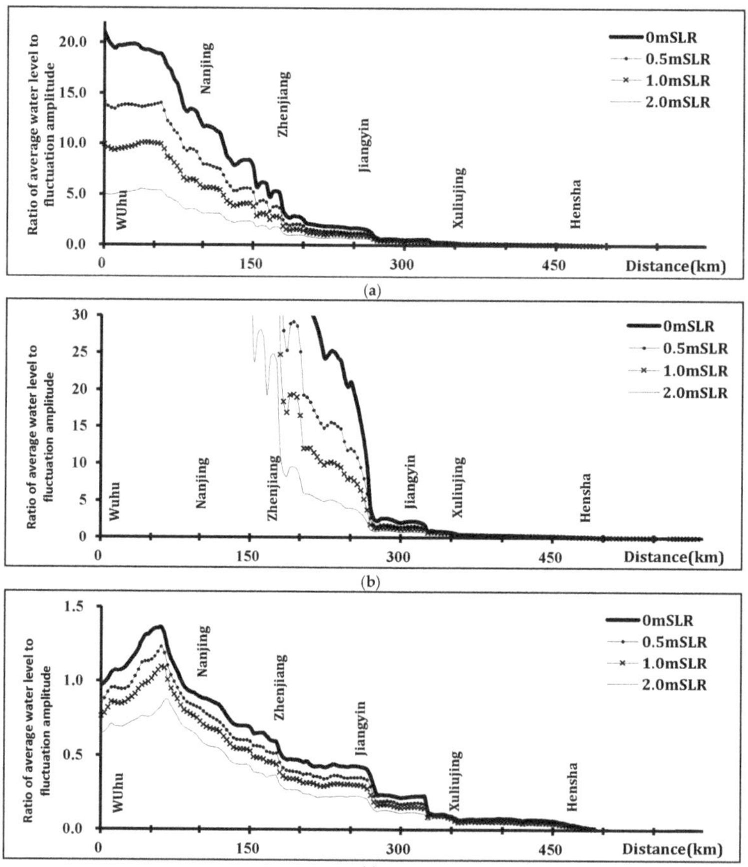

Figure 11. Ratio of average water level to maximum water level fluctuation amplitude under different discharges and SLR scenarios. (**a**) Ratio of average water level to maximum water level fluctuation amplitude under average annual discharge. (**b**) Ratio of average water level to maximum water level fluctuation amplitude under maximum peak discharge. (**c**) Ratio of average water level to maximum water level fluctuation amplitude under minimum dry discharge.

Under the multi-year average discharge, the ratio changes relatively smoothly in the river section downstream of Zhenjiang, while in the upstream section, it exhibits a fluctuating upward trend. The magnitude of this increase diminishes as sea level rises, as shown in Figure 11a.

Under maximum flood discharge conditions, this ratio is significantly affected by SLR at 270 km downstream from Wuhu. Under the original sea level, it rapidly increases from 2 times to 20. However, under the 2 m SLR scenario, the magnitude of change significantly narrows, increasing only from 1 to 4, as shown in Figure 11b. Changes in the ratio at 330 km downstream from Wuhu are relatively smaller and primarily influenced by river discharge.

Under minimum dry-flow conditions, this ratio within the study area is consistently less than 1.5, as shown in Figure 11c. The peak occurs at approximately 65 km downstream from Wuhu.

5.2. Changes in Tidal Current Velocity Within the Tidal Reach

Figure 12 shows the percentage change in tidal current velocity magnitude under different river discharge and SLR scenarios. The amplitude of tidal current change correlates positively with the magnitude of SLR, while the shape of the change curve is determined by river discharge conditions. Moving upstream from the estuary, the velocity magnitude increases slightly at 500 km mark but decreases slightly between approximately 400 and 430 km. SLR-induced increases in velocity magnitude generally remain below 60%.

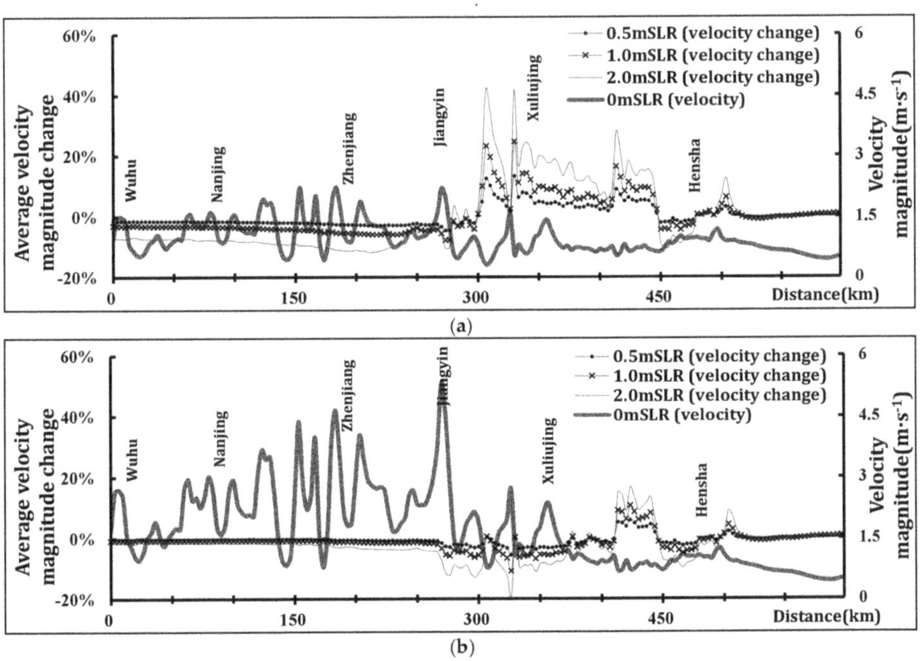

Figure 12. *Cont.*

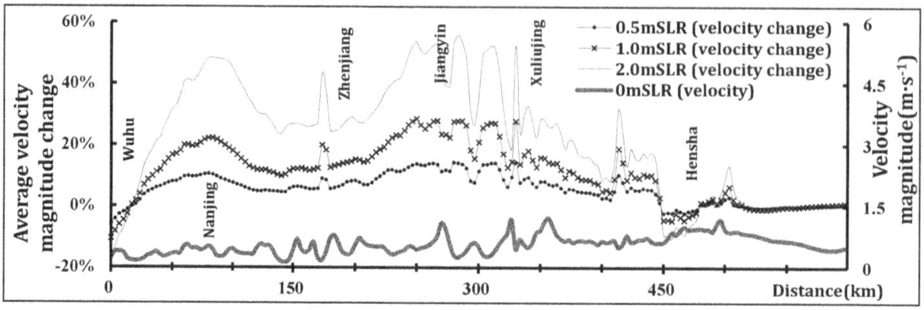

(c)

Figure 12. Changes in tidal current under different discharges and SLR scenarios. (**a**) Change in tidal current velocity under multi-year average discharge. (**b**) Change in tidal current velocity under maximum peak discharge. (**c**) Change in tidal current velocity under minimum dry discharge.

Under the annual average discharge scenario (Figure 12a), the change in velocity magnitude correlates positively with SLR, showing fluctuating increases along the 350–400 km reach. Within the subsequent 300–350 km reach, two peaks of 40% change occur, but the magnitude rapidly decreases to zero at 300 km downstream from Wuhu. Upstream of 300 km, velocity changes become negligible, except a maximum reduction of −12% can be observed at 225 km under 2 m SLR.

Under maximum flood flow conditions (Figure 12b), velocity changes upstream of 400 km are negative (0% to −20%) with irregular fluctuations. The magnitude of this decrease correlates negatively with SLR magnitude (larger SLR leads to larger velocity reduction). Changes gradually converge around 250 km and stabilize by 150 km, showing a 1.6% reduction in velocity under 2 m SLR.

During minimum dry-flow conditions, tidal velocity magnitude changes significantly with SLR. Across most of the reach, velocity change positively correlates with SLR (greater SLR increases velocity). Under 2 m SLR, velocity magnitude increases by over 20% between 100 and 350 km, peaking at 54%. However, after reaching its maximum at 100 km, velocity magnitude rapidly decays upstream, dropping to zero at 20 km downstream from Wuhu. Further upstream, velocity changes become negative, with the magnitude of reduction also positively correlated with SLR (larger SLR leads to larger velocity reduction).

5.3. Tidal Wave Attenuation

The attenuation of tidal wave amplitudes under different SLR conditions obtained from tidal harmonic analysis is shown in Table 3. Tidal energy within the river channel is attenuated by bottom friction and river discharge, while SLR reduces the bottom friction of the channel.

The tidal wave attenuation rate in the Wuhu–Xuliujing reach is significantly reduced by SLR. It is worth noting that the tidal wave attenuation rate increases in the Xuliujing–estuary area under 0.5 m and 1 m SLR, but decreases under 2 m SLR. In the Zhenjiang–Xuliujing River section, the decrease in attenuation rate caused by SLR is more obvious than that in the Wuhu–Zhenjiang River section. The decrease in the former reach is about 1.3×10^{-6}, 2.4×10^{-6}, and 4.2×10^{-6} under an SLR of 0.5 m, 1 m, and 2 m, respectively. In comparison, the decrease in the latter is only about 0.4×10^{-6}, 0.9×10^{-6}, and 2.1×10^{-6}. The reduction in the Wuhu–Zhenjiang reach is about half of that in the Zhenjiang–Xuliujing reach.

Table 3. The attenuation rate (D) of tidal wave under different SLR scenarios.

Reach	D ($\times 10^{-6}$, m^{-1})			
	0 m	0.5 m	1.0 m	2.0 m
Shanghai–Xuliujing	−0.06	−0.12	−0.12	−0.10
Xuliujing–Jiangyin	−8.76	−7.23	−6.05	−4.36
Jiangyin–Zhenjiang	−10.32	−9.24	−8.20	−6.41
Zhenjiang–Nanjing	−10.85	−10.34	−9.76	−8.60
Nanjing–Wuhu	−3.53	−3.28	−2.82	−1.49

6. Conclusions

Based on MIKE21 2015, a two-dimensional hydrodynamic numerical model of the YRE was established to study the impact of SLR on tidal waves in the tidal reach under different river discharges. The main conclusions are as follows:

(1) The influence of SLR on the study area can be divided into two parts: the Shanghai–Xuliujing and Xuliujing–Wuhu reaches. The impact undergoes a significant transition near Xuliujing.

(2) Under SLR, the change in the flow velocity and water level fluctuation amplitude first increases and then decreases in the Xuliujing–Shanghai area, with the greatest impact occurring in the channel between Chongming Island and Shanghai. Within the Wuhu–Xuliujing area, as the tidal wave propagates along the river, the water level increases while the water level fluctuation amplitude decreases. River water level fluctuation amplitude increases dramatically due to SLR, posing a threat to coastal environments.

(3) The tide limit shifts upstream due to SLR, extending approximately 5.71 km under a 2 m SLR. The tidal wave attenuation rate within the Xuliujing–Shanghai reach increases with SLR, while the attenuation rate within the Wuhu–Xuliujing reach decreases gradually.

Author Contributions: K.Y.: writing of original draft and conceptualization. W.D.: data analysis and visualization. Y.N.: data analysis, writing, and editing. W.C.: funding and conceptualization. M.J.: methodology, writing, and review. Z.D.: writing, review, and editing. All authors have read and agreed to the published version of the manuscript.

Funding: This research was financially sponsored by the National Natural Science Foundation of China (No. 42406155).

Data Availability Statement: The datasets supporting the conclusions of this article are available from the corresponding author upon reasonable request.

Conflicts of Interest: Author Kai Yu was employed by the company Zhejiang Dayang Bihai Planning and Design Co., Ltd. The remaining authors declare that the research was conducted in the absence of any commercial or financial relationships that could be construed as a potential conflict of interest.

References

1. Kuang, C.; Chen, W.; Gu, J.; He, L. Response of offshore sea level to river discharge in flood season in Yangtze Estuary. *J. Tongji Univ.* **2015**, *43*, 1266–1272.
2. Hou, C. *Numerical Study on the Tidal Current Limit and Tidal Limit in the Yangtze River and the Response of Saltwater Intrusion in the Changjiang Estuary to the River Discharge Change*; East China Normal University: Shanghai, China, 2013.
3. Huang, J.; Zhang, X.; Wu, Z.; Liu, F.; Cai, H.; Yang, Q. Investigation into the spatial and temporal tide-river dynamics and the underlying controlled factors along the tidal reach of the Changjiang River. *Acta Oceanol. Sin.* **2020**, *42*, 25–35.
4. Shi, S.; Cheng, H.; Xuan, X.; Hu, F.; Yuan, X.; Jiang, Y.; Zhou, Q. Fluctuations in the tidal limit of the Yangtze River estuary in the last decade. *Sci. China Earth Sci.* **2018**, *61*, 1136–1147. [CrossRef]

5. Lu, C.T.; Luo, X.F.; Chen, Z.C. Study of current limit causing by runoff and tidal range in Yangtze River. *Eng. J. Wuhan Univ.* **2016**, *49*, 201–205.
6. Xu, H.X.; Fan, L.F.; Gu, M.J. On tidal mark and tidal current mark in the Yangtze River. *Port Waterw. Eng.* **2012**, *6*, 15–20.
7. Buschman, F.A.; Hoitink, A.J.; Van Der Vegt, M.; Hoekstra, P. Subtidal water level variation controlled by river flow and tides. *Water Resour. Res.* **2009**, *45*, W10420. [CrossRef]
8. Nghiningwa, N.A.; Adelekan, O.I.; Mshelia, H.Z. Shoreline change, sea level rise and the impacts along the coastline of Walvis Bay, Namibia. *Ocean. Coast. Manag.* **2025**, *266*, 107690.
9. Lepesant, G. Between hard protection measures, nature-based solutions, and managed retreat: Adapting coastal areas to sea level rise in the Netherlands and France. *Z. Für Vgl. Polit.* **2024**, *18*, 141–156.
10. Mueller, W.; Zamrsky, D.; Essink, G.O.; Fleming, L.E.; Deshpande, A.; Makris, K.C.; Wheeler, B.W.; Newton, J.N.; Narayan, K.V.; Naser, A.M.; et al. Saltwater intrusion and human health risks for coastal populations under 2050 climate scenarios. *Sci. Rep.* **2024**, *14*, 15881. [CrossRef]
11. Feist, S.E.; Hoque, M.A.; Islam, M.A.; Dewan, A.; Fowler, M. Sea-level rise drives changes in salinisation patterns in low-lying Bangladesh. *Ecol. Indic.* **2025**, *176*, 113543. [CrossRef]
12. Ballantyne, C.K.; Dawson, S.; Dick, R.; Fabel, D.; Kralikaite, E.; Milne, F.; Sandeman, G.F.; Xu, S. The coastal landslides of Shetland. *Scott. Geogr. J.* **2018**, *134*, 71–96. [CrossRef]
13. Dai, Z.L.; Li, X.F.; Lan, B.S. Three-dimensional modeling of tsunami waves triggered by submarine landslides based on the smoothed particle hydrodynamics method. *J. Mar. Sci. Eng.* **2023**, *11*, 2015. [CrossRef]
14. Cazenave, A.; Llovel, W. Contemporary sea level rise. *Annu. Rev. Mar. Sci.* **2010**, *2*, 145–173. [CrossRef]
15. Horton, B.P.; Rahmstorf, S.; Engelhart, S.E.; Kemp, A.C. Expert assessment of sea-level rise by AD 2100 and AD 2300. *Quat. Sci. Rev.* **2014**, *84*, 1–6. [CrossRef]
16. Li, Q.; Mao, X.; Huthnance, J.; Cai, S.; Kelly, S. On internal waves propagating across a geostrophic front. *J. Phys. Oceanogr.* **2019**, *49*, 1229–1248. [CrossRef]
17. Vivier, F.; Kelly, K.A.; Harismendy, M. Causes of large-scale sea level variations in the Southern Ocean: Analyses of sea level and a barotropic model. *J. Geophys. Res. Ocean.* **2005**, *110*, C9014.
18. Zhou, J.J.; Zhang, M. Effect of dams on the regime of the mid-lower Yangtze River runoff and countermeasures. *J. Lake Sci.* **2018**, *30*, 1471–1488. [CrossRef]
19. Holleman, R.C.; Stacey, M.T. Coupling of sea level rise, tidal amplification, and inundation. *J. Phys. Oceanogr.* **2014**, *44*, 1439–1455. [CrossRef]
20. Chen, M.X. *Dynamics of Sea Level Variation in the North Pacific Ocean Kuroshio Region of East Sea of China and Kuroshio Extension*; Ocean University of China: Qingdao, China, 2009.
21. Kuang, C.; Chen, W.; Gu, J.; Su, T.C.; Song, H.; Ma, Y.; Dong, Z. River discharge contribution to sea-level rise in the Yangtze River Estuary, China. *Cont. Shelf Res.* **2017**, *134*, 63–75. [CrossRef]
22. Wang, J.; Yi, S.; Li, M.; Wang, L.; Song, C. Effects of sea level rise, land subsidence, bathymetric change and typhoon tracks on storm flooding in the coastal areas of Shanghai. *Sci. Total Environ.* **2018**, *621*, 228–234.
23. Liu, Z.H.; Hong, B. Influence of sea level rise on the saltwater intrusion of the east four outlets of the pearl river estuary. *Pearl River* **2019**, *40*, 43–49+81.
24. Liu, Z.H. *Numerical Study on the Effects of Sea Level Rise on Saltwater Intrusion and Substances Transport in the Pearl River Estuary*; South China University of Technology: Guangzhou, China, 2019.
25. Chen, Y.P.; Huang, H.M.; Wang, Y.G.; Tang, S.T. Impact of sea level rising on flood and ebb duration difference of the Yangtze River estuary. *Port Waterw. Eng.* **2019**, *5*, 42–47.
26. Yan, B.Y. *Assessment of Socio-Economic Vulnerability of the Megacity of Shanghai to Sea-Level Rise and Associated Storm Surges*; East China Normal University: Shanghai, China, 2016.
27. DHI. *MIKE21 and MIKE3 Flow Model FM-Hydrodynamic and Transport Module Scientific Documentation*; DHI Water and Environment: Hørsholm, Demark, 2014.
28. Jiangsu Maritime Safety Administration of the People's Republic of China. *Regulation on Ship Routing System in Jiangsu Section of Yangtze River*; Jiangsu Maritime Safety Administration of the People's Republic of China: Nanjing, China, 2013.
29. Casulli, V.; Cheng, R.T. Semi-implicit finite difference methods for three-dimensional shallow water flow. *Int. J. Numer. Methods Fluids* **1992**, *15*, 629–648. [CrossRef]
30. Willmott, C.J. On the validation of models. *Phys. Geogr.* **1981**, *2*, 184–194. [CrossRef]
31. Moriasi, D.N.; Arnold, J.G.; Van Liew, M.W.; Bingner, R.L.; Harmel, R.D.; Veith, T.L. Model evaluation guidelines for systematic quantification of accuracy in watershed Simulations. *Trans. ASABE* **2007**, *50*, 885–900. [CrossRef]
32. Department of Marine Early Warning and Monitoring, Ministry of Natural Resources. China Sea Level Communique 2019. Available online: https://coast.hhu.edu.cn/_upload/article/files/f8/60/c2d410124827ab4d83e4f5be3147/75fb99bc-64f9-4b0a-a4cf-6cb1b0baf0e4.pdf (accessed on 17 May 2025).

33. Cheng, H.Q.; Wang, D.M.; Chen, J.Y. Study and prediction of the relative sea level rise in 2030 in Shanghai area. *Progress. Inquisitiones de Mutat. Clim.* **2015**, *11*, 231–238.
34. Pachauri, R.K.; Allen, M.R.; Barros, V.R.; Broome, J.; Cramer, W.; Christ, R.; Church, J.A.; Clarke, L.; Dahe, Q.; Dasgupta, P.; et al. *Climate Change 2014: Synthesis Report. Contribution of Working Groups I, II and III to the Fifth Assessment Report of the Intergovernmental Panel on Climate Change*; IPCC: Geneva, Switzerland, 2014.
35. Rahmstorf, S. A semi-empirical approach to projecting future sea-level rise. *Science* **2007**, *315*, 368–370. [CrossRef]
36. Zhang, X.N.; Chen, C.Y.; Liu, H. Studying on dividing and connecting of the hydrology standard between river and sea navigation engineering. *Adv. Water Sci.* **2015**, *26*, 707–713.
37. Zhang, X.Y.; Huang, J.Z.; Yang, H.; Ou, S.Y.; Liu, F.; Cai, H.Y.; Yang, Q.S. The governing mechanism of tidal wave propagation and threshold effect in the Yangtze river estuary. *Oceanol. Limnol. Sin.* **2019**, *50*, 788–798.

Journal of
Marine Science and Engineering

MDPI

Article

Mobilization of PAHs by Wave-Induced Resuspension and Liquefaction in Silty Sediment

Fang Lu [1,2], Qian Song [1] and Wenquan Liu [3,*]

1 College of Ocean Science and Engineering, Shandong University of Science and Technology, Qingdao 266590, China; fanglu@sdust.edu.cn (F.L.)
2 Laboratory for Marine Geology, Qingdao Marine Science and Technology Center, Qingdao 266237, China
3 Key Laboratory of Marine Sedimentology and Environmental Geology, First Institute of Oceanography, Ministry of Natural Resources, Qingdao 266061, China
* Correspondence: liuwq@fio.org.cn

Abstract: Silty seabed sediments in the subaqueous delta of the Yellow River are heavily contaminated with petroleum-derived polycyclic aromatic hydrocarbons (PAHs). Storm-induced sediment resuspension and liquefaction are key mechanisms responsible for the remobilization of PAHs into the overlying water column. In this study, laboratory-scale wave flume experiments were conducted to simulate PAH release under three hydrodynamic scenarios: (i) static diffusion (Stage I), (ii) low-intensity wave action (5 cm wave height, Stage II), and (iii) high-intensity wave action (12 cm wave height, Stage III). Results revealed a strong positive correlation between suspended particulate matter (SPM) and PAH concentrations in the aqueous phase during sediment disturbance. In particular, sediment liquefaction significantly enhanced PAH release, with concentrations up to five times higher than those under static conditions. Furthermore, liquefaction facilitated vertical migration of PAHs within sediments, resulting in reductions in PAH levels below the original background concentrations. The release dynamics varied notably among PAH species: low-molecular-weight (2–3 ring) PAHs, with lower hydrophobicity, were primarily detected in the aqueous phase, while medium- and high-molecular-weight PAHs remained predominantly associated with sediment particles. These findings underscore the critical role of hydrodynamic disturbances—especially sediment liquefaction—in influencing PAH mobility and offer important implications for pollution risk assessment and coastal management in storm-impacted deltaic environments.

Keywords: polycyclic aromatic hydrocarbons (PAHs); sediment liquefaction; wave-induced resuspension; hydrodynamic disturbance

1. Introduction

The combustion of spilled oil significantly reduces its volume on the water surface, thereby decreasing the concentration of toxic compounds such as polycyclic aromatic hydrocarbons (PAHs). As major constituents of petroleum hydrocarbons, PAHs are predominantly formed through incomplete combustion processes and are well-documented for their teratogenic, carcinogenic, and mutagenic effects [1,2]. In marine environments, PAHs exist in three principal forms: dissolved in water, adsorbed onto suspended particulate matter, and predominantly deposited in sediments [3]. The distribution of PAHs between the aqueous and particulate phases is governed by dynamic exchange processes, including deposition, resuspension, and molecular diffusion between porewater and overlying water.

Marine sediments function as both sinks and secondary sources of PAHs, playing a pivotal role in their environmental fate [4,5]. Physical disturbances such as tidal movements, wind-driven currents, and wave activity can resuspend bottom sediments, leading to the reintroduction of associated pollutants into the water column [6–8]. Previous studies have demonstrated that sediment resuspension significantly influences PAH migration across the sediment–water interface, with desorption from resuspended particles being a dominant release pathway [9,10]. This process is largely driven by wave-induced pressure gradients, which facilitate porewater flow from high- to low-pressure zones within sediments, promoting PAH transport into the overlying water [11]. In addition, salinity has been shown to enhance the accumulation of both dissolved and particle-bound PAHs in the water column during the early stages of sediment resuspension and deposition [12]. To isolate the effects of wave action, our experimental conditions maintained constant temperature and salinity.

The transport and transformation of PAHs at the sediment–water interface are influenced by a range of environmental parameters, including sediment characteristics (e.g., grain size, organic matter content, mineral composition), physicochemical conditions (e.g., temperature, pH, coexisting ions, colloids), and dynamic factors such as hydrodynamic forces and bioturbation [13,14]. Furthermore, the physicochemical properties of individual PAH compounds—namely, molecular weight, number of aromatic rings, and hydrophobicity—affect their mobility and partitioning behavior [14–16]. Wave-induced shear stress has been identified as a critical factor influencing both the magnitude and rate of PAH release. Elevated shear stress not only increases overall PAH mobilization but also enhances the release rate, particularly of low-molecular-weight PAHs (e.g., two- to three-ring structures), which are more susceptible to desorption under turbulent conditions [17,18].

The Yellow River subaqueous delta, situated within the semi-enclosed Bohai Sea, is subject to significant oil pollution resulting from anthropogenic activities such as offshore oil extraction, transportation, and accidental spills. The seabed sediments in this region are particularly susceptible to liquefaction, a process that mobilizes fine-grained subsurface sediments and facilitates their transport into the overlying water column. While the release of oil-derived pollutants during sediment resuspension has been extensively studied in freshwater environments—such as lakes, rivers, and estuaries—limited attention has been given to the behavior and migration of polycyclic aromatic hydrocarbons (PAHs) during wave-induced sediment liquefaction in marine settings.

This study aims to investigate the release and migration dynamics of PAHs from marine sediments under varying wave conditions, employing a controlled wave flume system capable of simulating sediment liquefaction. The findings are expected to provide a scientific basis for evaluating the potential environmental risks and quantifying the release of PAHs in the Yellow River subaqueous delta.

2. Materials and Methods

2.1. Experiment Material

The sediments were collected from the surface layer of the Diao Kou Pile tidal flat at the Yellow River Estuary as presented in Figure 1.

Sediment samples were collected manually using a box sampler and stored in nylon bags. The pretreatment involved crushing the samples, removal of impurities (e.g., grass and branches), air-drying at room temperature in a ventilated, shaded environment, and thorough homogenization by manual mixing. The homogenized sediment was then sieved through an 80-mesh sieve (180 μm) to remove coarse debris while retaining the fine fraction most relevant for PAH adsorption and binding. This approach is consistent with standard sediment preparation methods in PAH studies, such as U.S. EPA Methods 8260D [19] and

8275A [20], which recommend sieving to remove coarse debris before analysis, and USGS sediment protocols, which commonly use an 80-mesh screen to retain the fine, PAH-binding fraction. Although a very small portion of the finest particles may have been lost during sieving, the impact on total PAH concentrations is expected to be minimal. The primary components of the soil samples were silt, sand, and clay, with a clay content of 15%, organic matter content of 1.39%, and PAHs content of 13.81 mg/kg. Artificial seawater with a salinity of 35 was prepared by dissolving sea crystals in clean water, which was used as overlying water for the experiment.

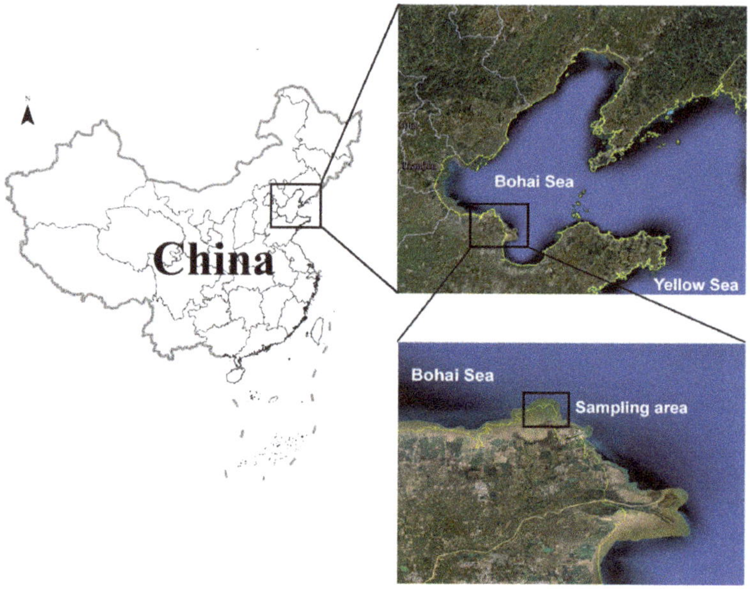

Figure 1. Sampling location at Diao Kou Pile tidal flat in the Yellow River Estuary.

The wave flume (4 m × 0.4 m × 1.0 m) used in this study is shown in Figure 2. The sediment tank (0.6 m × 0.4 m × 0.3 m) was positioned at the center of the flume bottom. A wave generator is located on the right side of the flume, which reciprocally pushes the water to form waves. A wave-absorbing device is installed on the left side of the flume to reduce the interference caused by wave reflection. The capacitive wave gauge (RBR, Ottawa, ON, Canada) is installed in the middle of the flume to collect wave height and wave period during the experiment, and the entire system is automatically controlled by a computer.

The turbidity meter (JFE Advantech Co., Ltd., Nishinomiya, Hyogo, Japan), an infrared backscattering-type sensor with a measurement range of 0–1000 FTU, a resolution of 0.03 FTU, and a measurement uncertainty of ±2% was mounted on a custom bracket at the top of the flume. Its probe was positioned 5 cm above the sediment bed, aligned with the water sampling point, to monitor suspended sediment concentration dynamics. Calibration was performed using standard suspensions, prepared by mixing known soil masses with 1 L of water to achieve target concentrations. For each concentration, three repeated measurements were taken and averaged, yielding an excellent linear relationship between measured and theoretical values (R^2 = 0.998).

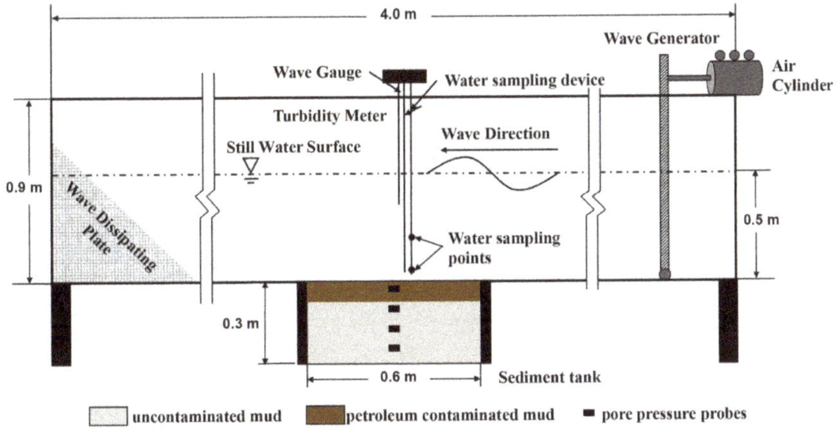

Figure 2. Diagram of wave flume device.

Porewater pressure was measured using a DEWE-43 Transient Data Acquisition System (DEWESoft d.o.o., Trbovlje, Slovenia), with CYY2 sensors (measuring range: 0–100 kPa; resolution: 0.005 kPa, measurement uncertainty: 0.2% FS) connected to the data logger to convert voltage signals into porewater pressure data, which were then transmitted to a computer. The sensor probes were embedded at depths of 2, 8, 14, and 20 cm within the sediment to monitor porewater pressure variations, as illustrated in Figure 3. The pore pressure sensor is immersed in clear water for 24 h before burial and continuously shakes sufficiently to ensure that the internal air is completely discharged. The porewater pressure probes were calibrated at depths of 0–50 cm in 10 cm intervals by comparing probe readings with theoretical hydrostatic pressures. The calibration showed excellent agreement, with $R^2 = 0.999$.

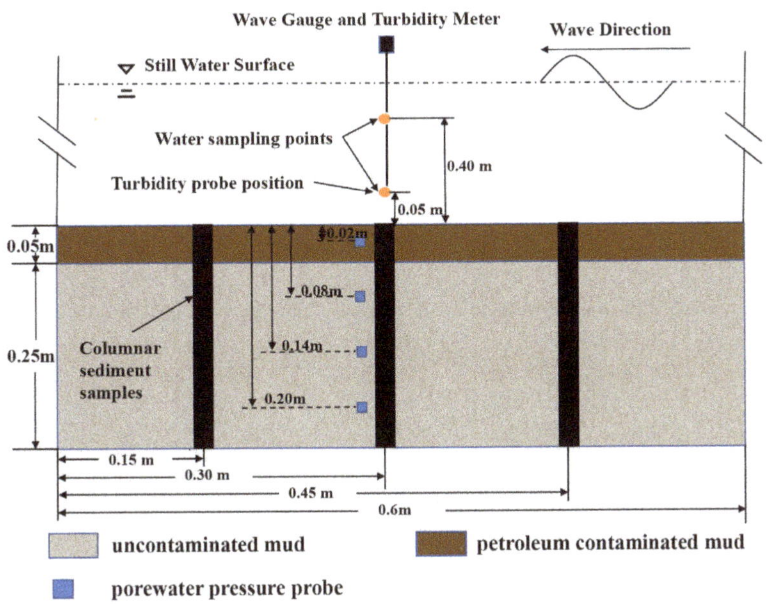

Figure 3. Schematic diagram of the bottom sediment deposition.

2.2. Configuration of Flume Experiment

The oil-saturated solution was prepared by placing approximately half of the artificial seawater in a plastic bucket, followed by the dropwise addition of excess light crude oil onto the water surface. The mixture was thoroughly stirred and allowed to stand for 1–2 days to achieve saturation. The oil-saturated solution was then collected from the bottom of the bucket using a rubber hose, and the oil concentration was determined to be 15.25 mg/L by ultraviolet spectrophotometry.

Subsequently, 8.5 kg of the oil-saturated seawater was mixed with 30.0 kg of dry soil in a sealed vessel using a low-speed twin-screw mixer for 60 min to ensure uniform contaminant distribution while minimizing volatilization losses. The resulting slurry was sealed, stored in darkness at low temperature, and allowed to equilibrate for 1 day before being used for sediment preparation.

The clean soil slurry was prepared by mixing 100.0 kg dry soil with 30.0 kg artificial seawater, and has the same water content as the contaminated soil slurry. Transfer the prepared clean soil slurry to the bottom of the flume until the slurry height reaches 0.25 m, and evenly spread a 5 cm thick layer of oil-containing contaminated soil slurry on top of it (Figure 3), resulting in a total thickness of 0.30 m for the bottom bed sediment. After the bottom bed is paved, slowly add artificial seawater into the flume until the water depth reaches 0.50 m.

2.3. Flume Experiment Procedure

The flume experiment was divided into three stages: Static diffusion stage (Stage I), 5 cm wave height stage (Stage II), and 12 cm wave height stage (Stage III). The selection of wave heights in this study was based on previous experiments conducted in the same flume [21], which ensured the chosen wave parameters were able to meet the requirements of different experimental stages.

In Stage I, the overlying water was static, and the bottom sediments were in a state of static diffusion, lasting for 40 h. Stage II and III were wave action states, with each wave height lasting 120 and 150 min, respectively. During Stage I, two bottles of 500 mL water samples were taken at 0.30 m on the left side of the bed, 0.05 m and 0.20 m above the sediment (Figure 3) with a self-made water sample sampler at intervals of 8 h. One of the bottles was filtered to obtain the filtrate, which was acidified with sulfuric acid and stored at low temperature (4 °C) to determine the concentration of PAHs. In Stage II and III, water samples were taken at 10-min intervals for the first 60 min and 20-min intervals for the next 120 min after the wave height had stabilized. The sampling locations and methods were the same as in Stage I. The experimental wave parameters and experimental process are shown in Table 1.

Table 1. Wave parameters and sampling time.

Test Stage	Wave Parameters		Duration	Sampling Time
	Wave Height (cm)	Period (s)		
Static diffusion (Stage I)	-	-	84 h	12 h/session
5 cm wave height (Stage II)	5.0	2.4	120 min	30 min/session
12 cm wave height (Stage III)	12.0	1.2	150 min	30 min/session

At the end of each stage, sediment core samples of 11, 11.5, and 15 cm in length were collected using a self-made PVC tube sampler at distances of 15 cm, 30 cm, and 45 cm from the left edge of the sediment flume (Figure 3). The core samples were then sectioned to determine the PAHs content at different depths within the bed sediment.

2.4. Analytical Methods of the Samples

The method for determining PAHs content in water samples refers to National Ecological and Environmental Standards of the People's Republic of China GBT 26411-2010 [22]. The water samples obtained from the sink were subjected to solid phase extraction on an SPE device (Supelco, Bellefonte, PA, USA) using an HLB solid phase extraction column (Waters Corporation, Milford, MA, USA). Before passing through the column, the sample was activated and equilibrated by 5 mL of n-hexane, dichloromethane, methanol, and ultrapure water, and the SPE column was eluted with 10 mL of a mixture of dichloromethane and n-hexane (v:v = 3:7) after extraction. The eluent was concentrated using a quantitative concentrator (LabTech, Inc., Sarasota, FL, USA), concentrated to about 2 mL, and the eluent was exchanged to n-hexane. Then, it continued to be concentrated and quantified to 0.9 mL. After adding 0.1 mL of decachlorobiphenyl as an internal standard, 7890A-5975C Gas Chromatography-Mass Spectrometry System (Agilent Technologies, Santa Clara, CA, USA) was used for analysis and testing.

The method for determining PAHs content in sediment samples refers to National Ecological and Environmental Standards of the People's Republic of China HJ 805-2016 [23]. Sediment samples obtained from the soil tanks were lyophilized by a Freeze Dryer (Martin Christ Gefriertrocknungsanlagen GmbH, Osterode am Harz, Germany), 20 g of the sample were accurately weighed, and they were extracted using an Accelerated Solvent Extractor ASE 300 (Thermo Fisher Scientific, Sunnyvale, CA, USA). The extraction conditions were as follows: extraction solvent acetone and dichloromethane (v:v = 1:1), heating temperature of 100 °C, extraction pressure of 1500 psi (1 psi = 6.895 kPa), static cycle 3 times. The extract was collected, concentrated by quantitative concentrate, and the solvent was converted to n-hexane, and then purified by chromatography column. A 70 mL eluent of dichloromethane and n-hexane (v:v = 3:7) was prepared to elute and purify the PAHs components on the purification column. The eluate was concentrated to about 2 mL by a quantitative concentrator, then the solvent was converted to n-hexane and continued to be concentrated to 0.9 mL. Then, 0.1 mL of decachlorobiphenyl (PCB) was added as the internal standard, and analyzed and tested by GC/MS.

Polycyclic aromatic hydrocarbons (PAHs) in water and sediment samples were analyzed using gas chromatography–mass spectrometry (GC/MS). The instrument was equipped with a HP-5MS capillary column (30 m × 0.25 mm × 0.25 μm). Helium was used as the carrier gas at a constant flow rate of 1.0 mL/min. The injector temperature was maintained at 280 °C under splitless mode, with an injection volume of 1.0 μL. The oven temperature program was set as follows: initial temperature at 80 °C held for 2 min; ramped at 20 °C/min to 180 °C, held for 5 min; then ramped at 10 °C/min to 290 °C, and held for another 5 min. Mass spectrometric detection was performed using an electron ionization (EI) source at an ionization energy of 70 eV and an ion source temperature of 230 °C. The mass scanning range was m/z 45–500. Qualitative identification of PAHs was achieved through spectral library matching and retention time comparison under full-scan mode. Quantification was performed using the internal standard method.

The method quantification limits (LOQs) and detection limits (LODs) for individual PAHs in sediment and water are listed in Table 2.

Table 2. Method quantification limits (LOQs) and detection limits (LODs) for individual PAHs in sediment and water.

Individual PAHs	Sediment		Water	
	LOD (mg/kg, dw)	LOQ (mg/kg, dw)	LOD (ng/L)	LOQ (ng/L)
Naphthalene	0.09	0.36	20.00	80.00
Acenaphthylene	0.09	0.36	3.00	12.00
Acenaphthene	0.12	0.48	2.90	11.60
Fluorene	0.08	0.32	6.30	25.20
Phenanthrene	0.10	0.40	12.00	48.00
Anthracene	0.12	0.48	2.20	8.80
Fluoranthene	0.14	0.56	3.20	12.80
Pyrene	0.13	0.52	3.20	12.80
Benzo[a]anthracene	0.12	0.48	2.40	9.60
Chrysene	0.14	0.56	4.40	17.60
Benzo[b]fluoranthene	0.17	0.68	2.60	10.40
Benzo[k]fluoranthene	0.11	0.44	4.10	16.40
Benzo[a]pyrene	0.17	0.68	2.90	11.60
Indeno[1,2,3-cd]pyrene	0.13	0.52	2.40	9.60
Dibenzo[a,h]anthracene	0.13	0.52	3.10	12.40
Benzo[g,h,i]perylene	0.12	0.48	4.10	16.40

2.5. Data Statistical Analysis

Statistical analyses were performed using SPSS 26.0 software. The normality of the parameters was examined using the Shapiro–Wilk test, which is more suitable for relatively small sample sizes. The Spearman rank correlation coefficient was employed to assess the monotonic associations between parameters. Unlike Pearson correlation, which assumes linearity and normality, Spearman correlation is designed to capture monotonic rather than strictly linear relationships, making it more robust for datasets with non-normal distributions or potential outliers.

2.6. Quality Control

All reagents were of HPLC grade, and all analytical procedures were conducted in accordance with the GB/T 26411-2010 for seawater Chinese [22] and national standards HJ 805-2016 for sediments [23]. Quality assurance was maintained through the analysis of blanks and duplicate samples after every 10 runs. Certified reference materials (No. GBW08737 for water and No. GBW07352 for sediments) were employed to verify method accuracy and ensure the reliability of analytical results. Calibration curves for PAHs analysis were established using a mixed standard solution of 16 PAHs (GBW08737) at six concentrations (10 μg/L, 25 μg/L, 50 μg/L, 100 μg/L, 250 μg/L, and 500 μg/L, prepared in n-hexane), with coefficients of determination (R^2) consistently $\geq$0.995. Recoveries of 65.4–103.0% for aqueous samples and 73.0–98.5% for sediments were obtained, confirming acceptable analytical precision and accuracy. Consequently, the raw data were used directly for subsequent analysis.

3. Results

3.1. Characteristics of Porewater Pressure Changes in Sediments

The temporal variation in porewater pressure within sediments under wave action is illustrated in Figure 4. When subjected to wave forcing with a height of 5 cm, the porewater pressure exhibited a trend of initial increase followed by gradual dissipation. This phenomenon can be attributed to the attenuation of wave-induced pressure gradients during wave propagation, particularly under small wave disturbances (wave height < 8 cm).

In these conditions, the upward pressure gradient becomes notably weakened—especially in the region of wave breaking—resulting in a lower rate of pressure buildup compared to the dissipation rate [24]. Consequently, no significant accumulation of porewater pressure occurs within the sediment matrix [25]. During this stage, only a limited number of suspended particles were observed above the sediment surface through the transparent side wall of the wave flume, and no evidence of sediment liquefaction was detected.

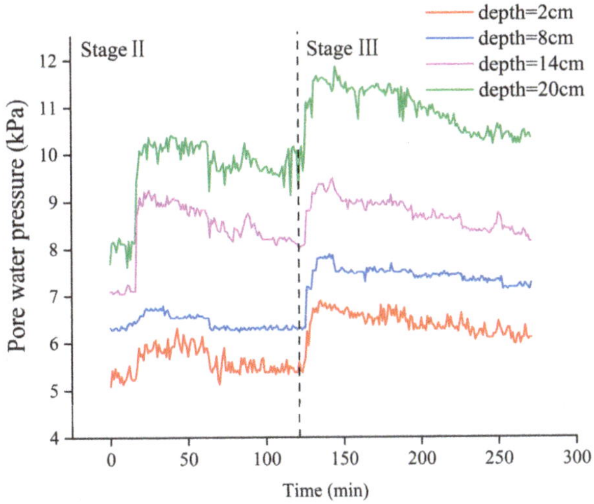

Figure 4. Variation in porewater pressure under different wave heights.

In contrast, under the influence of a 12 cm wave height, the porewater pressure increased rapidly, reaching pronounced peaks that exceeded the dissipation capacity of the sediment. This pressure buildup coincided with sharp rises in suspended sediment concentration and visual observation of fluid-like oscillation of the upper sediment layer. The sediments oscillated synchronously with wave motion, indicating a transition into a fluidized state. Such reciprocating motion and arc-shaped oscillatory interfaces of fluidized soils have also been reported in earlier flume studies on silty and sandy beds, where sustained excess pore pressure growth and strong sediment suspension under wave forcing were recognized as key indicators of liquefaction onset [24,26–28]. Accordingly, in this study, the simultaneous occurrence of excess porewater pressure buildup and significant increases in suspended solids was adopted as the criterion for identifying the onset of sediment liquefaction. Visual observations also confirmed that liquefaction occurred, and the sediment entered a state of dynamic instability [29,30].

3.2. Variation Characteristics of Suspended Sediment Concentration and Median Particle Diameter of Suspended Particles in Overlying Water

Figure 5 illustrates the temporal variation in suspended sediment concentration (SSC) and the median particle diameter (D_{50}) of suspended sediment measured 5 cm above the sediment surface under continuous wave forcing. Experimental results in this study and field observations consistently demonstrate a strong positive correlation between wave height and suspended sediment concentration (SSC) in coastal environments. This relationship is particularly evident in the subaqueous Yellow River delta, where in situ measurements by Liu et al. (2022) [31] recorded SSC values exceeding 30 g/L during storm events with significant wave heights (Hs) of 2.73 m. The study documented the formation of wave-supported fluid mud (WSFM) layers with thicknesses > 60 cm under such conditions, confirming that hydrodynamic energy is a primary driver of sediment resuspension. These field observations are consistent with the results of controlled wave flume experiments presented in this study. Under wave conditions with a height of 5 cm, SSC values ranged from 0.027 to 0.193 g/L. In contrast, under a 12 cm wave regime, SSC increased significantly, ranging from 0.375 to 0.867 g/L. This represents that suspended sediment concentrations (SSCs) increase by a factor of 4 to 10 as wave heights rise from 5 cm to 12 cm. The mechanistic link between wave energy and sediment mobilization involves both direct shear stress at the sediment–water interface and indirect effects through wave-induced pore pressure accumulation that reduces sediment cohesion [26]. SSCs increased sharply in the initial stage of wave action at a height of 12 cm and reached a near-steady state within 30 to 40 min, reflecting an adjustment period driven by sediment resuspension and vertical redistribution. This behavior is attributed to the intensification of turbulence and bed shear stress induced by wave action, which facilitates the detachment and suspension of particles from the sediment.

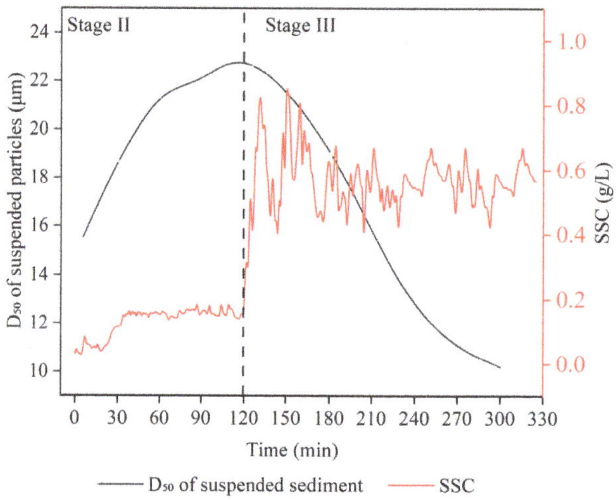

Figure 5. Temporal variation in SSC and D_{50} of suspended sediment measured 5 cm above the sediment surface under different wave height conditions.

The variation in the median particle diameter (D_{50}) of suspended sediment in the overlying water following wave action is presented in Figure 5. Under wave forcing, the D_{50} of suspended particles increased continuously during Stage II, reaching its peak value at the end of this stage. In Stage III, the median particle diameter gradually decreased over time. Notably, approximately 220 min after the onset of Stage III, the D_{50} dropped below the initial value recorded at the beginning of Stage II. Overall, the median particle diameter of suspended sediments was larger during Stage II. This is primarily attributed to the dominance of coarser particles resuspended from the surface sediment layer under the influence of low wave energy. In contrast, Stage III was characterized by sediment liquefaction, during which finer particles, often associated with pollutants, were transported upward through seepage channels and dispersed into the overlying water [25]. Meanwhile, larger particles rapidly settled out due to the oscillatory motion of the water column.

3.3. Variation Characteristics of PAHs Concentration and Composition in Overlying Water

Figure 6 presents the temporal variation in dissolved polycyclic aromatic hydrocarbons (PAHs) concentrations in the overlying water during the wave flume experiment. In the static diffusion phase (Stage I), the concentration of dissolved PAHs at 5 cm above the sediment ranged from 337.390 to 882.510 ng/L, characterized by an initial rapid decrease followed by a gradual increase. At 40 cm above the sediment, concentrations ranged from 567.346 to 852.245 ng/L and exhibited a steady upward trend. Overall, dissolved PAH concentrations at 40 cm were consistently higher than those near the sediment–water interface.

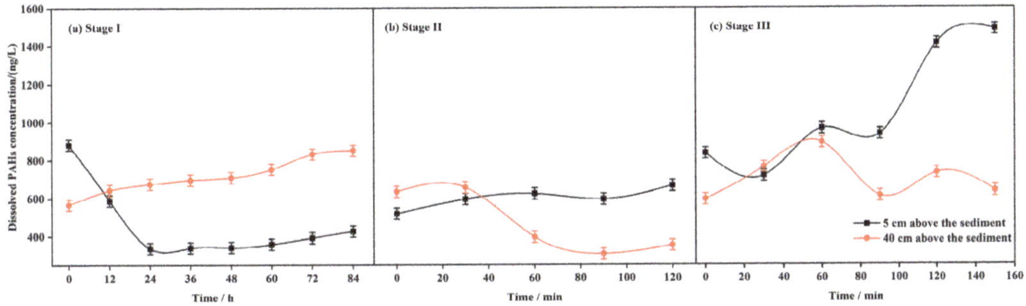

Figure 6. Characteristics of the concentration of dissolved PAHs in the overlying water at 5 cm and 40 cm above the sediment at different experimental stages: (**a**) static diffusion stage (Stage I), (**b**) wave action at 5 cm wave height (Stage II), and (**c**) wave action at 12 cm wave height (Stage III) (Error bars denote the mean absolute deviation (MAD) calculated from parallel samples).

In Stage II, the concentration of PAHs at 5 cm above the sediment exhibited a generally steady upward trend, increasing from 518.722 ng/L to 664.390 ng/L, with only minor fluctuations observed during the 60–90 min period. At 40 cm above the sediment, the PAH concentration remained nearly constant during the initial 30 min, followed by a rapid decline over the next 30 min, ultimately dropping below 400 ng/L.

In Stage III, the dissolved PAH concentration at 5 cm above the sediment showed pronounced temporal variation, characterized by multiple phases of increase and decrease. The overall concentration ranged from 835.966 to 1486.063 ng/L, indicating that intense wave action significantly influenced the release and migration of PAHs. In contrast, at 40 cm above the sediment, the concentration increased steadily during the first 60 min, followed by a slight decline and a subsequent rebound, reaching a peak of 892.626 ng/L. These results indicate that enhanced wave energy promotes the release and vertical redistribution of dissolved PAHs, particularly in near-bed waters.

A total of six dissolved PAH congeners were detected in the overlying water. Figure 7 illustrates the variations in the concentrations of individual dissolved PAH congeners in the overlying water during the wave flume experiments. Among them, the majority were low-molecular-weight PAHs consisting of 2–3 rings. In contrast, among the 4–6 ring PAHs, only fluoranthene, pyrene, and benzo[g,h,i]perylene were detected at relatively high frequencies, while other congeners were either detected at extremely low concentrations or remained undetectable due to their low abundance and the limited sampling volume. Overall, the release of 2–3 ring PAH congeners was significantly higher than that of 4-ring congeners.

During Stage I (no wave action), the initial concentrations of PAH congeners in the overlying water were relatively high. As the experiment progressed, the concentration of naphthalene showed a continuous increasing trend, whereas the concentrations of fluorene, phenanthrene, fluoranthene, pyrene, and other 3–4 ring congeners remained largely unchanged (Figure 7a). In Stage II (5 cm wave action), the concentrations of 2–3 ring PAHs decreased compared to Stage I, while the concentrations of 3–6 ring congeners remained relatively stable (Figure 7b). In Stage III (12 cm wave action), the concentrations of 2–3 ring congeners increased significantly. Although 4–6 ring PAHs exhibited an upward trend in concentration, their increase remained lower than that observed for 2–3 ring congeners (Figure 7c).

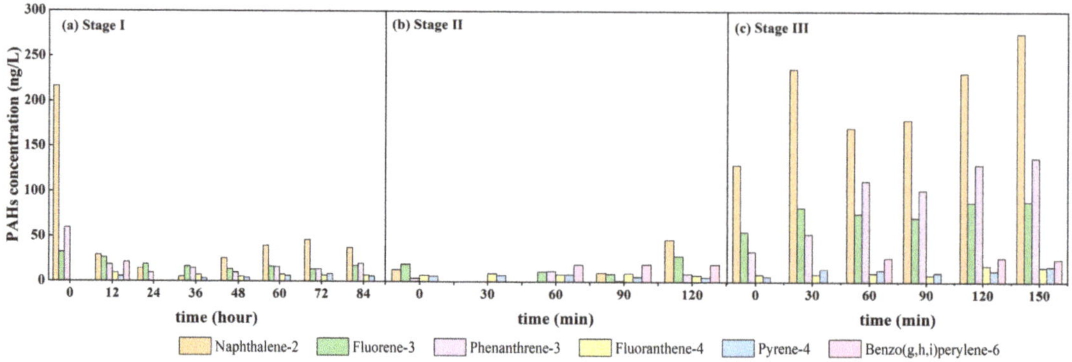

Figure 7. Temporal variation in the concentrations of individual dissolved PAH monomers in the overlying water at different experimental stages: (**a**) static diffusion stage (Stage I), (**b**) wave action at 5 cm wave height (Stage II), and (**c**) wave action at 12 cm wave height (Stage III).

Overall, the dissolved PAH profile indicates higher mobility and release activity among certain 2–3 ring compounds, such as naphthalene and fluorene, likely due to their greater water solubility and weaker sediment binding affinities. This is consistent with Sun et al. (2019) [32], who reported that 2–3 ring PAHs dominate vertical migration in porous media due to their higher aqueous solubility and lower K_{ow} values.

3.4. Variation Characteristics of PAHs Concentration and Composition in Sediments

The vertical distribution of PAHs in the sediment following different experimental stages is illustrated in Figure 8.

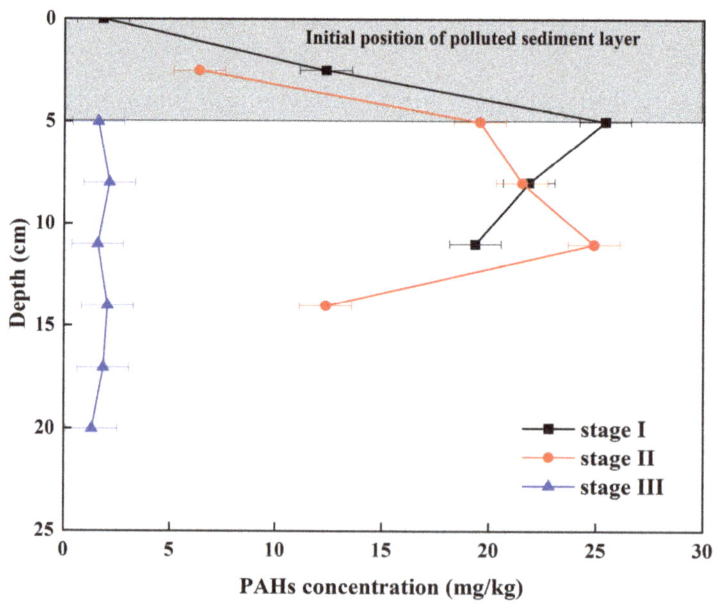

Figure 8. Vertical distribution of PAHs content in sediments at the end of Stage I, II, and III (Error bars denote the mean absolute deviation (MAD) calculated from parallel samples).

After Stage I, a sediment core measuring 11 cm in length was collected. During this static diffusion stage, petroleum-derived PAHs in the surface sediment were observed to migrate both upward into the overlying water and downward into the deeper sediment layers. The maximum PAH concentration was detected at a depth of 5 cm, reaching 25.404 mg/kg, with an overall concentration range of 1.795–25.404 mg/kg. Below 5 cm, the PAH content gradually decreased with depth, indicating limited vertical migration in the absence of hydrodynamic forcing.

After Stage II with a wave height of 5 cm, a sediment core with a length of 11.5 cm was collected. Wave-induced disturbance facilitated deeper penetration of pollutants, with the highest PAH concentration (24.899 mg/kg) detected at a depth of 12 cm. The PAH content in this stage ranged from 6.400 to 24.899 mg/kg, suggesting enhanced vertical transport due to sediment resuspension and porewater advection.

A sediment core with a length of 15 cm was obtained at the end of Stage III. The PAH content in the sediment was substantially reduced, with values ranging from 1.314 to 2.128 mg/kg across all depths. The vertical profile showed minimal variation, suggesting that stronger wave action may have facilitated the desorption and release of PAHs from the sediment matrix into the overlying water column.

The variations in the concentrations of particulate-phase PAH congeners in the sediment during the experiments are shown in Figures 9–11.

During Stage I, the concentrations of three- to four-ring PAHs exhibited marked fluctuations (Figure 9). Phenanthrene exhibited the highest concentration among all detected PAH congeners across all sediment depths, reaching up to 6.5 mg/kg. This was followed by pyrene, fluoranthene, and fluorene, all of which are classified as 3–4 ring PAHs and were consistently distributed across the sediment profile. In contrast, PAH congeners such as benzo[a]fluoranthene and benzo[b]fluoranthene, both 5-ring compounds, were detected at relatively low concentrations.

The proportional distribution of individual PAH congeners remained relatively stable across different depths. Notably, although there were variations in the concentrations of different congeners, nearly all were detected at multiple depths, indicating a widespread vertical presence. The only exception was anthracene, which was not detected at the 2.5 cm depth.

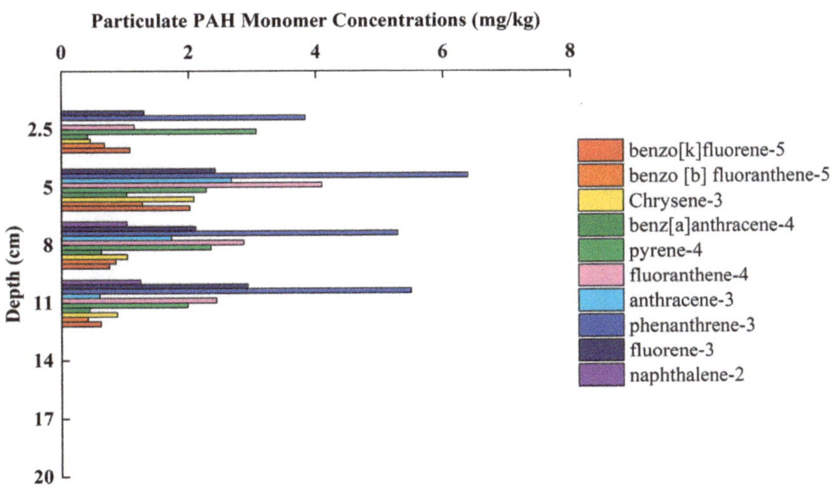

Figure 9. Particulate PAH monomer concentrations at the end of Stage I.

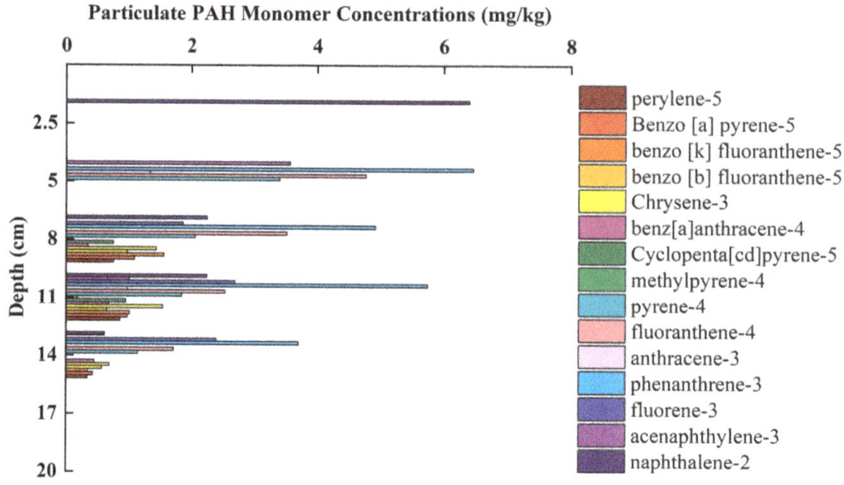

Figure 10. Particulate PAH monomer concentrations at the end of Stage II.

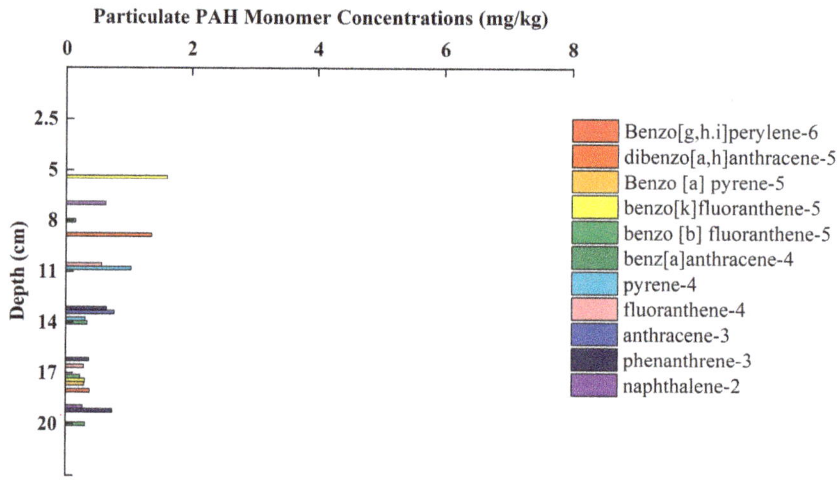

Figure 11. Particulate PAH monomer concentrations at the end of Stage III.

In Stage II, significant vertical migration toward deeper sediment layers was observed for two- to four-ring PAHs, such as naphthalene, phenanthrene, fluoranthene, and pyrene. Compared to Stage I, Stage II was characterized by a greater diversity of migrating PAH compounds and more pronounced changes in their concentrations (Figure 10).

In Stage III, due to sediment liquefaction and the subsequent release of PAHs from the sediment matrix, the concentrations of PAH congeners detected at various depths were generally low, all below 2 mg/kg (Figure 11).

The congeners that migrated to greater depths primarily included low-molecular-weight PAHs such as phenanthrene, naphthalene, and anthracene (2–3 rings), as well as the higher molecular weight compound benzo[a]anthracene. This deeper migration of benzo[a]anthracene may be attributed to its unique molecular structure, which potentially alters its adsorption and diffusion behavior in sediment compared to other five-ring PAHs. Its relatively easier desorption may have facilitated deeper transport within the bed sediment.

4. Discussion

4.1. Correlation Analysis Between Porewater Pressure and Suspended Sediment Concentration Variations

Under wave action with a height of 5 cm, the porewater pressure exhibited an initial increase followed by a gradual decline. During this stage, it was observed that the natural dissipation rate of pore pressure within the sediment matrix exceeded the rate of pressure buildup induced by the waves, resulting in no significant increase in overall pore pressure. Visual observations through the side glass wall of the flume revealed only a small amount of suspended particles present in the upper layer of the sediment bed.

When subjected to waves with a height of 12 cm, the porewater pressure underwent a pronounced dynamic change, rapidly reaching a peak value. This phenomenon was attributed to the accelerated increase in pore pressure driven by wave loading, which far outpaced the natural dissipation rate of pore pressure within the sediment. During this phase, the sediment exhibited fluid-like behavior, oscillating laterally under wave action, which is an intuitive manifestation of sediment liquefaction.

These results indicate that changes in wave conditions led to corresponding variations in porewater pressure and suspended sediment concentration. In Stage II, under 5 cm wave height, pore pressure remained relatively stable and suspended sediment concentration exhibited minor fluctuations. Upon increasing the wave height to 12 cm in Stage III, both the amplitude of pore pressure variation and suspended sediment concentration increased markedly, with the latter reaching values several times higher than those in Stage II.

To further analyze the relationship between porewater pressure variations and suspended sediment concentration changes, a correlation analysis was conducted between these two parameters during different stages. As indicated by the normality test results in Table 3, at the significance level of $\alpha = 0.05$, all variables were found to deviate significantly from a normal distribution (indicated by $p < 0.05$), indicating clear non-normality.

Table 3. Normality test of porewater pressure and SSC in Stage II and Stage III.

	Shapiro–Wilk		
	Statistic	**Degrees of Freedom**	**Sig. (*p*-Value)**
PWP_2CM_II	0.961	120	0.001
PWP_8CM_II	0.849	120	0.000
PWP_14CM_II	0.861	120	0.000
PWP_20CM_II	0.764	120	0.000
SSC_II	0.805	120	0.000
PWP_2CM_III	0.935	152	0.000
PWP_8CM_III	0.777	152	0.000
PWP_14CM_III	0.978	152	0.016
PWP_20CM_III	0.944	152	0.000
SSC_III	0.969	152	0.002

Note: PWP_2CM_II denotes the porewater pressure at a depth of 2 cm during Stage II; SSC_II denotes the suspended sediment concentration during Stage II. Other variables are defined in the same manner.

The results of the Spearman correlation analysis between porewater pressure and SSC during stages II and III are presented in Figures 12 and 13. In the figures, Spearman's rho values are represented by colored circles accompanied by numerical labels. The size of each circle is proportional to the absolute value of Spearman's rho, with larger circles indicating stronger correlations.

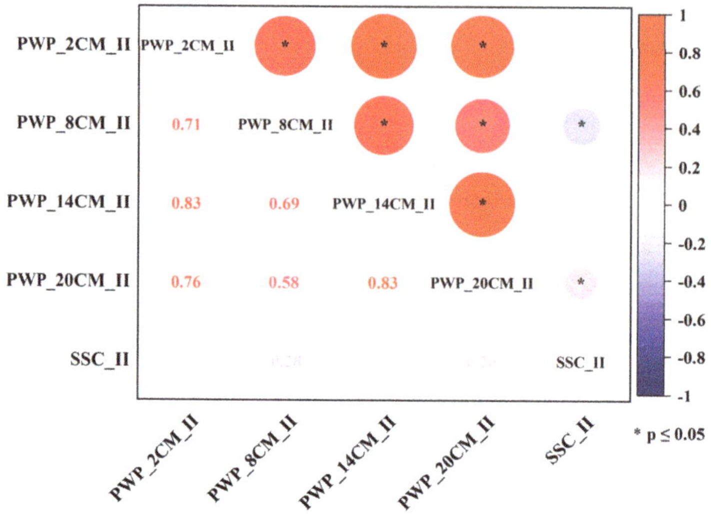

Figure 12. Spearman correlation between porewater pressure and SSC during Stage II (Note: PWP in this figure denotes porewater pressure).

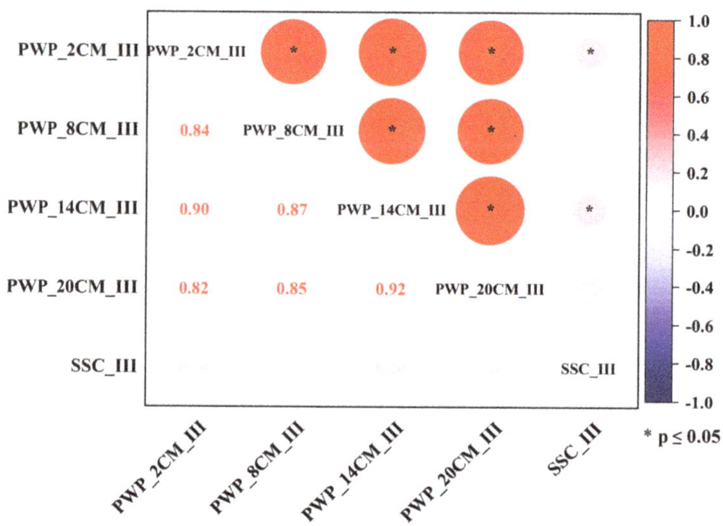

Figure 13. Spearman correlation between porewater pressure and SSC during Stage III (Note: PWP in this figure denotes porewater pressure).

The statistical results from the Spearman rank correlation analysis reveal a clear shift in the sediment dynamic regime between non-liquefied (Stage II) and liquefied (Stage III) conditions, providing important insights into the mechanisms of sediment resuspension under wave action.

In the absence of liquefaction (Stage II), the correlation structure between porewater pressure (PWP) and suspended sediment concentration (SSC) was complex and indicative of a stable seabed response. The strong, highly significant positive correlations among PWPs at all depths ($\rho > 0.70$, $p < 0.01$ for all pairs) confirm the effective transmission of wave-induced pressures through an intact sediment skeleton. Notably, a significant negative correlation was found between PWP_8CM_II and SSC_II ($\rho = -0.28$, $p = 0.002$).

This finding supports the mechanism of wave-induced cyclic compaction [33], whereby wave loading rearranges sediment grains, increasing pore pressure while enhancing seabed stability and resistance to erosion, thereby suppressing sediment resuspension. The absence of any strong positive PWP-SSC correlations suggests that direct shear erosion by wave orbital velocities, rather than pressure-driven processes, was the main contributor to SSC during this stage.

In contrast, under liquefied conditions (Stage III), the correlation pattern shifted markedly, highlighting the role of liquefaction as a primary driver of sediment resuspension. While the strong inter-PWP correlations persisted ($\rho > 0.81$, $p < 0.01$), significant positive correlations emerged between SSC and PWPs at 2 cm and 14 cm depths (PWP_2CM_III: $\rho = 0.20$, $p = 0.013$; PWP_14CM_III: $\rho = 0.20$, $p = 0.011$). These results provide statistical evidence that the sharp rise in pore pressure, which indicates sediment liquefaction and loss of effective stress, triggered large-scale sediment entrainment [28]. The transformation of the seabed surface into a liquefied layer substantially reduced its critical shear stress, enabling resuspension by wave-induced turbulence and leading to the observed increase in SSC.

However, the relatively weak positive correlations and their absence at certain depths are equally informative. They indicate that while pore pressure buildup is a necessary trigger for liquefaction, it is not the sole determinant of SSC. The magnitude and persistence of suspension are strongly modulated by post-liquefaction processes, including the spatial patchiness of liquefaction, the thickness of the liquefied layer, and the advective transport and mixing induced by ambient currents and turbulence. This decoupling between the initiation of erosion and the sustenance of suspension explains the modest correlation coefficients and reflects the inherent complexity of sediment dynamics.

4.2. Correlation Analysis Between Porewater Pressure and Suspended Sediment Grain Size Variations

After wave loading, the median grain size of suspended sediments increased during the low-wave stage but decreased during the high-wave stage. In Stage III, the median grain size of suspended particles decreased rapidly over time, and overall, the median grain size in Stage II was larger than that in Stage III. This trend indicates that the enhanced wave intensity in Stage III increased the porewater pressure, eventually triggering soil liquefaction. Soil liquefaction is a process that occurs under dynamic loading, during which soil particles tend to rearrange into a denser configuration. This leads to a rapid rise in porewater pressure, a sharp reduction in effective stress, and ultimately a loss of shear strength. Under high wave heights, the shear action of waves on the sediment bed becomes stronger, promoting particle rearrangement and pore space reduction, which accelerates the buildup of porewater pressure. As inter-particle interactions and friction decrease, finer particles (e.g., clay and silt), due to their smaller size and lower settling velocity, are more easily entrained and transported by the flow. In contrast, coarser particles (e.g., sand) may settle due to gravity or accumulate around local flow structures, forming residual layers [26,30,31].

Therefore, the observation that the median grain size of suspended sediments in Stage II is generally larger than that in Stage III supports the conclusion that increased wave intensity leads to greater pore pressure fluctuations, and the progressive accumulation of pore pressure in Stage III results in sediment liquefaction.

4.3. Effects of Wave-Induced Disturbance on the Release of PAHs into the Overlying Water

The conceptual diagram of release processes of PAHs from sediment into seawater is illustrated in Figure 14.

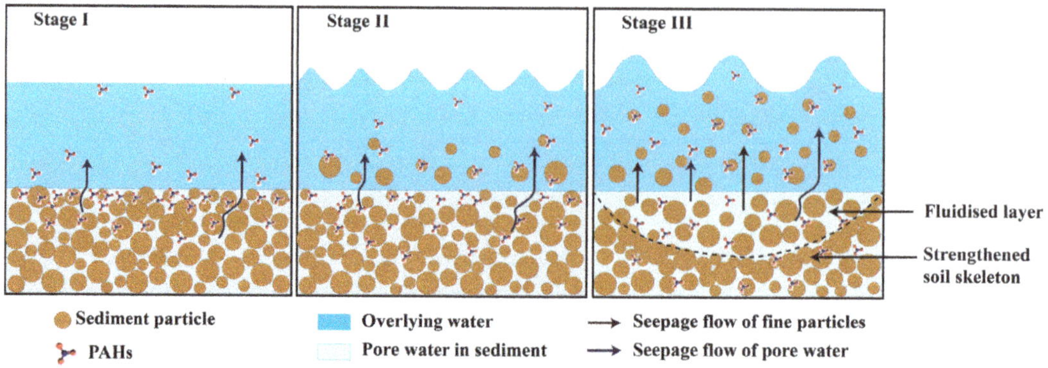

Figure 14. Conceptual diagram of release processes of PAHs from sediment into seawater.

In Stage I, during the initial period of the experiment, the concentration of dissolved polycyclic aromatic hydrocarbons (PAHs) in the overlying water was relatively high. Interestingly, PAH concentrations at 40 cm above the sediment surface were consistently higher than those at 5 cm. This pattern is attributed to the hydrodynamic disturbance generated during the water-filling process, which scoured the surface sediment and promoted the diffusion of PAHs into the overlying water. Furthermore, because PAHs are less dense than artificial seawater, they exhibited an upward migration tendency, contributing to elevated concentrations in the upper water column. Over a 24 h period, the concentration of dissolved PAHs gradually increased. This increase is primarily driven by two mechanisms: (1) a concentration gradient between the contaminated surface sediment and the overlying water, which facilitated molecular diffusion of PAHs; and (2) the consolidation of water-rich remolded sediments, which, through porewater discharge, mobilized PAHs from the sediment into the water column [4,34].

In Stage II, under wave forcing with a wave height of 5 cm, the concentration of suspended sediments exhibited a slight upward trend (Figure 5). Simultaneously, a distinct divergence in the distribution of polycyclic aromatic hydrocarbons (PAHs) was observed: the concentration of dissolved PAHs decreased in the upper water column but increased near the sediment–water interface. This phenomenon aligns with the findings of Li et al. (2016) [35], who demonstrated that sediment resuspension triggers phase redistribution of PAHs, with low-molecular-weight compounds preferentially partitioning into the dissolved phase. This phenomenon can be attributed to three primary mechanisms. First, the oscillatory pressure gradients generated by alternating wave crests and troughs induced advective porewater flow, facilitating the upward transport of PAHs from the sediment matrix [11,35]. Second, the elevated pore pressure triggered partial resuspension of surface sediments, promoting the release of low-molecular-weight PAHs. Due to their higher water solubility and stronger affinity for the dissolved phase, these compounds contributed to an increase in dissolved PAH concentrations [36–39]. Third, in the upper water column, the presence of fine suspended particles with high specific surface areas provided ample vacant adsorption sites. This facilitated the adsorption of dissolved PAHs onto particulates, thereby reducing their concentrations in the aqueous phase [40,41].

In Stage III, the wave height increased to 12 cm, inducing a progressive buildup of porewater pressure within the sediment bed. When the pressure surpassed the sediment's effective stress threshold, liquefaction ensued, as visually confirmed by the transition of the sediment from a cohesive solid to a fluidized state. This phenomenon triggered a sharp increase in suspended sediment concentrations in the water column, accompanied by a pronounced surge in dissolved PAH levels, increasing approximately threefold compared to pre-liquefaction conditions. As summarized by Berríos-Rolón et al. (2025) [42], sediment liquefaction under extreme hydrodynamic conditions is a critical yet understudied pathway for PAH remobilization in deltaic environments. The liquefaction process significantly intensified sediment–water mixing, facilitating the rapid release of both dissolved and particle-bound PAHs [43]. Concurrently, fine sediment particles remaining preferentially suspended in the upper water column acted as efficient PAH carriers, leading to the enrichment of particulate-bound PAHs [44]. Due to sorption processes, this resulted in a relative decline in the dissolved PAH fraction in the upper layer [16].

To quantitatively evaluate the linkage between SSC and PAHs during these two stages, the normality of the datasets (n = 11) was assessed and results are presented in Table 4. Shapiro–Wilk tests indicated marginal normality for SSC (p = 0.091) but significant non-normality for PAHs (p = 0.029). Given the non-normality of PAHs, Spearman's rank correlation was employed, revealing a significant positive relationship between SSC and PAHs across Stages II and III (ρ = 0.673, p = 0.023), as illustrated in Table 5. This statistical evidence corroborates the mechanistic interpretation that elevated SSC promotes PAH mobilization via adsorption and co-transport processes.

Table 4. Normality test of SSC and PAHs concentration.

Shapiro–Wilk Tests of Normality			
	Statistic	**Degrees of Freedom**	**Sig. (p-Value)**
SSC	0.876	11	0.091
PAHs	0.837	11	0.029

Table 5. Spearman correlation of SSC and PAHs concentration.

Spearman Correlations			
		SSC	**PAHs**
SSC	Correlation Coefficient ρ	1.000	0.673 *
	Sig. (2-tailed)		0.023
	N	11	11
PAHs	Correlation Coefficient ρ	0.673 *	1.000
	Sig. (2-tailed)	0.023	
	N	11	11

* Correlation is significant at the 0.05 level (2-tailed).

Together, these results demonstrate that both moderate resuspension and intense liquefaction can mobilize PAHs, but through distinct mechanisms. Stage II is characterized by phase redistribution governed by adsorption–desorption dynamics, whereas Stage III liquefaction enhances large-scale sediment–water exchange, resulting in a stronger coupling between SSC and PAHs. A similar relationship has been reported in resuspension experiments with sediments from Toulon Bay (northwestern Mediterranean Sea, France), where the remobilization of PAH-enriched sediments led to marked increases in dissolved PAH concentrations [44].

The differential release patterns of PAH congeners observed in Figure 7 can be directly attributed to their fundamental physicochemical properties, which govern their partitioning behavior between sediment and water phases. The significantly higher aqueous concentrations of 2–3 ring PAHs (e.g., naphthalene, fluorene, and phenanthrene) throughout all experimental stages are a direct consequence of their higher aqueous solubility and lower octanol–water partition coefficients (K_{ow}). For instance, naphthalene (2-ring) has a solubility of 31.5 mg/L and log K_{ow} of 3.37 [45], facilitating its rapid desorption and dissolution into the overlying water, even under static conditions (Stage I). This explains its continuous concentration increase during Stage I and its pronounced response to hydrodynamic disturbances in Stage III.

In contrast, the limited detection and lower concentrations of 4–6 ring PAHs (exemplified by fluoranthene, pyrene, and benzo[g,h,i]perylene) are fundamentally constrained by their low solubility and high hydrophobicity. Pyrene (4-ring) has a solubility of 0.13 mg/L and a log K_{ow} of 4.88 [45], indicating a strong tendency to adsorb onto sedimentary organic matter. Their large molecular size and planar structure further enhance their affinity for sediment matrices, resulting in resistant sorption and limited mobility. This is also proved by findings of PAHs. Consequently, even under the energetic conditions of Stage III (12 cm wave action), the release of these high-molecular-weight PAHs was markedly suppressed compared to their low-molecular-weight counterparts.

The wave-induced release dynamics further highlight this physicochemical control. In Stage II (5 cm waves), the decrease in 2–3 ring PAHs may reflect dilution or mild aerobic degradation enhanced by mixing. The dramatic resurgence of low-molecular-weight PAHs in Stage III underscores their responsiveness to physical perturbation; their weak sorption allows them to be readily mobilized by shear stress and advective flow. The subdued response of high-molecular-weight PAHs, which exhibits only a slight concentration increase, reinforces their strong sequestration by sediments. Their release likely requires more intense energy inputs to overcome desorption kinetic limitations or to resuspend sediment particles to which they are tightly bound.

4.4. Migration and Diffusion Processes of PAHs in Sediments

Following each experimental stage, sediment cores were collected and analyzed for PAH concentrations at 2.5 cm depth intervals.

At the conclusion of Stage I, PAH concentrations exhibited a progressive increase from the sediment surface (0 cm), reaching a distinct peak at 5 cm depth (Figure 8). This vertical distribution pattern resulted from two concurrent transport processes: (1) gravity-driven downward migration of PAHs through porewater transport, and (2) upward diffusion along the concentration gradient between surface sediments (high concentration) and overlying water (low concentration). The primary diffusion zone occurred within the 0–5 cm layer, where continuous PAH release into the water column gradually depleted concentrations in surface sediments. Below this depth, diminishing concentration gradients progressively weakened the diffusion driving force, resulting in reduced PAH mobility.

After the cessation of wave forcing in Stage II, PAHs originally concentrated in the contaminated layer began to diffuse into the surrounding sediments. As observed by Li et al. (2016) [35] in successive suspension experiments, such diffusion is driven by porewater advection and particle rearrangement under hydrodynamic forcing. Intense mixing between sediment particles and porewater facilitated the downward migration and dispersion of PAHs, resulting in the concentration peak shifting to greater depths [44,46,47]. This was reflected in the experimental results, which showed a continuous increase in PAH concentrations within the 2.5–12 cm sediment depth range. This general pattern of vertical migration and partitioning is also consistent with broader reviews on PAH deposition and distribution in environmental matrices [48]. The review highlights that low-molecular-weight PAHs (2–3 rings) tend to exhibit higher mobility and diffuse more readily through sediment layers due to their greater aqueous solubility, whereas high-molecular-weight PAHs (4–6 rings) are more strongly adsorbed to organic matter and fine particles, limiting their vertical transport. These general observations align with our experimental findings, where 2–3 ring PAHs migrated deeper into the sediment under wave-induced disturbance, while 4–6 ring congeners remained more confined near the initial contaminated layer.

In Stage III, sediment liquefaction occurred, with a maximum liquefaction depth of 15 cm. During this phase, the overall PAH concentration in the sediment decreased and became more uniformly distributed, with the migration depth extending to 20 cm. This enhanced migration was primarily driven by the rapid accumulation of porewater pressure under high-intensity wave conditions, which disrupted the sediment structure and triggered liquefaction. The resulting vigorous mixing between the sediment and overlying water transformed the sediment into a fluidized state, promoting the release of PAHs into the water column and leading to an overall reduction in residual PAH concentrations in the sediment. Simultaneously, the expansion of seepage channels facilitated the further downward migration of deeply buried contaminants [49].

5. Conclusions

Through systematic wave flume experiments, this study elucidates the complex interplay between hydrodynamic forces and PAH mobilization in coastal sediments. Under static conditions (Stage I), PAH release was primarily driven by molecular diffusion along concentration gradients, with upward migration enhanced by density differences between PAHs and seawater, resulting in higher dissolved PAH concentrations in the upper water column. The sediment profile developed a distinct concentration peak at 5 cm depth, reflecting the balance between upward diffusion and downward porewater transport.

The introduction of moderate wave action (Stage II, 5 cm wave height) revealed nuanced transport dynamics. While wave-induced resuspension enhanced porewater advection, PAH release remained limited due to competing processes. Fine suspended particles with high specific surface areas effectively scavenged dissolved PAHs from the water column, as evidenced by the temporal variations in SSC and D50. Simultaneously, the development of a deeper concentration peak at 12 cm sediment depth indicated active downward migration through the sediment matrix.

High-intensity waves (12 cm wave height) triggered sediment liquefaction, leading to complete homogenization of PAH concentrations throughout the sediment column, a fivefold increase in near-bed dissolved PAHs, and contaminant migration extending to 20 cm depth, highlighting liquefaction's capacity to fundamentally alter sediment–pollutant interactions.

A wave flume system was employed to enable synchronized monitoring of porewater pressure, suspended sediment concentration (SSC), and PAH release, with high-frequency multi-depth sampling providing detailed characterization of PAH mobilization dynamics. Liquefaction was found to significantly enhance the mobility of low-ring PAHs, and a strong positive correlation between SSC and PAH release, particularly during liquefaction events, identified fine-grained, organic-rich sediment deposits as critical zones for pollution release, emphasizing the need for storm-resilient pollution control strategies. Nevertheless, laboratory-scale flumes may not fully reproduce natural turbulence, and remolded sediments may differ from naturally aged deposits in adsorption–desorption behavior, potentially leading to overestimation of PAH release. Short-duration wave stages also do not capture the full temporal dynamics of storm events, particularly for slow-release high-ring PAHs. Building on these findings, future research should include multi-factor experiments incorporating bio-physical interactions, co-migration with other contaminants, and sediment aging effects, alongside field validation in the Yellow River Delta to test liquefaction thresholds and quantify PAH fluxes under storm conditions. These results further highlight the importance of targeted risk management strategies, such as depth-specific capping or sediment stabilization, to mitigate PAH release in liquefaction-prone areas.

Author Contributions: Conceptualization, F.L. and W.L.; methodology, F.L.; investigation, F.L.; data curation, F.L. and Q.S.; writing—original draft preparation, Q.S.; writing—review and editing, F.L.; visualization, F.L. and Q.S.; supervision, F.L.; project administration, F.L.; funding acquisition, F.L. and W.L. All authors have read and agreed to the published version of the manuscript.

Funding: This research was funded by the National Natural Science Foundation of China (No. 41807247).

Data Availability Statement: The raw data supporting the conclusions of this article will be made available by the authors on request.

Acknowledgments: The authors gratefully acknowledge the invaluable technical and financial support provided by the Observation and Research Station of Seawater Intrusion and Soil Salinization, Laizhou Bay, Ministry of Natural Resources (Qingdao, China). We also extend our sincere thanks to the Shandong Provincial Key Laboratory of Marine Environment and Geological Engineering (Qingdao, China) for their essential technical assistance throughout this research.

Conflicts of Interest: The authors declare no conflicts of interest.

References

1. Ghandourah, M.A. An insightful overview of the distribution pattern of polycyclic aromatic hydrocarbon in the marine sediments of the Red Sea. *Open Chem. J.* **2022**, *20*, 777–784. [CrossRef]
2. Abdel-Shafy, H.I.; Mansour, M.S.M. A review on polycyclic aromatic hydrocarbons: Source, environmental impact, effect on human health and remediation. *Egypt. J. Pet.* **2016**, *25*, 107–123. [CrossRef]
3. Montuori, P.; De Rosa, E.; Di Duca, F.; De Simone, B.; Scippa, S.; Russo, I.; Sarnacchiaro, P.; Triassi, M. Polycyclic Aromatic Hydrocarbons (PAHs) in the Dissolved Phase, Particulate Matter, and Sediment of the Sele River, Southern Italy: A Focus on Distribution, Risk Assessment, and Sources. *Toxics* **2022**, *10*, 401. [CrossRef]
4. Lu, Z.; Tian, W.; Chen, Z.; Chu, M.; Zhang, S.; Liu, B.; Zhao, J.; Zou, M.; Huo, B.; Xu, G. Release of PAHs from sediments to seawater under wave: Indoor microcosms and level IV fugacity models. *J. Hazard. Mater.* **2024**, *474*, 134799. [CrossRef]
5. Maletić, S.P.; Beljin, J.M.; Rončević, S.D.; Grgić, M.G.; Dalmacija, B.D. State of the art and future challenges for polycyclic aromatic hydrocarbons in sediments: Sources, fate, bioavailability and remediation techniques. *J. Hazard. Mater.* **2019**, *365*, 467–482. [CrossRef]
6. Dong, J.; Xia, X.; Wang, M.; Xie, H.; Wen, J.; Bao, Y. Effect of recurrent sediment resuspension-deposition events on bioavailability of polycyclic aromatic hydrocarbons in aquatic environments. *J. Hydrol.* **2016**, *540*, 934–946. [CrossRef]
7. Li, Y.; Tang, C.; Wang, J.; Acharya, K.; Du, W.; Gao, X.; Luo, L.; Li, H.; Dai, S.; Mercy, J.; et al. Effect of wave-current interactions on sediment resuspension in large shallow Lake Taihu, China. *Environ. Sci. Pollut. Res.* **2017**, *24*, 4029–4039. [CrossRef]
8. Ding, Y.; Sun, L.; Qin, B.; Wu, T.; Shen, X. Characteristics of sediment resuspension in Lake Taihu, China: A wave flume study. *J. Hydrol.* **2018**, *561*, 702–710. [CrossRef]
9. Ma, Y.; Wang, P.; Hua, Z.; Dong, Y.; Yu, L.; Huang, S. Field study on endogenous perfluoroalkyl acid release and their spatiotemporal distribution processes induced by inland navigation. *Sci. Total Environ.* **2024**, *916*, 170394. [CrossRef] [PubMed]
10. Wang, J.; Chai, J.; Xu, R.; Pang, Y. The effects of wind-wave disturbances on sediment resuspension and phosphate release in Lake Chao. *Sci. Total Environ.* **2024**, *912*, 169254. [CrossRef] [PubMed]
11. Liu, X.; Liu, S.; Zhao, H.; Jeng, D.S.; Guo, L.; Zheng, T. Numerical investigation of solute migration and release from sediments driven by wave-induced accumulation of porewater pressure. *Adv. Water Resour.* **2023**, *179*, 104508. [CrossRef]
12. Li, R.; Feng, C.; Wang, D.; Li, B.; Hu, L.; Shen, Z. Role of salinity in the multiphase redistribution of polycyclic aromatic hydrocarbons (PAHs) in sediment suspension. Environ. *Earth Sci.* **2016**, *75*, 116. [CrossRef]
13. Yang, X.; Yu, L.; Chen, Z.; Xu, M. Bioavailability of Polycyclic Aromatic Hydrocarbons and their Potential Application in Eco-risk Assessment and Source Apportionment in Urban River Sediment. *Sci. Rep.* **2016**, *6*, 23134. [CrossRef]
14. Duran, R.; Cravo-Laureau, C. Role of environmental factors and microorganisms in determining the fate of polycyclic aromatic hydrocarbons in the marine environment. *FEMS Microbiol. Rev.* **2016**, *40*, 814–830. [CrossRef] [PubMed]
15. Hu, W.; Zhang, J.; Li, D.; Yuan, Y.; Tang, Y.; Hui, K.; Jiang, Y.; Tan, W. Study on factors influencing the transport and transformation of polycyclic aromatic hydrocarbons in soil-groundwater systems. *Emerg. Contam.* **2025**, *11*, 100472. [CrossRef]
16. Vijayanand, M.; Ramakrishnan, A.; Subramanian, R.; Issac, P.K.; Nasr, M.; Khoo, K.S.; Rajagopal, R.; Greff, B.; Wan Azelee, N.I.; Jeon, B.-H.; et al. Polyaromatic hydrocarbons (PAHs) in the water environment: A review on toxicity, microbial biodegradation, systematic biological advancements, and environmental fate. *Environ. Res.* **2023**, *227*, 115716. [CrossRef]

17. Wang, D.; Feng, C.; Huang, L.; Niu, J.; Shen, Z. Historical deposition behaviors of PAHs in the Yangtze River Estuary: Role of the sources and water currents. *Chemosphere* **2013**, *90*, 2020–2026. [CrossRef]
18. Ma, J.; Ailijiang, N.; Mamat, A.; Wu, Y.; Luo, X.; Li, M. Fugacity-based multimedia transport modeling and risk assessment of PAHs in Urumqi. *Sci. Rep.* **2025**, *15*, 4063. [CrossRef]
19. *Method 8260D (SW-846)*; Volatile Organic Compounds by Gas Chromatography/Mass Spectrometry (GC/MS), Revision 3. U.S. EPA: Washington, DC, USA, 2006.
20. *Method 8275A (SW-846)*; Semivolatile Organic Compounds in Soil/Sludges and Solid Wastes Using Thermal Extraction/Gas Chromatography/Mass Spectrometry (TE/GC/MS). U.S. EPA: Washington, DC, USA, 1996.
21. Lu, F.; Zhang, H.Q.; Jia, Y.G.; Liu, W.; Wang, H. Migration and Diffusion of Heavy Metal Cu from the Interior of Sediment during Wave-Induced Sediment Liquefaction Process. *J. Mar. Sci. Eng.* **2019**, *7*, 449. [CrossRef]
22. *GBT 26411-2010*; Marine Water Quality—Determination of 16 Polycyclic Aromatic Hydrocarbons—Gas Chromatography-Mass Spectrometry. General Administration of Quality Supervision, Inspection and Quarantine of the People's Republic of China, Standardization Administration of the People's Republic of China: Beijing, China, 2011.
23. *HJ 805-2016*; Water Quality—Determination of 16 Polycyclic Aromatic Hydrocarbons in Sediments—Gas Chromatography-Mass Spectrometry. Ministry of Environmental Protection of the People's Republic of China: Beijing, China, 2016.
24. Liu, X.; Jia, Y.; Zheng, J.; Wen, M.; Shan, H. An Experimental Investigation of Wave-induced Sediment Responses in a Natural Silty Seabed: New Insights into Seabed Stratification. *Sedimentology* **2016**, *64*, 508–529. [CrossRef]
25. Tong, L.L.; Zhang, J.S.; Chen, N.; Lin, X.F.; He, R.; Sun, L. Internal solitary wave-induced soil responses and its effects on seabed instability in the South China Sea. *Ocean Eng.* **2024**, *310*, 118697. [CrossRef]
26. Liu, X.; Jia, Y.; Zheng, J.; Hou, W.; Zhang, L.; Zhang, L.; Shan, H. Experimental evidence of wave-induced inhomogeneity in the strength of silty seabed sediments: Yellow River Delta, China. *Ocean. Eng.* **2013**, *59*, 120–128. [CrossRef]
27. Tzang, S.; Ou, S. Laboratory flume studies on monochromatic wave-fine sandy bed interactions: Part 1. Soil fluidization. *Coast. Eng.* **2006**, *53*, 965–982. [CrossRef]
28. Tzang, S.; Ou, S.; Hsu, T. Laboratory flume studies on monochromatic wave-fine sandy bed interactions Part 2. Sediment suspensions. *Coast. Eng.* **2009**, *56*, 230–243. [CrossRef]
29. Niu, J.; Xu, J.; Dong, P.; Li, G. Porewater pressure responses in silty sediment bed under random wave action. *Sci. Rep.* **2019**, *9*, 11685. [CrossRef]
30. Xu, X.B.; Xu, G.H.; Yang, J.J.; Xu, Z.Q.; Ren, Y.P. Field observation of the wave-induced pore pressure response in a silty soil seabed. *Geo-Mar. Lett.* **2021**, *41*, 13. [CrossRef]
31. Liu, X.; Lu, Y.; Yu, H.; Ma, L.; Li, X.; Li, W.; Zhang, H.; Bian, C. In-Situ Observation of Storm-Induced Wave-Supported Fluid Mud Occurrence in the Subaqueous Yellow River Delta. *J. Geophys. Res. Ocean.* **2022**, *127*, e2021JC018190. [CrossRef]
32. Sun, Y.; Zhang, S.; Lan, J.; Xie, Z.; Pu, J.; Yuan, D.; Yang, H.; Xing, B. Vertical migration from surface soils to groundwater and source appointment of polycyclic aromatic hydrocarbons in epikarst spring systems, southwest China. *Chemosphere* **2019**, *230*, 616–627. [CrossRef]
33. Baas, J.H.; Manica, R.; Puhl, E.; de Oliveira Borges, A.L. Thresholds of intrabed flow and other interactions of turbidity currents with soft muddy substrates. *Sedimentology* **2016**, *63*, 2002–2036. [CrossRef]
34. Dong, J.; Xu, J.; Li, G.; Li, A.; Zhang, S.; Niu, J.; Xu, X.; Wu, L. Experimental Study on Silty Seabed Liquefaction and Its Impact on Sediment Resuspension by Random Waves. *J. Mar. Sci. Eng.* **2022**, *10*, 437. [CrossRef]
35. Li, R.; Feng, C.; Wang, D.; Li, B.; Shen, Z. Multiphase redistribution differences of polycyclic aromatic hydrocarbons (PAHs) between two successive sediment suspensions. *Front. Environ. Sci. Eng.* **2016**, *10*, 381–389. [CrossRef]
36. Li, C.; Wang, Y.; Gao, F.; Yang, L. Spatiotemporal evolution of excess pore pressures in a silty seabed under progressive waves during residual liquefaction. *Appl. Ocean Res.* **2022**, *129*, 103401. [CrossRef]
37. Wang, L.; Zhang, J.; Jeng, D.; Zhang, Q.; Chen, T. Experimental study on the dynamic response of a silty seabed under waves. *Ocean. Eng.* **2023**, *269*, 113554. [CrossRef]
38. Yang, Z.; Feng, J.; Niu, J.; Shen, Z. Release of polycyclic aromatic hydrocarbons from Yangtze River sediment cores during periods of simulated resuspension. *Environ. Pollut.* **2008**, *155*, 366–374. [CrossRef]
39. Lu, Z.; Tian, W.; Zhang, S.; Chu, M.; Zhao, J.; Jiang, J.; Zou, M.; Liu, B. A study combining a sediment-seawater microcosm with multimedia fugacity model to evaluate the effect of tidal cycles on polycyclic aromatic hydrocarbon release from sediments. *Sci. Total Environ.* **2023**, *891*, 164340. [CrossRef] [PubMed]
40. Liu, K.; Wang, X.; Zhang, H.; Wei, Y.; Zhao, G.; Liu, X.; Cui, H.; Han, J. Sources, transport and fate of polycyclic aromatic hydrocarbons (PAHs) in a typical river-estuary system in the North China: From a new perspective of PAHs loading. *Mar. Pollut. Bull.* **2025**, *117*, 692. [CrossRef]
41. Usanase, G.; Azema, N.; El Bitouri, Y.; Souche, J.-C.; Gonzalez, C. Contribution of settling measurements to the study of polycyclic aromatic hydrocarbons' (PAHs) mobilisation during resuspension of PAHs-associated sediment. *Environ. Sci. Pollut. Res.* **2021**, *28*, 68349–68363. [CrossRef] [PubMed]

42. Berríos-Rolón, P.J.; Cotto, M.C.; Márquez, F. Polycyclic Aromatic Hydrocarbons (PAHs) in Freshwater Systems: A Comprehensive Review of Sources, Distribution, and Ecotoxicological Impacts. *Toxics* **2025**, *13*, 321. [CrossRef]
43. Yan, D.; Meng, L.; Li, H.; Song, T.; Sun, P.; Bao, M.; Li, X. Petroleum hydrocarbon release behavior study in oil-sediment aggregates: Turbulence intensity and chemical dispersion effect. *RSC Adv.* **2019**, *9*, 7922–7931. [CrossRef]
44. Guigue, C.; Tedetti, M.; Dang, D.; Mullot, J.; Garnier, C.; Goutx, M. Remobilization of polycyclic aromatic hydrocarbons and organic matter in seawater during sediment resuspension experiments from a polluted coastal environment: Insights from Toulon Bay (France). *Environ. Pollut.* **2017**, *229*, 627–638. [CrossRef]
45. Mackay, D.; Shiu, W.Y.; Ma, K.C.; Lee, S.C. *Handbook of Physical-Chemical Properties and Environmental Fate for Organic Chemical*; CRC Press: Boca Raton, FL, USA, 2006.
46. Castillo-Ilabaca, C.; Gutiérrez, M.H.; Aranda, M.; Henríquez-Aedo, K.; Pereira, A.; Salamanca, M.; Galand, P.E.; Jessen, G.L.; Pantoja-Gutiérrez, S. PAH contamination in coastal surface sediments. *Sci. Rep.* **2024**, *15*, 12345. [CrossRef]
47. Wang, L.; Yang, Z. Simulation of polycyclic aromatic hydrocarbon remobilization in typical active regions of river system under hydrodynamic conditions. *J. Soils Sediments* **2010**, *10*, 1380–1387. [CrossRef]
48. Shubhankar, B.; Ambade, B. A Review on Deposition, Distribution of Polycyclic Aromatic Hydrocarbons in Different Environmental Matrix and Study Its Toxicity and Carcinogenic Effect. *Asian J. Chem.* **2016**, *28*, 2341–2345. [CrossRef]
49. Minick, D.J.; Anderson, K.A. Diffusive flux of PAHs across sediment-water and water-air interfaces at urban Superfund sites. *Environ. Toxicol. Chem.* **2017**, *36*, 2281–2289. [CrossRef]

Journal of
Marine Science and Engineering

MDPI

Article

Integration of the CEL and ML Methods for Landing Safety Prediction and Optimization of Full-Scale Track Design in a Deep-Sea Mining Vehicle

Yifeng Zeng [1], Zongxiang Xiu [1,*], Lejun Liu [1], Qiuhong Xie [1], Yongfu Sun [2], Jianghui Yang [3] and Xingsen Guo [4,5,6,7,*]

1 First Institute of Oceanography, Ministry of Natural Resources, Qingdao 266061, China
2 National Deep Sea Base Management Center, Qingdao 266237, China
3 China Offshore Oil Engineering Corp CNOOC Ltd., Tianjin 300461, China
4 Shandong Provincial Key Laboratory of Marine Environment and Geological Engineering, Ocean University of China, Qingdao 266100, China
5 Sanya Oceanographic Institution, Ocean University of China, Sanya 572024, China
6 Department of Civil, Environmental and Geomatic Engineering, University College London, London WC1E 6BT, UK
7 Department of Engineering, University of Cambridge, Cambridge CB2 1PZ, UK
* Correspondence: xzx@fio.org.cn (Z.X.); xg298@cam.ac.uk (X.G.)

Abstract: Ensuring the safe landing of deep-sea mining vehicles (DSMVs) on soft seabed sediments is critical for the stability and operational reliability of subsea mineral extraction. However, deep-sea sediments, particularly in polymetallic nodule regions, are characterized by low shear strength, high compressibility, and rate-dependent behavior, posing significant challenges for full-scale experimental investigation and predictive modeling. To address these limitations, this study develops a high-fidelity finite element simulation framework based on the Coupled Eulerian–Lagrangian (CEL) method to model the landing and penetration process of full-scale DSMVs under various geotechnical conditions. To overcome the high computational cost of FEM simulations, a data-driven surrogate model using the random forest algorithm is constructed to predict the normalized penetration depth based on key soil and operational parameters. The proposed hybrid FEM–ML approach enables efficient multiparameter analysis and provides actionable insights into the complex soil–structure interactions involved in DSMV landings. This methodology offers a practical foundation for engineering design, safety assessment, and descent planning in deep-sea mining operations.

Keywords: deep-sea soft sediments; deep-sea mining vehicle; large deformation finite element method; random forest algorithm; prediction method

1. Introduction

With the expansion of marine resource exploration into deep-sea territories, the demand for specialized underwater mining systems has increased globally [1–4]. Among these, deep-sea mining vehicles (DSMVs) play a pivotal role in harvesting valuable seabed minerals such as polymetallic nodules and sulfides [5–8]. However, their deployment is hindered by the unique mechanical traits of deep-sea sediments, namely, their low resistance to shear, high compressibility, and near-fluid behavior, all of which pose challenges for vehicle stability and control [9,10]. Understanding the settlement behavior of DSMVs on these sediments is critical to ensure their operational stability and minimize their impact on sensitive marine environment. Tracked mining vehicles (as shown in Figure 1) have emerged as the most viable solution for these applications because of their ability to distribute loads and maintain mobility across soft, yielding substrates [11,12]. Therefore, the settlement mechanism of tracked mining vehicles in deep-sea sediments has received increasing attention.

Figure 1. Mining vehicles of polymetallic sulfides developed by Nautilus Minerals Commence (Source: https://dsmobserver.com/2017/07/nautilus-png-submerged-trials/ accessed on 5 December 2024).

The traction and settlement behavior of tracked DSMVs are intrinsically linked, especially under conditions involving significant soil deformation [13]. The classical pressure–sinkage relationship proposed by Bekker remains a foundational framework for understanding terrain interactions, particularly highlighting how soil cohesion, fraction, and saturation levels influence deformation behavior [13–15]. Nevertheless, its applicability is often limited under the large-deformation and low-strength scenarios typical in deep-sea environments [16,17], and seabed soil heterogeneity presents challenges for these models. Variations in sediment properties complicate predictions, as noted in studies on seabed heterogeneity [18–20]. Research on DSMV ground interactions has historically been constrained by the difficulty of recreating authentic deep-sea conditions in laboratory settings. The unique mechanical behavior of abyssal soils—particularly their rate-dependent and strain-softening responses—renders traditional experimental methodologies inadequate [4,21–25]. Moreover, the complexity and scale of full-track systems further complicate physical replication, thereby necessitating advanced computational methods that can simulate realistic seabed interactions under varying environmental and operational parameters. Prior work by Einav and Randolph [26] emphasized the significance of strain-softening behavior in large-deformation problems, highlighting the necessity of incorporating more advanced constitutive formulations in such contexts.

Although widely adopted in terramechanics, Bekker's model exhibits limitations when applied to highly heterogeneous and viscoelastic deep-sea sediments [15,16], prompting the need for more nuanced numerical approaches in extreme geotechnical conditions. Numerical simulations, particularly finite element method (FEM), have emerged as a powerful tools for addressing these challenges. Early work by Benson [27] laid the foundation for CEL-based FEM techniques applied to problems involving extreme deformation and contact separation, providing critical support for their adoption in geotechnical applications. Unlike traditional experimental approaches, FEM allows for the detailed modeling of complex geotechnical phenomena, such as large deformations, sediment–fluid interactions, and the dynamic response of sediments under loading [28–32]. Janarthanan &

Kuttikrishnan [33] examined the sinkage behavior of deep-sea soil under mining vehicle operations using numerical simulations and experimental studies, emphasizing the importance of optimizing the design configuration to ensure stability under varying soil conditions. Amudha et al. [34] analyzed breakout and sinkage forces of tracked vehicles on soft marine clays using finite element methods and field-calibrated models, further supporting the need for integrated predictive approaches in seabed mobility assessments. More recently, Janarthanan et al. [35] performed multi-body dynamic simulations to compare inline and offset track configurations for DSMVs, revealing how geometric layout influences traction efficiency and sinkage in soft seabed conditions. Sun et al. [36,37] investigated the sinking pattern of a single track plate under different conditions with the Coupled Eulerian–Lagrangian (CEL) method, indicating the necessity of considering different track parameters and soil mechanical parameters. Luan et al. [38] analyzed the dynamic stresses generated in the seabed by moving deep-sea mining vehicles, highlighting the critical influence of seabed properties and vehicle parameters on the stress distribution using numerical case studies. Xia et al. [39] developed a multibody dynamics model for a four-track deep-sea mining vehicle and simulated its straight-line motion, demonstrating the impact of vehicle dynamics and seabed conditions on travel performance and stability. Liu et al. [40] investigated the bearing capacity of consolidated clay subjected to track loads from deep-sea mining vehicles, stressing the importance of soil consolidation and stress distribution patterns for assessing vehicle-induced deformation. Similar approaches have also been explored internationally. Tho et al. [41] applied a three-dimensional Eulerian finite element method to simulate penetration of spudcans into seabed soils, successfully capturing large soil deformations and validating the approach against centrifuge tests. Agarwal et al. [17] analyzed wheel–granular media interactions using the Material Point Method (MPM) and Resistive Force Theory (RFT), providing insights into wheel/track–soil dynamics relevant to deep-sea vehicles. Gonzalez and Iagnemma [42] employed deep learning techniques to estimate terrain slip and classify terrain types for ground vehicles, illustrating the potential of ML for terrain–vehicle interaction prediction, while Chavez-Garcia et al. [43] demonstrated that convolutional neural networks trained on simulated terrains can robustly predict ground traversability, reinforcing the strategy of using simulated datasets to inform ML-based safety or mobility models. Zhang et al. [44] studied soil failure patterns caused by vehicle motion trajectories to evaluate the traction performance of deep-sea mining vehicles. Zande et al. [45] applied a coupled numerical–experimental framework to examine track–soil interaction for planetary rovers, offering insights relevant to DSMV track design. White and Randolph [46] investigated seabed–structure interaction using strain-softening models to predict penetration resistance under cyclic loading.

However, most existing studies have focused on simplified track geometries or small-scale models, neglecting the critical influences of full-scale DSMV operations. Moreover, the mechanical behavior of deep-sea soft sediments is often approximated using basic constitutive models, which fail to capture key factors such as strain softening and rate dependency. Several recent studies have demonstrated that finite element software combined with modified constitutive soil models can be employed to characterize this property of deep-sea sediments more accurately. Xiao et al. [47] studied the effects of strain softening on the penetration resistance of offshore bucket foundations in nonhomogeneous clay, emphasizing the importance of accounting for soil heterogeneity and strain-softening mechanisms in design and analysis. Xie et al. [48] developed a convenient method for simulating large-deformation submarine landslides, highlighting the need to incorporate shear-softening and strain rate correlation effects for accurate numerical predictions. Guo et al. [49] conducted a comprehensive study on the bearing capacity factors of T-bar penetrometers during penetration into deep-sea sediments, providing valuable insights into the soil resistance characteristics under various penetration stages. Their findings underscore the necessity of considering the complex soil behavior and penetration dynamics in deep-sea environments.

Although these methods enhance the fidelity of modeling full-scale DSMV subsidence under soft seabed conditions, simulating the entire track system remains a nontrivial task because of its intricate structural and mechanical interdependencies. For instance, Kim and Hossain [19] developed numerical strategies for anchor and soil interaction under strain-rate-dependent seabed conditions, which provide useful parallels for DSMV–seabed interaction analysis. The complexity arises from the interplay between multiple factors, including track–soil contact mechanics, the geometric and material properties of the undercarriage, and the effects of transient dynamic loads during operation. Consequently, such simulations demand considerable computational resources and processing time. To address these limitations, recent research has increasingly explored predictive modeling techniques and algorithmic optimization strategies that aim to reduce the computational burden without compromising model integrity. Among these, machine learning (ML) has gained significant traction in geomechanics and engineering applications [50–52], owing to its ability to learn nonlinear relationships and adapt across various data domains [50–55]. Ensemble-based frameworks such as random forest (RF) have shown strong potential for improving predictive robustness. By integrating the outputs of numerous base learners, these models enhance both accuracy and generalizability under diverse conditions [56–58]. The RF algorithm constructs a collection of decision trees trained on different subsets of the data and combines their predictions to form a consensus output [59]. Its simplicity of deployment, versatility across different input feature spaces, and resilience to irregular data distributions make it a valuable tool for engineering prediction tasks [60].

To address the challenges associated with predicting the landing behavior of DSMVs on ultra-soft seabed sediments, this study establishes a hybrid computational framework that integrates a CEL finite element model with an ensemble-based machine learning algorithm. The proposed FEM model incorporates an improved strain-softening and rate-dependent constitutive law to simulate the full-scale dynamic track–seabed interaction under various sediment strengths, sensitivity levels, and loading conditions. The quantitative influence of these parameters on the dimensionless settlement depth is systematically analyzed, and the underlying mechanisms are discussed. To overcome the high computational cost of large-deformation FEM analyses and enable efficient prediction, the simulation outputs are further used to train a RF regression model. This multiparameter prediction approach achieves high accuracy while offering greater generalizability than conventional empirical formulas.

This study contributes a dual-method evaluation framework that combines high-fidelity numerical simulation with data-driven learning for DSMV landing safety assessment. The framework not only enhances predictive capability in soft marine geology but also provides practical insights for the structural design and operational planning of DSMVs in complex deep-sea environments.

2. FEM Method for Obtaining the DSMV Sinking Process

2.1. CEL Numerical Method

To effectively represent the substantial deformation and material flow that occurs when vehicles interact with the seabed, this research adopts the CEL method [27]. By integrating Eulerian meshes that remain stationary with Lagrangian meshes that move with the material, CEL is capable of accurately simulating the severe distortions typical of soft seafloor environments and scenarios often beyond the reach of traditional modeling approaches [61]. In the ABAQUS 2022 (Dassault Systèmes, Waltham, MA, USA) software framework, this hybrid strategy assigns Eulerian treatment to regions undergoing extreme deformation, whereas relatively stable zones are handled through Lagrangian modeling. A volume fraction tool facilitates the representation of material distribution by generating a field where the Eulerian volume fraction (EVF) varies between 0 and 1, indicating the presence of materials. Figure 2 illustrates the comparative behavior of the two methods. The system ensures accurate tracking of material flow through remeshing procedures and adheres to conservation laws, whereas dynamic interaction at the Eulerian–Lagrangian boundaries (where EVF equals 1) is managed via contact mechanics. CEL is extensively utilized in explicit dynamic simulations, including scenarios such as track–terrain interactions, making it a robust approach for handling complex deformation and contact phenomena.

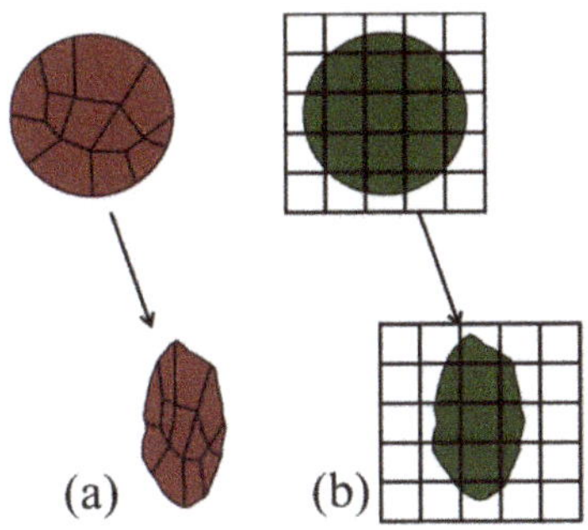

Figure 2. Continuum deformation using the (**a**) Lagrangian and (**b**) Eulerian methods.

2.2. A Constitutive Soil Model That Integrates Strain Softening and Strain Rates

To realistically represent the mechanical behavior of soft deep-sea soils, this study adopts a strain-softening and rate-sensitive constitutive model, which was originally proposed by Einav and Randolph [26]. This model enables a dynamic update of the undrained shear strength based on the cumulative plastic strain and applied shear strain rate. Under undrained conditions, soft seabed soils exhibit post-peak stress–strain softening, quantified by the cumulative plastic strain [26,62]. Under compression by an applied load, strain rate effects increase the soil's undrained shear strength [63–67]. These effects can be expressed using semilogarithmic or power-law functions [68–71].

$$S_u = S_{u0}\alpha\beta = S_{u0}[\delta_{rem} + (1-\delta_{rem})\exp(-3\xi/\xi_{95})]\left[1+\eta\left(\dot{\gamma}/\dot{\gamma_{ref}}\right)^n\right] \quad (1)$$

where S_u is the current undrained strength and S_{u0} is the undrained shear strength of the undisturbed soil. δ_{rem} is the inverse of the soil sensitivity S_t, denoting the ratio between the fully remolded and initial shear strengths; ξ_{95} is the accumulated shear strain when 95% soil strength degradation occurs between the intact and fully remolded conditions; accumulated absolute plastic shear strain, ξ; η is the viscosity coefficient; where n is the shear-thinning index; $\dot{\gamma}_{ref}$ is the reference shear strain rate; and $\dot{\gamma}$ is shear strain rate. Each increment (dt) with respective to $\dot{\gamma}$ in Equation (2) can be calculated from

$$\dot{\gamma} = \frac{(\Delta\varepsilon_1 - \Delta\varepsilon_3)}{\Delta t} \quad (2)$$

The process of realizing the modified constitutive soil model through numerical modeling is as follows: Shear strength is initially defined by S_{u0} and updated dynamically using cumulative strain and rate effects through the VUSDFLD and VGETVRM subroutines [47,48]. The shear strength evolves iteratively, as calculated from the cumulative plastic strains and updated coefficients (α and β), enabling the precise simulation of soil deformation under dynamic loading.

To assess the feasibility of the proposed model and verify its simulation accuracy, a falling cone test previously reported in the literature was numerically reproduced. The experimental setup was conducted under atmospheric pressure conditions and is widely used in geotechnical laboratories to estimate soil shear strength based on penetration depth. The geometrical parameters of the simulation model were based on a falling cone apparatus manufactured by GEONOR Inc., Oslo, Norway, as shown in Figure 3a. In the test, the cone tip contacted the soil surface and was released from a magnetic holder, with shear strength calculated using Hansbo's empirical equation [72]. Although the original falling cone test was performed in air under atmospheric conditions, the soil sample used was saturated clay. Therefore, in the corresponding numerical simulation, the soil density was adjusted to its submerged density by subtracting the density of water. This buoyant unit weight assumption is commonly employed in marine geotechnical modeling to simulate soil behavior under effective stress conditions. Accordingly, the revised simulation more realistically captures the large-deformation behavior of saturated cohesive soils while remaining consistent with the reference experimental setup.

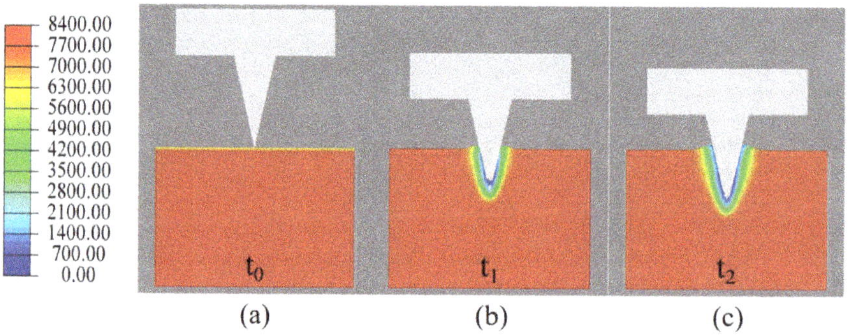

Figure 3. Distribution and variation in the shear strength (Pa) of Sample III at different moments of the numerical modeling process: (**a**) Initial moment; (**b**) Penetration process; (**c**) Final moment.

Table 1 lists the soil sample parameters. A 100 g cone with a 30° tip angle was used in the simulation. As shown in Figure 4, the numerical simulation results show good agreement with the experimental data, indicating that the CEL-based numerical model combined with a modified soil constitutive model is capable of capturing the large-deformation behavior during dynamic penetration. This validation, although conducted under simplified conditions, lays the groundwork for the subsequent application of the model to simulate DSMV–seabed interaction under submerged conditions.

Table 1. Physical and mechanical characteristics of the sample soil.

Sample No.	Submerged Density (kg/m^3)	Cohesion (kPa)	Elastic Modulus (MPa)	Soil Sensitivity	Poission's Ratio
I	590	15	7.5	10	0.49
II	520	4.2	2.1	42	0.49
III	660	8.2	4.1	16.4	0.49
IV	180	11	5.5	8.5	0.49

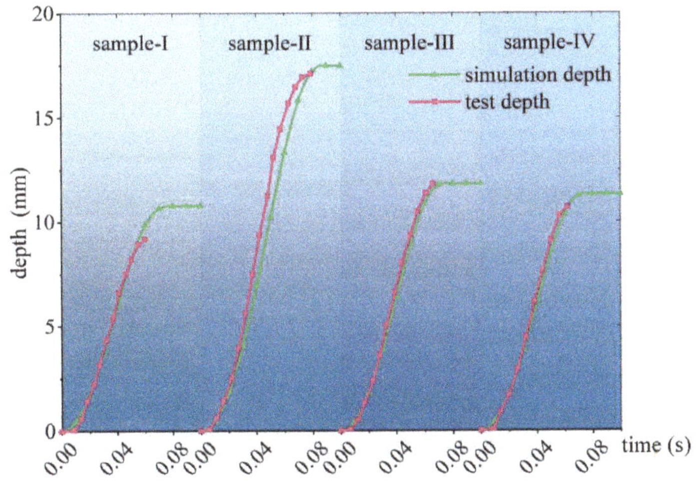

Figure 4. Comparison between the numerical simulation results and the fall cone test results.

2.3. Numerical Model of the Track–Soil Sinkage Interaction

The full-scale track–soil interaction model developed in this study is inspired by the "Pioneering-I" DSMV prototype designed by Sun et al. [36]. The track system modeled includes detailed mechanical components such as single-pin plates, sprockets, and support assemblies. Figure 5a shows a gear radius of 0.5 m and a spacing of 3.12 m between gears. The single-pin track plate design (Figure 5b) has dimensions of 0.6 m × 0.15 m × 0.01 m.

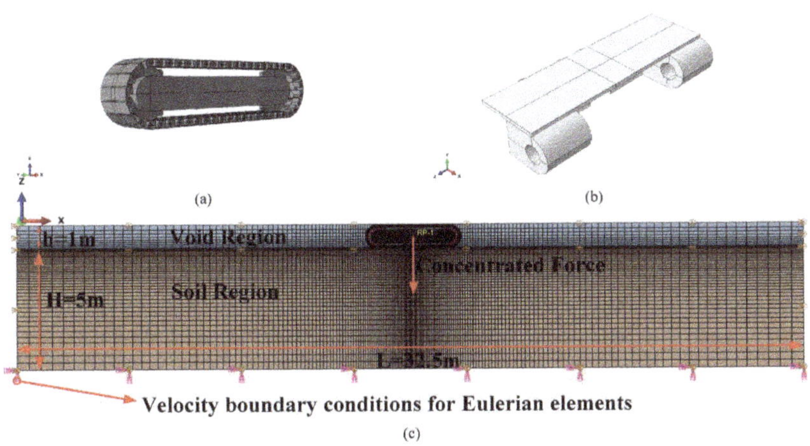

Figure 5. Geometric parameters of the track and track plate: (**a**) Full-size track geometry model; (**b**) Track plate geometry model; (**c**) Overall diagram of the model.

The seabed model, constructed with Eulerian elements, consists of material and void regions to simulate sediment displacement and ensure convergence. Figure 5c shows the parameters: the soil height is five times the track height, with the length and width scaled to 10 and 20 times the track size, respectively. The void region is 1 m high. To approximate the submerged conditions typical of deep-sea environments, the soil material was assigned a buoyant density, obtained by subtracting the density of seawater from the total soil density. This assumption enables the simulation to reflect effective stress conditions and is commonly used in marine geotechnical modeling. Owing to sediment softness, the track is modeled as a rigid body with constraints applied at its center of mass for velocity and load control.

Separate Eulerian and Lagrangian meshes are used for the simulation. The C3D8R and EC3D8R elements model the Lagrangian and Eulerian regions, respectively, providing accurate representations of seabed–track interactions. This validated framework ensures reliable analysis of DSMV operations in deep-sea environments.

2.4. Sensitivity Analyses of Mesh Parameters

To ensure numerical stability and computational efficiency, a mesh sensitivity analysis was conducted on the Eulerian domain, with the mesh division information detailed in Table 2. First, the sediment resistance acting on the DSMV was tested under different mesh sizes for the same sinking depth. The relative errors between models with different mesh sizes were subsequently calculated, and the results are shown in Figure 6, with the simulations performed on a 12th Gen Intel(R) Core(TM) i7-12700 CPU. Overall, the calculation results with mesh sizes of 0.05 m, 0.04 m, and 0.03 m have relatively good convergence. Within certain depth ranges, the results for the 0.05 m mesh size exhibited larger deviations than those for the 0.03 m and 0.04 m mesh sizes. In contrast, the models with mesh sizes of 0.03 m and 0.04 m maintained better consistency with the 0.02 m mesh model across all depth ranges. This consistency indicates that the model achieves a high level of accuracy at these mesh resolutions, making 0.03 m and 0.04 m suitable choices for balancing computational efficiency and accuracy. This agreement indicates that the model maintains a high simulation accuracy at these mesh resolutions. Considering both the accuracy of numerical convergence and the need for computational efficiency, this study selects a central mesh size of 0.04 m for the densest Eulerian elements. Beyond the core region, the mesh size is gradually increased toward the periphery, with the largest elements reaching 0.5 m at the boundaries. This hierarchical mesh design balances precision in critical areas with reduced computational costs in less critical regions, ensuring an efficient and reliable simulation framework.

Table 2. Mesh sensitivity parameter setting.

	Case 1	Case 2	Case 3	Case 4	Case 5	Case 6
Center region	0.02	0.03	0.04	0.05	0.06	0.08
Edge region	0.5	0.5	0.5	0.5	0.5	0.5
Number of elements	1,905,120	1,503,663	1,248,912	1,091,475	940,212	787,626
Stable time increment (s)	4.89×10^{-5}	7.34×10^{-5}	9.78×10^{-5}	1.22×10^{-4}	1.47×10^{-4}	1.96×10^{-4}

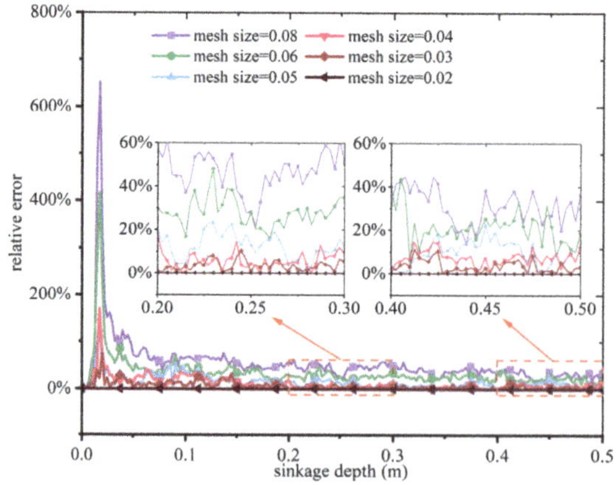

Figure 6. Mesh sensitivity analysis results.

3. Results and Discussion

This section provides an analytical discussion of the sinking behavior of the DSMV in deep-sea soft sediments, dividing the analysis into three parts using the proposed numerical model. The first part focuses on the undrained shear strength of the sediment, including its magnitude and spatial distribution. The second section examines the influence of strain-softening and strain rate parameters on the mechanical response of sediment. Finally, the discussion addresses the dynamic factors of the DSMV, such as its weight, bottom velocity, and size of the contact area.

3.1. Effect of the Sediment Undrained Shear Strength

To analyze the effect of the undrained shear strength of the sediment on the maximum sinkage depth, Figure 7a shows the sediment surface shear strength S_{u0} and its rate of change with depth as a function of the sinkage depth. Under the same softening effect parameter and DSMV dynamic parameter conditions, the maximum sinkage depth of the DSMV exhibits the opposite trend with respect to the sediment surface shear strength, i.e., the maximum sinkage depth of the DSMV decreases with increasing sediment shear strength. The rate of change of the curve decreases as the shear strength increases. The difference in sinking depth is close to 20 cm for the five strength conditions under these conditions. Figure 7b shows the relationship between the maximum sinking depth and the rate of change of shear strength with depth for a shear strength of 5 kPa. Under the same surface shear strength condition, the higher the rate of change in shear strength, the greater the shear strength at the same depth, resulting in a smaller sinking depth of the DSMV. In conclusion, the figure showed that the inverse proportional functional relationship is reflected between the shear strength and the maximum sinkage depth, which is consistent with the relationship described in Equation (3).

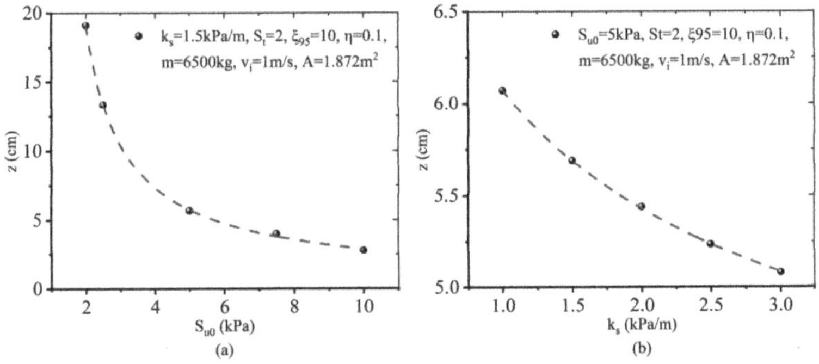

Figure 7. Variation in the maximum DSMV sinkage depth with increasing undrained shear strength magnitude and distribution: (**a**) Different surface shear strengths; (**b**) Different shear strength enhancement factors.

This trend reflects the fundamental role of shear strength in resisting local failure. A higher undrained shear strength enhances the mobilized bearing capacity, leading to a more stable stress distribution beneath the track. As the stress path approaches failure, lower shear strength sediments mobilize plastic strain earlier, resulting in deeper penetration. As shown in Figure 8, the cumulative shear strain occurring in the smaller deposits of k_s under the action of the tracks is greater, and the shear zone is deeper.

3.2. Effect of the Softening Effect Parameter

To further investigate the influence of softening-related parameters on the settlement behavior of the DSMV, we conducted a parametric study focusing on the soil sensitivity

(S_t), strain at 95% strength degradation (ξ_{95}), and viscosity coefficient (η). These parameters reflect the post-yield strength degradation, softening evolution rate, and sediment resistance to deformation under shear loading, respectively.

Figure 8. Accumulated plastic shear strain under low vs. high k_s.

As shown in Figure 9a, increasing the soil sensitivity results in a greater maximum sinkage depth. This can be attributed to the reduced post-disturbance shear strength in highly sensitive sediments. Once disturbed, such soils rapidly lose their strength, leading to localized failure zones that facilitate deeper track penetration. The rate of increase in settlement tends to diminish as St continues to increase, indicating a nonlinear sensitivity–sinkage relationship, possibly due to the saturation of strain localization effects. Figure 9b illustrates the effect of ξ_{95}. A higher ξ_{95} implies that the sediment undergoes more cumulative strain before reaching significant strength degradation. This delays the formation of shear bands, thus improving the load-bearing capacity and reducing settlement. In contrast, lower ξ_{95} values lead to rapid strength loss and early onset of localized failure. In Figure 9c, the role of viscosity is evident. Sediments with higher viscosity coefficients exhibit a more pronounced resistance to flow, thereby restricting the plastic deformation of soil particles. This dampens the velocity-induced inertial effects during landing and limits the formation of failure surfaces, resulting in reduced sinkage. Figure 10 visually confirms this mechanism, with the higher viscosity leading to a broader strain-hardened region within the sediment below the track, where an increase in shear strength occurs.

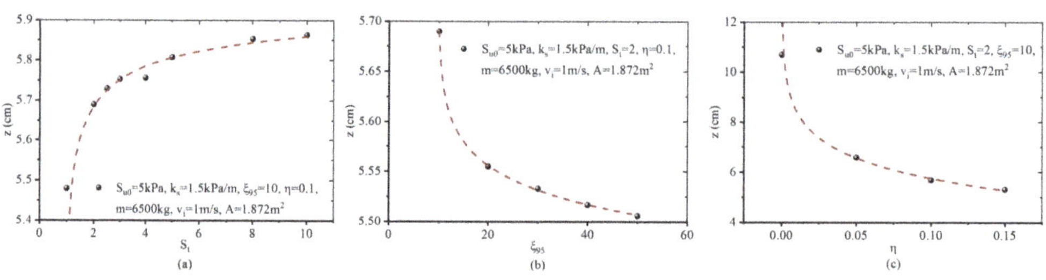

Figure 9. Relationship between the softening effect parameter and the maximum depth of sinkage of the DSMV under the same load and shear strength conditions: (**a**) Different soil sensitivities; (**b**) Different plastic strain parameters; (**c**) Different soil viscosities.

These results highlight the critical role of softening characteristics in shaping the post-yield behavior of soil and the transition from elastic-plastic to viscoplastic deformation modes.

3.3. Effect of the Dynamic Parameter Situation

The dynamic conditions of the DSMV, particularly bottom velocity, track–sediment contact area, and vehicle mass, significantly affect its interaction with deep-sea sediments.

Figure 11 presents the relationship between these parameters and the corresponding maximum sinkage depth.

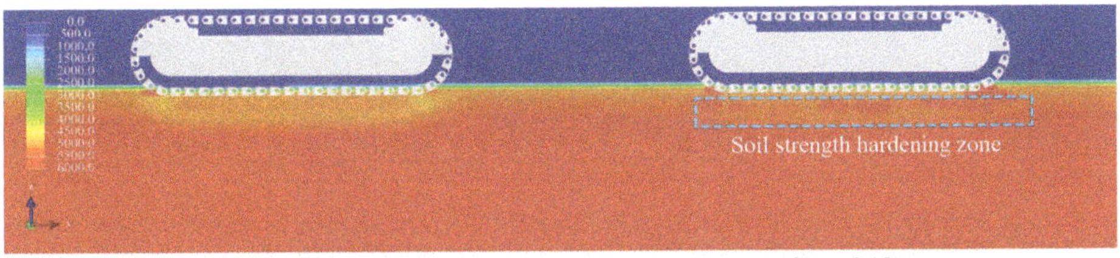

(a)$\eta = 0.05$ (b)$\eta = 0.15$

Figure 10. Shear strength contours under different viscosity coefficients (η).

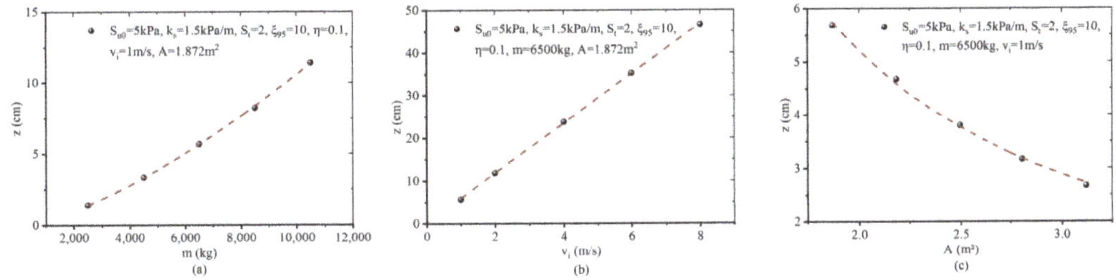

Figure 11. Relationship between the maximum sinkage of the DSMV and its dynamic parameters under the same soil parameter conditions: (**a**) Different wet weights of mining vehicles; (**b**) Different bottom contact speeds; (**c**) Different contact areas.

The observed trends indicate that higher landing velocities and larger vehicle masses result in increased settlement depths. This is due to the kinetic energy transferred to the soil upon contact, which increases the shear strain rate and accelerates the deformation process. The additional energy input not only induces localized failure but also expands the plastically deformed region beneath the track, as evidenced by Figure 12, which shows the soil velocity field contours. These contours reveal high-velocity zones around the failure fronts, indicating fluid-like flow behavior in low-viscosity sediment.

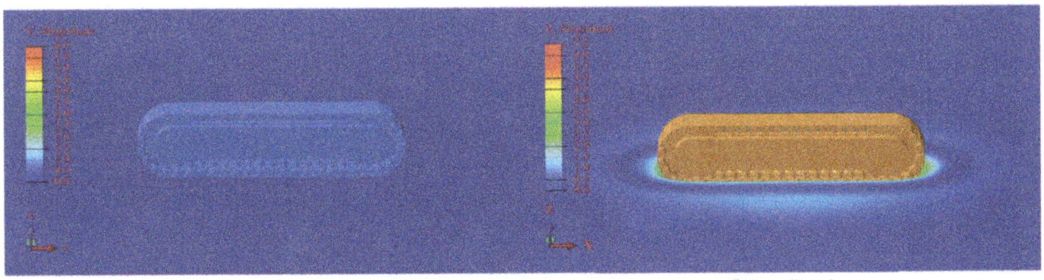

(a)$v_i = 1$ m/s (b)$v_i = 8$ m/s

Figure 12. Soil velocity field at low and high DSMV bottoming velocities.

Conversely, an increase in the track contact area leads to a reduction in pressure applied per unit area, thereby limiting the stress concentration and delaying failure initiation. The stress is more evenly distributed, reducing the severity of plastic deformation and resulting in shallower penetration depths. Among all the dynamic parameters, the bottom velocity has the most pronounced effect on settlement, underscoring the importance of controlling the descent speed during vehicle deployment to avoid excessive embedment.

3.4. Interplay Between Strain-Softening and Rate Effects

The combined influence of strain-softening and rate effects introduces a complex, competing dynamic that governs the overall sinkage behavior of DSMVs. While strain softening reduces the mobilized shear strength with progressive deformation, strain rate effects can lead to transient hardening, enhancing soil resistance under rapid loading conditions. The numerical results illustrate this interplay: at low bottom velocities, strain rate effects are minimal, allowing softening to dominate. In this regime, the shear strength rapidly decreases once yielding begins, resulting in early shear band formation and deep penetration. However, as the landing velocity increases, the rate-dependent hardening partially offsets the softening-induced strength loss, delaying localization and reducing settlement compared with the softening-only case. This antagonistic relationship is further complicated by sediment viscosity. In high-viscosity sediments, the mobilized strength is preserved for longer durations, reducing the softening effects even under slow loading. In contrast, low-viscosity sediments facilitate flow and strength degradation, making the rate effect insufficient to prevent deep penetration even under rapid descent.

Therefore, the final sinkage depth reflects the net outcome of this interaction: the balance between the rate-enhanced stiffness and the degradation-driven weakening. A clear understanding of this coupled mechanism is essential for the optimal design of DSMV landing strategies in heterogeneous deep-sea environments. While the FEM simulations provide detailed insights, they are computationally intensive. To overcome this limitation and support rapid assessments, we developed a data-driven surrogate model using the random forest algorithm, trained on the numerical results.

4. Prediction Model Using the Random Forest Algorithm

4.1. Random Forest Approach

Given the nonlinear and multivariable nature of the DSMV settlement behavior, a random forest regression model is adopted. Random forest is a typical ensemble machine learning model [59], which aggregates the outputs of multiple base models constructed on the dataset to produce a final prediction. This algorithm is well-known for its relative simplicity, its efficiency in handling large-scale datasets, and its ability to maintain predictive accuracy even when the data distribution exhibits significant variability [60]. This ensemble approach builds multiple decision trees and integrates their predictions to capture complex patterns within the high-dimensional input space, thus enhancing robustness and generalization, as depicted in Figure 13. The overall prediction of the random forest is derived by averaging the results of all decision trees in the model [52].

In a random forest model, the construction of individual decision trees adheres to a methodical and iterative framework. Each training set is defined as $D = \{(x_1, y_1), (x_2, y_2), \ldots, (x_N, y_N)\}$, where each sample x_i comprises n features representing the independent input variables and N denotes the number of training instances. At each node of the decision tree, the algorithm performs recursive data segmentation, dividing the input space into two distinct partitions, R_1 and R_2, based on candidate feature thresholds. To determine the most effective splitting criterion, all the variables and their potential values are evaluated exhaustively. This iterative assessment mechanism enables the construction of an ensemble where each decision tree captures unique patterns, thereby enhancing the overall accuracy and generalization capacity of the model.

4.2. Dimensionless Sinkage Depth

Machine learning requires a certain number of features (i.e., attributes) to train a model. One of the main challenges is determining the correct number and type of such features from the attributes of a given dataset [73]. The machine learning process generally applies the available feature values directly for model training. This approach makes the process vulnerable to overfitting, prediction errors, bias, and poor generalizability. The key to improving the accuracy and fit of ML is to determine the best set (i.e., grouping) of the correct set of features to best match the feature values. In this paper, we propose a new approach to enhance the feature engineering and selection optimization process for DSMV sinking.

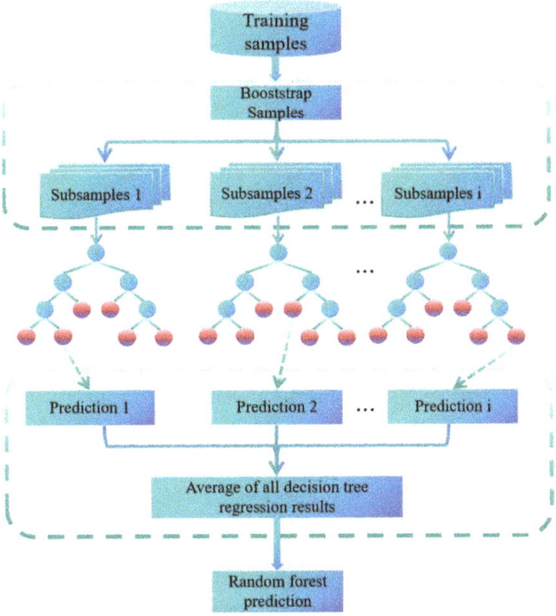

Figure 13. Conceptual diagram of the random forest modeling principle.

To develop the RF input structure, a dimensionless representation of the sinkage depth is adopted. This method builds upon the traditional energy approach [74]. The total energy E_{total} is defined as the sum of the total kinetic and potential energies, and then the soil strength slope k_s and the anchor projection equivalent diameter D_p are normalized to obtain the empirical model equation for predicting the penetration $z/D_p \approx \left(\frac{E_{total}}{k_s D_p^4}\right)^{1/3}$. The improvement lies in the incorporation of strain-softening and strain rate effects, which play a significant role in characterizing the mechanical behavior of soft sediments. These enhancements enable a more accurate representation of sediment deformation under dynamic loading conditions. The improved energy method is further employed for feature selection and feature engineering, serving as a critical preprocessing step for data preparation in the DSMV predictive model. This comprehensive feature formulation enables the RF model to capture complex coupled influences in a compact, interpretable input format, enhancing both predictive accuracy and training convergence.

Using this finite element model, the maximum sinking depth of the unloaded DSMV(z) can be determined under varying conditions of undrained shear strength, strain-softening parameters, and track dynamic factors. The sum of the kinetic energy when the DSMV hits the seabed and the potential energy at the cessation of motion is defined as the total energy E_{total}, and is dimensionless for E_{total} and z by setting $\frac{E_{total}}{AB(S_{u0}+k_s z)}$ and $z/\sqrt{A}$. Furthermore, the strain rate, strain-softening effect, and soil strength inhomogeneity are considered. The improved empirical formula is defined by Equation (3):

$$z/\sqrt{A} = f\left(\delta_{rem}, \xi_{95}, \eta, \frac{E_{total}}{AB(S_{u0}+k_s z)}\right) = f\left(\delta_{rem}, \xi_{95}, \eta, \frac{\frac{1}{2}mv_i^2 + mgz}{AB(S_{u0}+k_s z)}\right) \quad (3)$$

where m is the wet weight of the DSMV, vi is the initial bottom velocity, A is the contact area of the track, and B is the track width.

In this study, the binary tree-based machine learning integration method, i.e., random forest, is used to predict $(z/\sqrt{A})$. The process of model building consists of the following main steps.

(I) Data definition and generation: The initial step involves identifying the key input features $(\delta_{rem}, \xi_{95}, \eta, \frac{\frac{1}{2}mv_i{}^2+mgz}{AB(S_{u0}+kz)})$ alongside the predicted output variable $(z/\sqrt{A})$. These output targets are derived from high-fidelity numerical simulations, supplemented by an improved energy-based analytical approach.

(II) Data inspection and partitioning: Preliminary data exploration was carried out to detect potential anomalies and understand variable distributions. Next, the dataset was randomly partitioned into a training subset (90%) and a testing subset (10%) to ensure reliable model development and performance validation.

(III) Model training and hyperparameter optimization: The training subset was employed to fit the random forest regression model. A comprehensive grid search strategy was implemented to fine-tune critical hyperparameters—such as tree count, depth, and minimum split size—based on internal validation metrics, with the goal of maximizing predictive performance.

(IV) Model validation and robustness analysis: To assess model accuracy, standard regression metrics (e.g., R^2, MAE, and RMSE) were computed using the test subset. Additionally, K-fold cross-validation and repeated Monte Carlo simulations were conducted to evaluate the model's generalization capacity and resilience across multiple randomized data splits.

4.3. Establishment of the Prediction Model

In step (I), the model input parameters include δ_{rem}, ξ_{95}, η, and E_{total}. The output parameter is $z/\sqrt{A}$. Based on the strain parameter characteristics of the DSMV 'Pioneering-I' by Shanghai Jiaotong University [36] and a typical deep-sea polymetallic nodule belt [68,69], Table 3 shows a summary of the statistical characteristics of the 257 samples. In step (II), to facilitate more efficient feature selection [75,76], we subsequently performed correlation analyses on the input and output parameters. Correlation coefficients are often used to quantify the degree of association between two sets of variables. The Kendall correlation coefficient (often denoted as τ) is a rank-based statistic that measures the monotonicity of a relationship between two variables. It assesses correlation by comparing the relative rank order of pairs of variables for consistency. The Kendall method does not require the relationship between variables to be linear and is suitable for small datasets or discrete data. It is calculated as shown in Equation (4):

$$\tau = \frac{C-D}{\frac{1}{2}n(n-1)} \quad (4)$$

where n is the total number of samples, and the C and D instances refer to pairs of samples in which the rank order of the two variables is consistent and inconsistent, respectively. The denominator $\frac{1}{2}n(n-1)$ is the total number of sample pairs. If two sample pairs (x_i, y_i) and (x_j, y_j), $x_i > x_j$ and $y_i > y_j$ or $x_i < x_j$ and $y_i < y_j$, they are consistent pairs. The opposite is an inconsistent sample pair. The Kendall correlation coefficient τ is in the range of $[-1, 1]$, $\tau = 1$ indicates a perfect positive correlation, and $\tau = -1$ indicates a perfect negative correlation.

Table 3. Parameter ranges in the sample data.

m (kg)	A (m^2)	vi (m/s)	η	St	ξ_{95}	Su0 (Pa)	K (Pa/m)
2500	1.87	1	0	1	10	2000	0
3000	2.18	2	0.01	2	15	2500	1000
4500	2.5	3	0.04	2.5	20	3000	2000
6500	2.81	4	0.05	3	25	5000	2500
8500	3.12	5	0.1	3.33	30	6000	3000
9000		6	0.15	4	40	7500	
10,500		7		5	50	9000	
		8		8		10,000	
				10			

To verify the accuracy of the numerical simulation results, eight representative parameter combinations were randomly selected from Table 3 and compared with the analytical solution proposed by Sun et al. [37]. The comparison results are illustrated in Figure 14. As shown, the predicted settlement depths from the current FEM model closely match the analytical values under the same loading conditions.

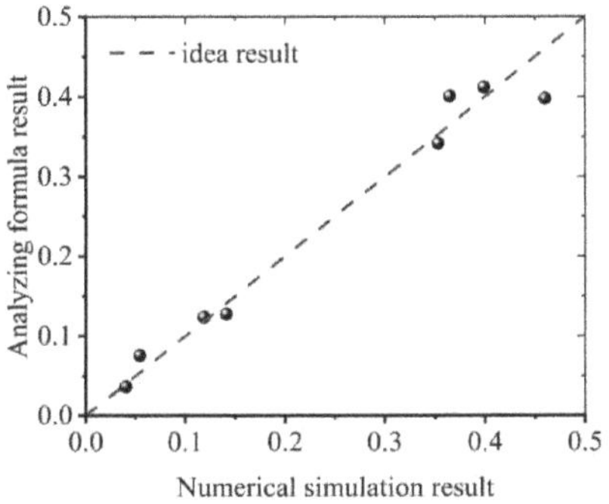

Figure 14. Comparison of the numerical simulation results with the analytical equation results.

Specifically, among the selected cases, the analytical results range from 0.047 m to 0.412 m, while the FEM results range from 0.040 m to 0.460 m. The absolute differences between paired values are generally less than 0.06 m. In terms of relative error, 6 out of 8 cases fall within ±15%. While two of the test cases exhibit slightly larger deviations (above 15%), the overall consistency remains acceptable for model validation purposes, especially considering the complexity of seabed–structure interactions and the absence of full-scale experimental data. This level of agreement confirms the fidelity of the proposed numerical model in capturing the key mechanisms governing DSMV–seabed interaction, thereby supporting its suitability for generating training data for the machine learning model.

Figure 15 shows a heatmap depicting the Kendall correlation between the input and output parameters. It is clear from the figure that E_{total} is positively correlated with $z/\sqrt{A}$. The correlation between E_{total} and $z/\sqrt{A}$ is very strong, with a correlation coefficient of 0.51. Based on the correlation analysis, the selected set of features is limited and the correlation between the input characteristic parameters and the target output parameters $z/\sqrt{A}$ is in accordance with the results of the analysis. Moreover, there is no evidence of multicollinearity between these features. Therefore, we can directly proceed with the training of the random forest model, which corresponds to step (III).

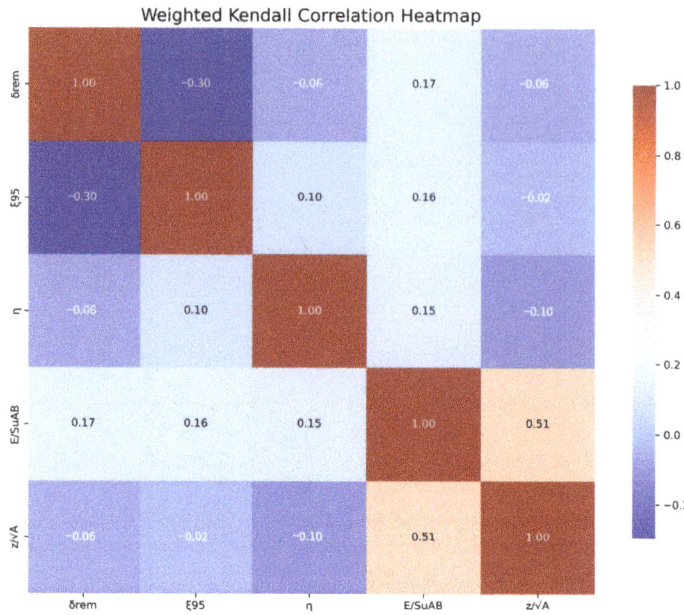

Figure 15. Heatmap of correlation coefficients.

Following the initial model development, a dedicated hyperparameter optimization phase was conducted to enhance the predictive performance of the random forest algorithm. The key hyperparameters subject to tuning included the total number of estimators (n_estimators), the maximum tree depth (max_depth), the upper bound on features considered per split (max_features), and the minimum thresholds for data splits (min_samples_split) and terminal leaf sizes (min_samples_leaf) [77]. Figure 16a–e presents the learning curves reflecting changes in model accuracy (R^2) as a function of these critical parameters. An increasing trend in predictive performance is observed with increasing values of (n_estimators), (max_depth), and (max_features), indicating that a more complex ensemble structure generally enhances model capability. In contrast, excessive constraints on internal node splitting or leaf node size tend to restrict model flexibility, thereby reducing predictive accuracy. Based on the comprehensive grid search and curve analysis, the final set of optimal hyperparameters—achieving a balance between accuracy and generalization—is summarized in Table 4.

Using the random forest model configured with optimal hyperparameters, predictive performance was assessed on the holdout dataset comprising the remaining 10% of the total samples. The evaluation results are visualized in Figures 17 and 18, which respectively illustrate the alignment between the predicted versus observed values, and the residual distributions for both the training and test sets. The strong convergence between the predicted outputs and true values across datasets indicates excellent model fidelity. The dimensionless penetration resistance coefficients predicted by the model exhibit high agreement with the simulation results, suggesting robust generalizability. Statistical performance metrics, presented in Table 5, further confirm this: the mean absolute error (MAE) remains consistently low across both subsets. Although a few outliers are observed, their impact is marginal, and overall predictive reliability is maintained. Therefore, the model is deemed effective for accurately estimating the maximum penetration depth experienced by the DSMV under varying geotechnical conditions.

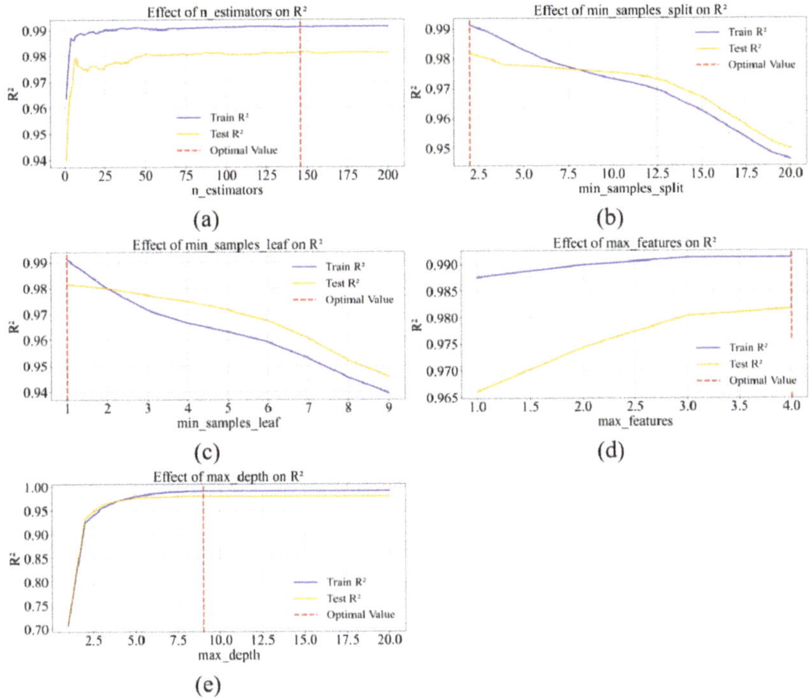

Figure 16. Variation in random forest accuracy versus key hyperparameters: (**a**) N_Estimators; (**b**) Max_Depth; (**c**) Min_Samples_Split; (**d**) Min_Samples_Leaf; (**e**) Max_Features.

Table 4. The corresponding values of the optimal hyperparameters.

Parameters	N_Estimators	Max_Depth	Min_Samples_Split	Min_Samples_Leaf	Max_Features
Value	146	9	2	1	4

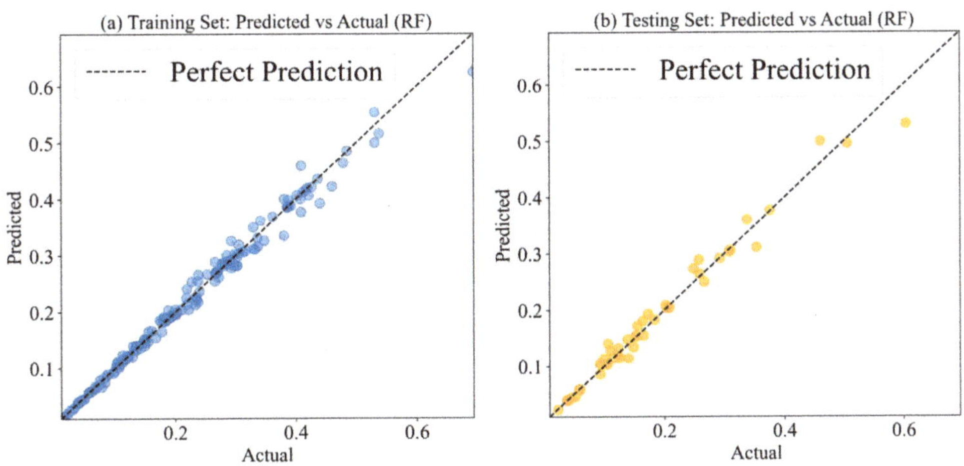

Figure 17. (**a**) Proximity of the training set actual and predicted values and (**b**) proximity of the testing set actual and predicted values.

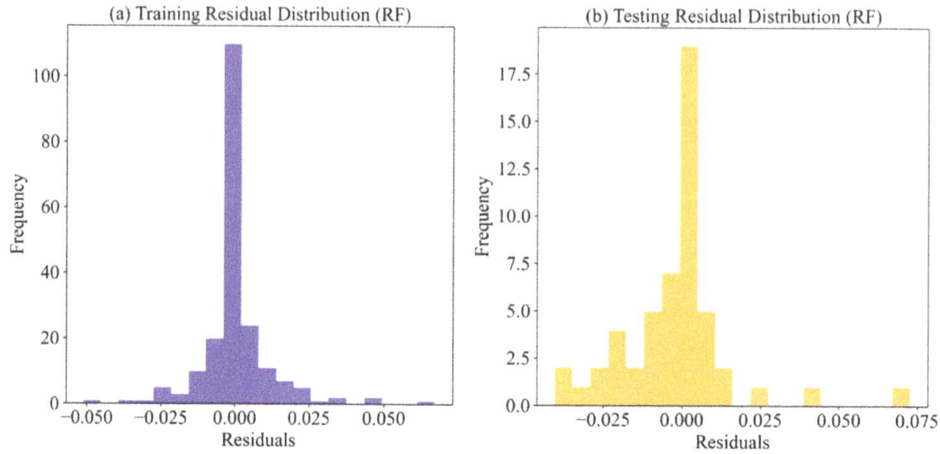

Figure 18. Error quantification analysis.

Table 5. Evaluation metrics and their statistical values for the established prediction models.

Evaluation Metrics	R^2	MSE	MAE	RMSE
Training set	0.9914	0.0001	0.0066	0.0120
Testing set	0.9807	0.0003	0.0104	0.0173

4.4. Evaluation of the Prediction Model

This section corresponds to step (IV) in the predictive modeling framework, which focuses on the comprehensive evaluation of model performance. The evaluation is structured into three dimensions: predictive accuracy, generalization capacity, and model robustness. To quantify accuracy, four statistical indicators were employed—the coefficient of determination (R^2), mean squared error (MSE), mean absolute error (MAE), and root mean squared error (RMSE)—as summarized in Table 5 [78]. These metrics were computed based on test set predictions to assess the model's ability to reproduce the dimensionless sinkage depth. The evaluation results confirm that the random forest algorithm achieves consistently high accuracy across all the metrics, indicating its strong capacity to capture the nonlinear relationships governing DSMV penetration resistance. Accordingly, the RF model is a suitable and reliable approach for predicting seabed interaction outcomes under varying geotechnical conditions.

To evaluate the model's generalizability, a k-fold cross-validation approach was adopted [79]. This technique involves partitioning the dataset into k equally sized subsets; during each iteration, one subset is reserved for validation while the remaining $k-1$ subsets are used for training. Repeating this process k times ensures that each sample is used for both training and validation, thereby reducing the variance caused by a single train–test split. For this study, a standard 10-fold cross-validation was applied, which is a widely accepted practice in machine learning applications [80]. The coefficient of determination (R^2) was employed as the primary metric to quantify model performance across all folds (Figure 19). The results demonstrate the strong generalizability of the random forest model: on the training set, the R^2 values ranged from 0.9919 to 0.9937, yielding an average of 0.9925; on the test set, the R^2 values ranged from 0.9168 to 0.9835, with a corresponding average of 0.9511. These consistently high R^2 scores confirm that the model maintains robust predictive accuracy, even when exposed to previously unseen data.

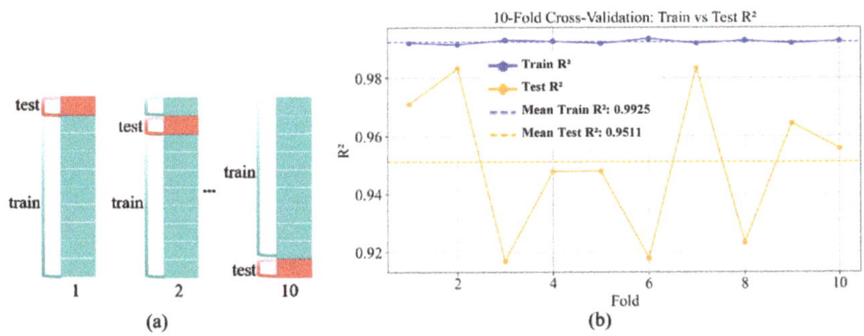

Figure 19. 10-fold cross-validation of the established prediction model: (**a**) schematic diagram of the 10-fold cross validation principle; (**b**) 10-fold cross validation results.

To further investigate the stability of the random forest prediction model under stochastic data conditions, a series of 200 independent Monte Carlo simulations was carried out [52,81]. In each iteration, the dataset was randomly partitioned into training and test subsets using a 90:10 split ratio, followed by the computation of normalized regression metrics—the coefficient of determination (R^2), root mean squared error (RMSE), and mean absolute error (MAE)—for both subsets. The simulation protocol follows the approach detailed in the literature [52]. The results from the training set show a high degree of consistency across iterations, indicating that the model's performance is largely unaffected by variations in data partitioning. While minor fluctuations are noted in the test set results, the normalized values of R^2, RMSE, and MAE exhibit strong convergence toward 1 as the number of runs approaches 200, as depicted in Figure 20. Across all the simulations, the predictive metrics derived from the training data consistently outperform those derived from the test data, although the discrepancy remains small. This performance pattern underscores the robustness and generalization strength of the proposed model, validating its applicability in diverse geotechnical scenarios.

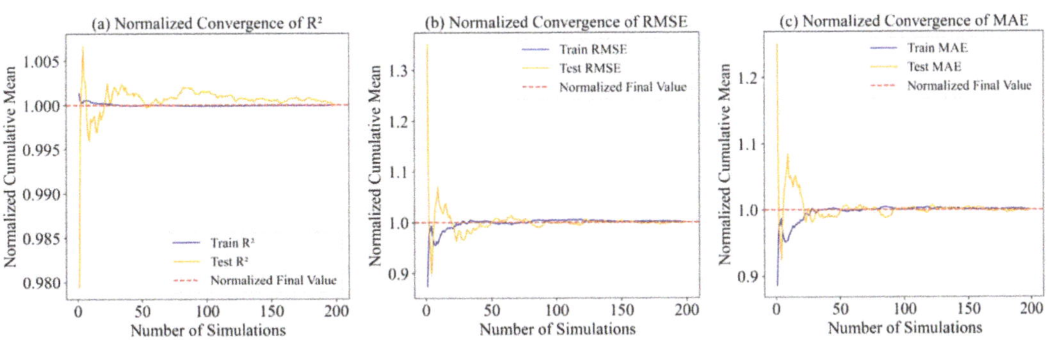

Figure 20. Normalized accuracy and errors from 200 runs of Monte Carlo simulation for the established prediction model: (**a**) normalized convergence values of R^2; (**b**) normalized convergence values of RMSE; (**c**) normalized convergence values of MAE.

5. Conclusions

To investigate the sinking patterns and assessment methods of an unloaded DSMV in deep ocean surface soft sediments under complex conditions, this study proposes a dual-method framework that integrates high-fidelity finite element simulations with data-driven machine learning predictions. The FEM model, based on the CEL approach and an improved constitutive law, enables accurate simulation of DSMV–seabed interactions under various sediment and loading conditions. The random forest algorithm is then trained using the FEM dataset to establish a fast and generalizable prediction model for dimensionless settlement depth. This methodological integration significantly improves the accuracy and efficiency of seabed landing safety assessments.

The main findings are summarized as follows:

(1) A high-fidelity CEL finite element model was developed to simulate the landing and penetration behavior of a full-scale DSMV track system under various sediment conditions. The track structure was modeled as a rigid assembly with detailed geometry and contact definitions, while the seabed was represented using a rate-dependent and strain-softening constitutive soil model. This configuration enables realistic capture of nonlinear soil–track interactions and large-deformation responses. The model predictions align well with those of existing studies, confirming its physical realism and engineering applicability.

(2) Parametric analyses revealed that the dimensionless settlement depth decreases with increasing undrained shear strength but increases significantly with higher sensitivity, faster softening rates (i.e., lower ξ_{95}), greater bottoming velocities, and stiffer contact conditions. Contour plots of the plastic strain and shear strain rate revealed that highly sensitive and rapidly loaded sediments, which are the primary contributors to deeper penetration, are more prone to shear localization and intense flow behavior.

(3) The incorporation of a strain-softening and rate-dependent constitutive model allowed for more accurate simulations of strength evolution in deep-sea sediments under dynamic loading. The results show that strain rate effects introduce apparent viscous hardening under fast landing scenarios, partially offsetting strength degradation. However, under high-sensitivity or low-rate conditions, softening dominates, leading to more extensive failure zones and greater settlement.

(4) An RF regression model based on ensemble learning was constructed to predict the dimensionless settlement depth using high-dimensional input features. With the test R^2 exceeding 0.94, the random forest model demonstrated strong alignment with simulation results. The feature importance analysis further highlighted strength, strain sensitivity, and descent dynamics as critical predictors validating the model's ability to identify key physical drivers in the DSMV landing process. Compared with conventional empirical approaches, the RF model demonstrates superior nonlinearity handling and generalization capability.

(5) The proposed integrated framework supports an efficient parameter screening strategy for evaluating settlement sensitivity across various landing scenarios. This approach provides a practical tool for optimizing the DSMV track structure design and descent planning. Future work may extend this framework by incorporating uncertainty quantification and real-time feedback control to support the development of a digital twin for deep-sea landing operations.

6. Limitations

Despite the encouraging results, this study has some limitations. First, due to the high cost and complexity of full-scale deep-sea field testing, the FEM results have not been validated by actual in situ DSMV landing data. Instead, consistency with prior empirical models and scaled laboratory tests has been used as a proxy for validation. Second, the sediment properties used in this study mainly represent ultra-soft clays found in polymetallic nodule regions, which may limit the generalizability of the findings to other seabed types. Future work will incorporate uncertainty quantification, real-time sensor feedback, and in situ data to further enhance the robustness of the proposed framework.

Author Contributions: Conceptualization, Y.Z., Z.X., L.L., Y.S. and X.G.; Data curation, Y.Z. and Q.X.; Formal analysis, Y.Z., Z.X. and J.Y.; Funding acquisition, Z.X., Q.X. and X.G.; Methodology, Z.X. and

L.L.; Resources, L.L., J.Y. and X.G.; Software, J.Y.; Supervision, L.L.; Validation, Y.Z.; Writing—original draft, Y.Z. and Z.X.; Writing—review and editing, Y.Z., Z.X., L.L., Q.X., Y.S., J.Y. and X.G. All authors have read and agreed to the published version of the manuscript.

Funding: This research was funded by the National Natural Science Foundation of China, grant numbers 41876066 and 41606084; the Opening Fund of State Key Laboratory of Geohazard Prevention and Geoenvironment Protection (Chengdu University of Technology), grant number SKLGP2019K023; and the National Key Research and Development Program of China, grant number 2024YFC2815400.

Data Availability Statement: Some data, models, or code generated or used during the study are available in a repository online in accordance with the funder's data retention policies. Other data, models, or code that support the findings of this study are available from the corresponding author upon reasonable request, including the results of the virtual element tests.

Conflicts of Interest: Author Jianghui Yang was employed by the company China Offshore Oil Engineering Corp CNOOC Ltd. The remaining authors declare that the research was conducted in the absence of any commercial or financial relationships that could be construed as a potential conflict of interest.

References

1. Du, K.; Xi, W.; Huang, S.; Zhou, J. Deep-sea mineral resource mining: A historical review, developmental progress, and insights. *Miner. Metall. Explor.* **2024**, *41*, 173–192. [CrossRef]
2. Sha, F.; Xi, M.; Chen, X.; Liu, X.; Niu, H.; Zuo, Y. A recent review on multi-physics coupling between deep-sea mining equipment and marine sediment. *Ocean Eng.* **2023**, *276*, 114229. [CrossRef]
3. Sharma, R. Deep-sea mining and the water column: An introduction. In *Deep-Sea Mining and the Water Column: Advances, Monitoring and Related Issues*; Springer: Cham, Switzerland, 2024; pp. 3–40.
4. Zhang, N.; Zhai, W.K.; Yin, S.Y.; Chen, X.G.; Xia, J.X.; Luan, L.B.; Zhai, A.S. Deep-sea soft soil and deep-sea mining vehicle interaction: From soil properties to locomotion performance. *Appl. Ocean Res.* **2025**, *154*, 104372. [CrossRef]
5. Atmanand, M.; Janardhanan, K.; Amuda, K.; Gnanaraj, A.; Rajesh, S.; Muthuvel, P.; Kuttikrishnan, G.; Ramadass, G. Contemporary technological progress in deepsea mining. In *Deep-Sea Mining and the Water Column: Advances, Monitoring and Related Issues*; Springer: Cham, Switzerland, 2024; pp. 67–89.
6. Leng, D.; Shao, S.; Xie, Y.; Wang, H.; Liu, G. A brief review of recent progress on deep sea mining vehicle. *Ocean Eng.* **2021**, *228*, 108565. [CrossRef]
7. Wu, Q.; Yang, J.; Lu, H.; Lu, W.; Liu, L. Effects of heave motion on the dynamic performance of vertical transport system for deep sea mining. *Appl. Ocean Res.* **2020**, *101*, 102188. [CrossRef]
8. Shen, Y.J.; Chen, M.F.; Du, Y.L.; Li, M. Key mechanical issues and technical challenges of deep-sea mining development system. *Mech. Eng.* **2022**, *44*, 1005–1020.
9. Liu, S.J.; Chang, L.; Yu, D. Status and progress on researches and developments of deep ocean mining equipments. *Chin. J. Mech. Eng.* **2014**, *50*, 8–18. [CrossRef]
10. Zhao, H.L.; Sun, Y.F.; Jia, C.; Wei, R.C.; Deng, H.; Wu, T. The sinkage by mining car traveling on deep-sea soft bottom based on Burger's creep model. *Mar. Geol. Quat. Geol.* **2024**, *44*, 179–190. [CrossRef]
11. Hong, S.; Kim, H.W.; Yeu, T.; Choi, J.S.; Lee, T.H.; Lee, J.K. Technologies for safe and sustainable mining of deep-seabed minerals. In *Environmental Issues in Deep-Sea Mining: Impacts, Consequences and Policy Perspectives*; Springer: Cham, Switzerland, 2019; pp. 95–143.
12. Kang, Y.J.; Liu, S.J. The development history and latest progress of deep-sea polymetallic nodule mining technology. *Minerals* **2021**, *11*, 1132. [CrossRef]
13. Rauf, O.; Yang, N.; Chen, M.; Ma, H.X. Evaluation of ground pressure, bearing capacity, and sinkage in rigid-flexible tracked vehicles on characterized terrain in laboratory conditions. *Sensors* **2024**, *24*, 1779. [CrossRef]
14. Laughery, S.; Gerhart, G.; Goetz, R. *Bekker's Terramechanics Model for Off-Road Vehicle Research*; US Army TARDEC: Hammond, WI, USA, 1990.
15. Bekker, M.G. *Introduction to Terrain-Vehicle Systems*; University of Michigan Press: Ann Arbor, MI, USA, 1969.
16. Bekker, M.G.; Kitano, M.; Kumagai, K. *Theory of Land Locomotion*; University of Michigan Press: Ann Arbor, MI, USA, 1969.
17. Agarwal, S.; Senatore, C.; Zhang, T.G.; Kingsbury, M.; Iagnemma, K.; Goldman, D.I.; Kamrin, K. Modeling of the interaction of rigid wheels with dry granular media. *J. Terramechanics* **2019**, *85*, 1–14. [CrossRef]
18. Wang, L.; Chen, X.G.; Li, Z.G.; Liu, X.L.; Liu, X.X.; Wang, L.Z.; Yang, W.D. Analysis of traction performance of deep-sea tracked mining vehicles under different grouser pitch-to-height ratios and traveling speeds. *Ocean Eng.* **2024**, *297*, 117112. [CrossRef]

19. Kim, Y.H.; Hossain, M.S. Dynamic installation of OMNI-Max anchors in clay: Numerical analysis. *Géotechnique* **2015**, *65*, 1037–1096. [CrossRef]
20. Kim, Y.H.; Hossain, M.S.; Wang, D. Effect of strain rate and strain softening on embedment depth of a torpedo anchor in clay. *Ocean Eng.* **2015**, *108*, 704–715. [CrossRef]
21. Kim, Y.H.; Hossain, M.S.; Wang, D.; Randolph, M.F. Numerical investigation of dynamic installation of torpedo anchors in clay. *Ocean Eng.* **2015**, *108*, 820–832. [CrossRef]
22. Zhao, J.; Jang, B.; Duan, M. On spudcan deep penetration using Eulerian finite element in multi-layer sediments: Soil plug mechanism. *Ships Offshore Struct.* **2017**, *12*, 715–721. [CrossRef]
23. Dai, X.; Xu, T.; Chen, J. Physical and mechanical properties of deep oceanic sediments cored from the bottom of Challenger Deep, Mariana Trench. *Geofluids* **2021**, *2021*, 9109132. [CrossRef]
24. Li, L.; Li, S.L. Simulation and mechanical characteristics of terramechanics of the surface soil on deep-sea bed. *Eng. Mech.* **2010**, *13*, 213–220.
25. Su, Q.; Lv, H.; Yang, J.M.; Sun, P.F. Analysis of the walking performance of tracked deep-sea mining vehicle on soft sediment. *Ocean Eng.* **2022**, *2*, 162–168.
26. Einav, I.; Randolph, M.F. Combining upper bound and strain path methods for evaluating penetration resistance. *Int. J. Numer. Meth. Eng.* **2005**, *63*, 1991–2016. [CrossRef]
27. Benson, D.J. Computational methods in Lagrangian and Eulerian hydrocodes. *Comput. Methods Appl. Mech. Eng.* **1992**, *99*, 235–394. [CrossRef]
28. Wang, J.Y.; Cao, W.G.; Zhai, Y.C. Experimental study of interaction between deep-sea sediments and tracks. *Rock Soil Mech.* **2011**, *S2*, 274–278.
29. Zhang, W.; Cassidy, M.J.; Tian, Y. Practical large-deformation finite-element method for 3D geotechnical problems involving free surface deformations. *Int. J. Geomech.* **2020**, *20*, 04020064. [CrossRef]
30. Liu, M.; Huang, W.X.; Yuan, W.H.; Zhang, W. Modelling the large deformation failure behaviour of unsaturated porous media with a two-phase fully-coupled smoothed particle finite element method. *SSRN Electron. J.* **2023**, *427*, 117013. [CrossRef]
31. Yuan, W.; Zhu, J.; Wang, N.; Zhang, W.; Dai, B.; Jiang, Y.; Wang, Y. A dynamic large-deformation particle finite element method for geotechnical applications based on Abaqus. *J. Rock Mech. Geotech. Eng.* **2023**, *15*, 1859–1871. [CrossRef]
32. Huang, C.; Qian, J. FEM-BEM modeling of dynamic responses of a fluid-filled lined tunnel under a water hammer. *J. Eng. Mech.* **2024**, *150*, 04022045. [CrossRef]
33. Janarthanan, C.; Kuttikrishnan, G.; Sundaramoorthi, V.; Chandran, V.; Ramadass, G.A. Deep Sea Soil Sinkage Simulation and Experimental Studies for Development of Deep Water Mining Machine. *Mar. Technol. Soc. J.* **2022**, *56*, 72–82. [CrossRef]
34. Amudha, K.; Reshma, K.V.; Ramesh, N.R.; Deepak, C.R.; Ramadass, G.A.; Atmanand, M.A. Estimation of sinkage and breakout forces for tracked vehicle for soft soils. In Proceedings of the OCEANS 2014—TAIPEI, Taipei, Taiwan, 7–10 April 2014; pp. 1–7. [CrossRef]
35. Janarthanan, C.; Muruganandhan, R.; Gopkumar, K. Design and multi-body dynamic analysis of inline and offset track configuration in deep-sea mining vehicles for enhanced traction in soft seabed. *J. Terramechanics* **2024**, *116*, 100999. [CrossRef]
36. Sun, P.; Lu, H.; Yang, J.; Deng, L.; Liu, M.; Li, S. Numerical study on multiple parameters of sinkage simulation between the track plate of the deep-sea mining vehicle and the seafloor soil. *J. Mar. Sci. Eng.* **2022**, *10*, 1680. [CrossRef]
37. Sun, P.; Lu, H.; Yang, J.; Xu, Z. A numerical investigation on the sinkage of the deep-sea mining vehicle based on a modified constitutive soil model. *ASME J. Offshore Mech. Arct. Eng.* **2025**, *147*, 022101. [CrossRef]
38. Luan, L.; Chen, X.; Kouretzis, G.; Ding, X. Dynamic seabed stresses due to moving deep-sea mining vehicles. *Comput. Geotech.* **2023**, *157*, 105356. [CrossRef]
39. Xia, M.; Lu, H.; Yang, J.; Sun, P. Multi-Body Dynamics Modeling and Straight-Line Travel Simulation of a Four-Tracked Deep-Sea Mining Vehicle on Flat Ground. *J. Mar. Sci. Eng.* **2023**, *11*, 1005. [CrossRef]
40. Liu, Y.Y.; Wan, Z.; Quan, D.G.; Liu, Y.F.; Jia, Y.G.; Zhu, N. Analysis of bearing capacity of consolidated clay under the track load of deep-sea mining vehicle. *Mar. Georesour. Geotechnol.* **2024**, *43*, 508–523. [CrossRef]
41. Tho, K.K.; Leung, C.F.; Chow, Y.K.; Swaddiwudhipong, S. Eulerian finite-element simulation of spudcan–pile interaction. *Can. Geotech. J.* **2013**, *50*, 595–608. [CrossRef]
42. Gonzalez, R.; Iagnemma, K. DeepTerramechanics: Terrain classification and slip estimation for ground robots via deep learning. *arXiv* **2018**, arXiv:1806.07379. [CrossRef]
43. Chavez-Garcia, R.O.; Guzzi, J.; Gambardella, L.M.; Giusti, A. Learning ground traversability from simulations. *IEEE Robot. Autom. Lett.* **2018**, *3*, 1695–1702. [CrossRef]
44. Zhang, N.; Zhai, W.; Yin, S.; Song, Y.; Chen, X.; Guo, C.; Ma, Y. Study of the tractive performance of a deep-sea mining vehicle based on the soil failure mode. *Int. J. Geomech.* **2025**, *25*, 04025013. [CrossRef]
45. Zande, M.; Smith, J.; Thompson, R. Coupled numerical–experimental framework for track–soil interaction in planetary rovers. *J. Terramechanics* **2024**, *101*, 35–43. [CrossRef]

46. White, D.; Randolph, M. Seabed characterisation and models for pipeline-soil interaction. *Int. J. Offshore Polar Eng.* **2007**, *17*, 193–204.
47. Xiao, Z.; Fu, D.F.; Zhou, Z.F.; Lu, Y.; Yan, Y. Effects of strain softening on the penetration resistance of offshore bucket foundation in nonhomogeneous clay. *Ocean Eng.* **2019**, *193*, 106594. [CrossRef]
48. Xie, Q.; Xu, Q.; Xiu, Z.; Liu, L.; Du, X.; Yang, J.; Liu, H. Convenient method for large-deformation finite-element simulation of submarine landslides considering shear softening and rate correlation effects. *J. Mar. Sci. Eng.* **2024**, *12*, 81. [CrossRef]
49. Guo, X.; Liu, Z.; Zheng, J.; Luo, Q.; Liu, X. Bearing capacity factors of T-bar from surficial to stable penetration into deep-sea sediments. *Soil Dyn. Earthq. Eng.* **2023**, *165*, 107671. [CrossRef]
50. Pham, B.T.; Hoang, T.A.; Nguyen, D.M.; Bui, D.T. Prediction of shear strength of soft soil using machine learning methods. *Catena* **2018**, *166*, 181–191. [CrossRef]
51. Phoon, K.K.; Zhang, W.G. Future of machine learning in geotechnics. *Georisk* **2023**, *17*, 7–22. [CrossRef]
52. Nguyen, M.; Pham, B.; Tuyen, T.; Yen, H.; Prakash, I.; Vu, T.; Chapi, K.; Shirzadi, A. Development of an artificial intelligence approach for prediction of consolidation coefficient of soft soil: A sensitivity analysis. *Open Constr. Build. Technol. J.* **2019**, *13*, 178–188. [CrossRef]
53. Shahani, N.M.; Kamran, M.; Zheng, X.G.; Liu, C.C.; Guo, X.W. Application of gradient boosting machine learning algorithms to predict uniaxial compressive strength of soft sedimentary rocks at Thar Coalfield. *Adv. Civ. Eng.* **2021**, *2021*, 2565488. [CrossRef]
54. Shao, W.; Yue, W.H.; Zhang, Y.; Zhou, T.X.; Zhang, Y.T.; Dang, Y.B.; Wang, H.Y.; Feng, X.H. The application of machine learning techniques in geotechnical engineering: A review and comparison. *Mathematics* **2023**, *11*, 3976. [CrossRef]
55. Guo, X.S.; Meng, X.S.; Han, F.H.; Zhang, H.; Liu, X.L. Assessing the strength of deep-sea surface ultrasoft sediments with T-bar penetration: A machine learning approach. *Eng. Geol.* **2024**, *338*, 107632. [CrossRef]
56. Sagi, O.; Rokach, L. Ensemble learning: A survey. *Wiley Interdiscip. Rev. Data Min. Knowl. Discov.* **2018**, *8*, e1249. [CrossRef]
57. Yin, J.N.; Li, N. Ensemble learning models with a Bayesian optimization algorithm for mineral prospectivity mapping. *Ore Geol. Rev.* **2022**, *145*, 104916. [CrossRef]
58. Zhang, D.M.; Zhang, J.Z.; Huang, H.W.; Qi, C.C.; Chang, C.Y. Machine learning-based prediction of soil compression modulus with application of 1D settlement. *J. Zhejiang Univ. Sci. A* **2020**, *21*, 430–444. [CrossRef]
59. Breiman, L. Random forests. *Mach. Learn.* **2021**, *45*, 5–32. [CrossRef]
60. Graw, J.H.; Wood, W.T.; Phrampus, B.J. Predicting global marine sediment density using the random forest regressor machine learning algorithm. *J. Geophys. Res. Solid Earth* **2021**, *126*, e2020JB020135. [CrossRef]
61. Benson, D.J.; Okazawa, S. Contact in a multi-material Eulerian finite element formulation. *Comput. Methods Appl. Mech. Eng.* **2004**, *193*, 4277–4298. [CrossRef]
62. Li, T.; Tang, X.W.; Wang, Z. Experimental Study on Unconfined Compressive and Cyclic Behaviors of Mucky Silty Clay with Different Clay Contents. *Int. J. Civ. Eng.* **2019**, *17*, 841–857. [CrossRef]
63. Aubeny, C.P.; Shi, H. Effect of rate-dependent soil strength on cylinders penetrating into soft clay. *IEEE J. Ocean. Eng.* **2007**, *32*, 49–56. [CrossRef]
64. Zhou, H.; Randolph, M.F. Computational techniques and shear band development for cylindrical and spherical penetrometers in strain-softening clay. *Int. J. Geomech.* **2007**, *7*, 287–295. [CrossRef]
65. Zhou, H.; Randolph, M.F. Numerical investigations into cycling of full-flow penetrometers in soft clay. *Géotechnique* **2009**, *59*, 801–812. [CrossRef]
66. Zhu, H.; Randolph, M.F. Numerical analysis of a cylinder moving through rate-dependent undrained soil. *Ocean Eng.* **2010**, *38*, 943–953. [CrossRef]
67. Boukpeti, N.; White, D.J.; Randolph, M.F.; Low, H.E. Strength of fine grained soils at the solid-fluid transition. *Géotechnique* **2012**, *62*, 213–226. [CrossRef]
68. Biscontin, G.; Pestana, J.M. Influence of peripheral velocity on vane shear strength of an artificial clay. *Geotech. Test. J.* **2001**, *24*, 423–429. [CrossRef]
69. Randolph, M.F. Characterisation of soft sediments for offshore applications. In Proceedings of the 2nd International Conference on Site Characterization, Rotterdam, The Netherlands, 19–22 September 2004; pp. 209–232.
70. Dutta, S.; Hawlader, B.; Phillips, R. Finite element modeling of partially embedded pipelines in clay seabed using coupled Eulerian-Lagrangian method. *Can. Geotech. J.* **2015**, *52*, 58–72. [CrossRef]
71. Dutta, S.; Hawlader, B. Pipeline soil-water interaction modelling for submarine landslide impact on suspended offshore pipelines. *Géotechnique* **2019**, *69*, 29–41. [CrossRef]
72. Hansbo, S. A new approach to the determination of the shear strength of clay by the fall-cone test. *Proc. Swed. Geotech. Inst.* **1957**, *14*, 1–46.
73. Uddin, M.F.; Lee, J.; Rizvi, S.; Hamada, S. Proposing Enhanced Feature Engineering and a Selection Model for Machine Learning Processes. *Appl. Sci.* **2018**, *8*, 646. [CrossRef]

74. O'Loughlin, C.D.; Richardson, M.D.; Gaudin, C.; Randolph, M.F. Penetration of dynamically installed anchors in clay. *Géotechnique* **2013**, *63*, 909–919. [CrossRef]
75. Wei, M.D.; Meng, W.Z.; Dai, F.; Wu, W. Application of machine learning in predicting the rate-dependent compressive strength of rocks. *J. Rock Mech. Geotech. Eng.* **2022**, *14*, 1356–1365. [CrossRef]
76. Zhang, W.G.; Wu, C.Z.; Zhong, H.Y.; Li, Y.Q.; Wang, L. Prediction of undrained shear strength using extreme gradient boosting and random forest based on Bayesian optimization. *Geosci. Front.* **2021**, *12*, 469–477. [CrossRef]
77. Abdi, Y.; Momeni, E.; Armaghani, D.J. Elastic modulus estimation of weak rock samples using random forest technique. *Bull. Eng. Geol. Environ.* **2023**, *82*, 185. [CrossRef]
78. Zhou, J.; Qiu, Y.G.; Armaghani, D.J.; Zhang, W.G.; Li, C.Q.; Zhu, S.L.; Tarinejad, R. Predicting TBM penetration rate in hard rock condition: A comparative study among six XGB-based metaheuristic techniques. *Geosci. Front.* **2021**, *12*, 101093. [CrossRef]
79. Yan, T.; Shen, S.L.; Zhou, A.N.; Chen, X.S. Prediction of geological characteristics from shield operational parameters by integrating grid search and K-fold cross validation into stacking classification algorithm. *J. Rock Mech. Geotech. Eng.* **2022**, *14*, 1292–1303. [CrossRef]
80. Zhang, P.; Yin, Z.Y.; Jin, Y.F.; Chan, T.H. A novel hybrid surrogate intelligent model for creep index prediction based on particle swarm optimization and random forest. *Eng. Geol.* **2020**, *265*, 105328. [CrossRef]
81. Ly, H.B.; Nguyen, T.A.; Pham, B.T. Estimation of soil cohesion using machine learning method: A random forest approach. *Adv. Civ. Eng.* **2021**, *2021*, 8873993. [CrossRef]

Journal of
Marine Science
and Engineering

MDPI

Article

Contrastive Analysis of Deep-Water Sedimentary Architectures in Central West African Passive Margin Basins During Late-Stage Continental Drift

Futao Qu [1,2], Xianzhi Gao [1,2,*], Lei Gong [3,*] and Jinyin Yin [4]

1 College of Geosciences, China University of Petroleum, Beijing 102249, China; ftqu07@163.com
2 State Key Laboratory of Petroleum Resources and Prospecting, China University of Petroleum, Beijing 102249, China
3 School of Highway and Construction Engineering, Yunnan Communications Vocational and Technical College, Kunming 650500, China
4 Petroleum Exploration and Production Research Institute, SINOPEC, Beijing 100083, China; yinjy.syky@sinopec.com
* Correspondence: gaoxz@cup.edu.cn (X.G.); leigong0607@126.com (L.G.); Tel.: +86-13801395531 (X.G.); +86-15101016957 (L.G.)

Abstract: The Lower Congo Basin (LCB) and the Niger Delta Basin (NDB), two end-member deep-water systems along the West African passive margin, exhibit contrasting sedimentary architectures despite shared geodynamic settings. The research comprehensively utilizes seismic reflection structure, root mean square amplitude slices, drilling lithology, changes in logging curves, and previous research achievements to elucidate the controlling mechanisms behind these differences. Key findings include: (1) Stark depositional contrast: Since the Eocene, the LCB developed retrogradational narrow-shelf systems dominated by erosional channels and terminal lobes, whereas the NDB formed progradational broad-shelf complexes with fan lobes and delta-fed turbidites. (2) Primary controls: Diapir-driven topographic features and basement uplift govern architectural variability, whereas shelf-slope break configuration and oceanic relief constitute subordinate controls. (3) Novel mechanism: First quantification of how diapir-induced seafloor relief redirects sediment pathways and amplifies facies heterogeneity. These insights establish a tectono-sedimentary framework for predicting deep-water reservoirs in diapir-affected passive margins, refine the conventional "source-to-sink" model by emphasizing salt-geomorphic features coupling as the primary driver. By analyzing the differences in lithofacies assemblages and sedimentary configurations among the above-mentioned different basins, this study can provide beneficial insights for the research on related deep-water turbidity current systems and also offer guidance for deep-water oil and gas exploration and development in the West African region and other similar areas.

Keywords: deep-water turbidites; sedimentary architecture; salt tectonics; diapir controls; passive margin; source-to-sink; lower Congo–Niger Delta basins

1. Introduction

Since Reading and Richards [1] proposed a classification of turbidite systems in deep-water basin margins, a comprehensive understanding has been achieved internationally regarding the deep-water sedimentary characteristics, the composition of architectural elements, and the hydrocarbon accumulation patterns in passive continental margin settings [2–4]. Wang et al. [5] identified various sedimentary units within deep-water turbidite systems of passive continental margins, considering sea-level changes as a primary factor influencing the development of these units. Zhao et al. [6] studied the composition of deep-sea turbidite fan lobe complexes in the Niger Delta. Li et al. [7] analyzed the structure of large-scale erosional channels in the Lower Congo Basin, suggesting that tectonic uplift, sea-level fluctuations, and climate change collectively control the development of deep-water channel complexes [5]. Deep-water turbidites on continental margins, as sedimentary systems driven by multiple factors, provide valuable insights into the formation, texture, and evolution of sediments within deep-water turbidites [8,9]. With the continuous accumulation of global deep-sea drilling and 3D seismic data, geologists have increasingly recognized the significant differences among deep-water turbidite systems formed in various sedimentary settings and tectonic regions [10], especially in passive continental margin areas with substantial variations in tilting styles, paleocurrent supply, paleogeomorphological changes, and shelf-slope types [11,12]. Therefore, in-depth analysis of deep-water sedimentation under multi-factor control, detailed examination of typical deep-water turbidite architectural elements, and comparison of diversified shelf-edge structural styles and morphological parameters are crucial foundations for further understanding of these sedimentary systems.

This study focuses on deep-water depositional systems within the Lower Congo Basin and Niger Delta Basin—passive margin petroleum basins in central West Africa. By integrating high-resolution seismic data, well logs, and previous research findings, we conduct qualitative and quantitative analyses to compare architectural elements of deep-water deposits. The aim is to reveal the characteristics of different sedimentary architectures in terms of basin geometry, internal structure, geological evolution, and sand bodies distribution. Furthermore, we elucidate the dominant controlling factors governing these architectural variations, providing guidance for deep-water hydrocarbon exploration and development in West Africa and analogous regions worldwide.

2. Regional Geological Overview

The basins on both sides of the South Atlantic are typical passive continental margin basins, resulting from the rifting of the African and South American plates of West Gondwana since the Late Jurassic [8,13]. These basins exhibit characteristics of segmented strike, transverse zoning, vertical layering, differential separation, and conjugate development [8,13–15]. The passive continental margin basins on the West African side of the South Atlantic (Figure 1) have undergone four evolutionary stages: cratonic, rift, transitional (sag), and continental drift [16]. They are divided into southern, central, and equatorial (including the Niger Delta) segments. The rift stage of the central segment basins (including Gabon and Lower Congo) occurred earlier than that of the equatorial segment basins, with widespread Aptian salt formations [17]. The Niger Delta Basin (Figure 1b) differs from other basins, representing a large-scale delta system that has been developing since the Cenozoic, lacking salt formations [18]. The equatorial segment basins (Côte d'Ivoire and Benin) are controlled by the Equatorial Atlantic transform fault, without salt development. Due to the gradual northward separation of Gondwana, the Lower Congo Basin formed earlier than the Niger Delta Basin, leading to differences in structural characteristics and stratigraphic infilling between the two basins [12] (Figure 1d,e).

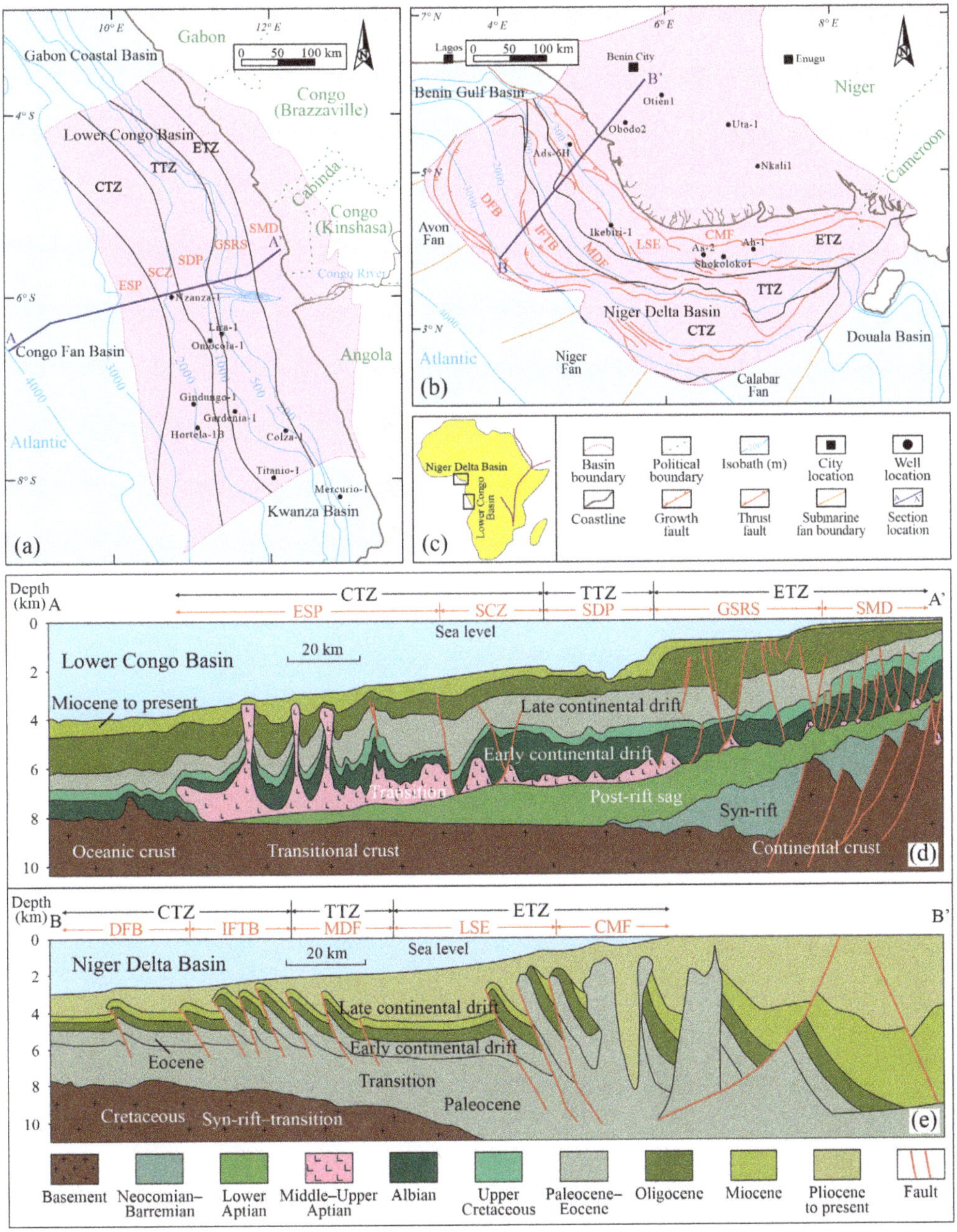

Figure 1. Location map (**c**), tectonic zonation, and structural cross-sections of the Lower Congo Basin (**a**,**d**) and the Niger Delta Basin (**b**,**e**) (modified from [16]). Abbreviations: CTZ, Compressional tectonic zone; TTZ, Transitional tectonic zone; ETZ, Extensional tectonic zone; ESP, Extensive salt province; SCZ, Salt canopy zone; SDP, Salt diapir province; GSRS, Giant salt raft system; SMD, Salt-raft mini-basin domain; DFB, Detachment fold belt; IFTB, Inner fold-and-thrust belt; MDF, Mud diapir field; LSE, Lower shoreface environment; CMF, Coastal marsh facies.

2.1. Tectonic Evolution and Basin Structure

The formation of both basins is closely related to the breakup of Gondwana. The Lower Congo Basin is a result of the separation of the African and South American plates, beginning in the Late Jurassic Tithonian stage and undergoing cratonic, rift, transitional (sag), and continental drift stages [19] (Figures 1d and 2a). The Niger Delta Basin was formed from the failed arm of a triple junction rift system in Nigeria, West Africa [20], developing since the Early Cretaceous Barremian stage and experiencing rift, transitional (sag), and continental drift stages [21] (Figures 1e and 2b).

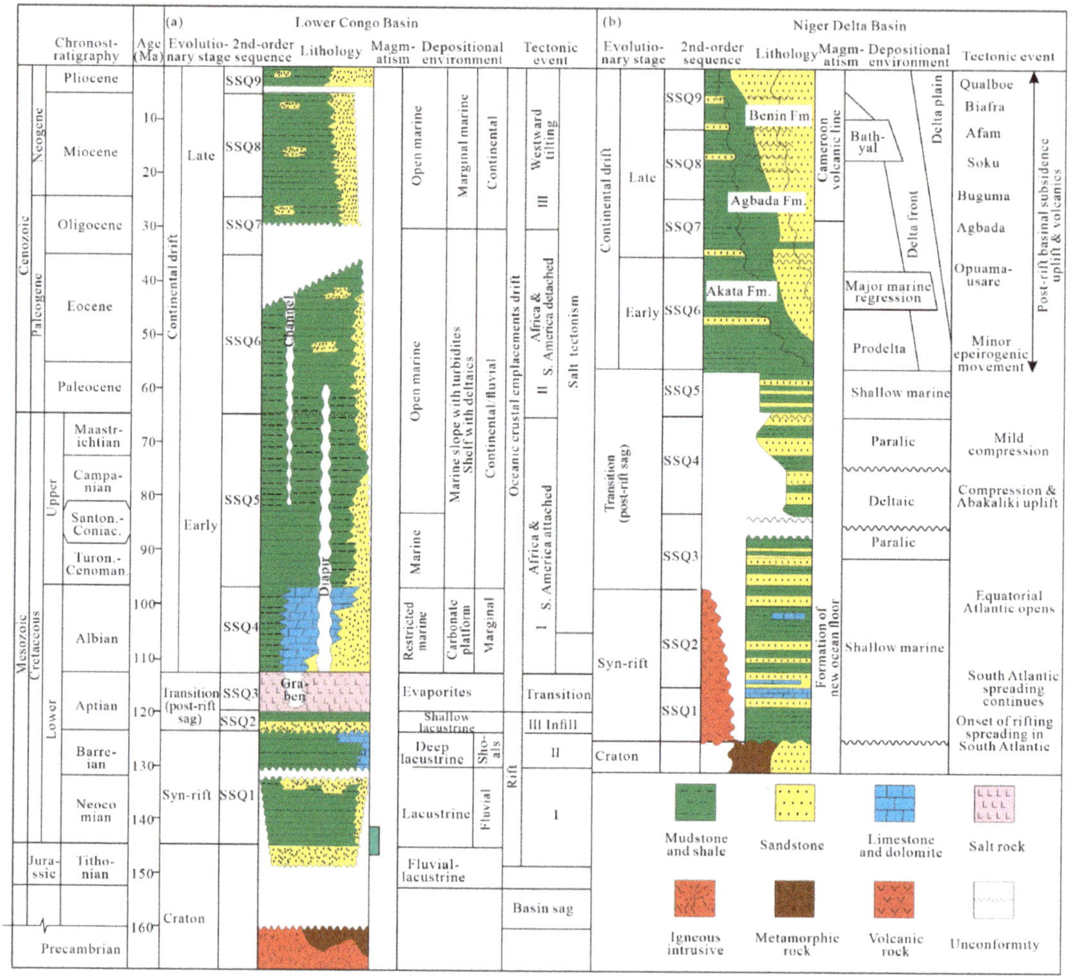

Figure 2. Composite chronostratigraphic columns of the Lower Congo Basin (**a**) and the Niger Delta Basin (**b**) (modified from [16]).

The Lower Congo Basin is located in central West Africa, covering a total area of approximately 170,000 km^2, with most (89%) located offshore. It is a typical West African passive continental margin basin [21,22]. The tectonic deformation and structural styles of the basin exhibit clear zonation in plan view, divided from east to west into a thin-skinned extension zone, a salt diapir transition zone, and a frontal thrust-fold zone [23]. The main body of the deep-sea fan is developed within the thrust-fold zone (Figure 1d).

The Niger Delta Basin is located in the Gulf of Guinea in the equatorial Atlantic, with a total area of approximately 210,000 km^2, with 62% located offshore. The basin originated in the Cretaceous and is mainly composed of Cenozoic deltas extending southwestward, with a sedimentary thickness exceeding 12 km [24]. Marine shales dominate the basin, with shale depositional activity significantly impacting hydrocarbon accumulation. Similar to the Lower Congo Basin, the tectonic features and structural styles of the Niger Delta Basin exhibit a typical tripartite zonation in plan view, divided from northeast to southwest into a thin-skinned extension zone, a mud diapir transition zone, and a thrust-fold zone (Figure 1e).

2.2. Stratigraphic Framework and Sedimentary Characteristics

The basement of the Lower Congo Basin is composed of Precambrian crystalline basement, overlain by Cretaceous to Neogene sedimentary strata, with local Jurassic formations. Aptian salt has a significant impact on the tectonics, sedimentation, and hydrocarbon accumulation of the basin [25]. Bounded by the salt layer, the basin's strata are divided into pre-salt, salt, and post-salt sequences, corresponding to the rift, transitional (sag), and drift stages, respectively [26]. The pre-salt strata include fluvial-lacustrine sandstones and shales of the Late Jurassic Tithonian to Early Cretaceous Barremian stages (SSQ1). The salt layer contains shallow lacustrine sandstones, mudstones, and evaporitic rocks deposited during the Aptian stage (SSQ2–SSQ3) of the transitional (sag) phase. The post-salt strata include the Early Cretaceous Albian to Quaternary stages (SSQ4–SSQ9), with Albian deposits comprising restricted marine carbonates, transitioning upward into marine clastic sediments. The deep-sea fan systems are mainly developed in strata since the Late Cretaceous (Figure 2a).

The basement of the Niger Delta Basin is similar to that of the Lower Congo Basin, overlain by transgressive sedimentary strata of the Early Cretaceous Aptian to Albian rift stages (SSQ1–SSQ2), extending upward to the transitional (sag) phase. The Late Cretaceous Turonian to Paleocene (SSQ3–SSQ5, transitional phase) represents a large-scale transgressive stage, with deposits of alternating terrestrial-marine and shallow-marine to semi-deep-marine facies. From the Eocene to the present (SSQ6–SSQ9, continental drift phase), a large-scale regressive phase occurred. Influenced by global sea-level fall, large-scale wave-dominated progradational deltas developed in the Niger Delta shelf area, while deep-sea fan deposits gradually developed below the continental slope. This stratigraphic sequence comprises three diachronic units arranged from base to top: the Akata (Eocene-Quaternary), Agbada (Oligocene-Quaternary), and Benin (Oligocene-Quaternary) Formations [20] (Figure 2b). The Akata Formation consists of thick-bedded shales of shelf-slope prodelta and shallow- to deep-marine facies, serving as the primary source rock in the area, with a maximum thickness exceeding 5000 m. The Agbada Formation consists of 3000–4500 m thick progradational delta front deposits, with shales in the lower part and gradually increasing sandstone content in the upper part, composed of multiple retrogradational–progradational sedimentary rhythms. The Benin Formation mainly consists of fluvial and back-barrier swamp sandstones and conglomerates, in parallel unconformable contact with the underlying Agbada Formation [27].

3. Datasets and Methods

Building on a comprehensive analysis of the regional geological setting, this study utilizes well-log data from 18 key wells (9 each in the Lower Congo Basin and Niger Delta Basin). The workflow proceeds as follows: First, lithology characteristics are calibrated and sedimentary facies sequences are classified for individual wells. Subsequently, integrating 4039 km of seismic lines (2560 km in the Lower Congo Basin and 1479 km in the Niger Delta Basin) with RMS amplitude attribute horizon slices and other relevant data, lithofacies types and architectural elements are described and analyzed. Seismic data processing is commonly performed using the Landmark software suite, developed by Landmark Graphics Corporation (a Halliburton subsidiary) in the United States.

RGB spectral decomposition blends low (red), medium (green), and high (blue) frequency magnitude volumes into a single-color image [28]. This enhances subtle stratigraphic and structural features invisible in conventional seismic amplitude data. By leveraging geological bodies' frequency responses, it improves resolution of thin beds, internal architecture variations, and depositional/erosional elements in complex systems like MTCs [29]. It is a valuable frontier tool in high-resolution seismic geomorphology and attribute analysis.

Shoreline and shelf-edge migration are primarily driven by imbalances among sediment supply, accommodation space, and climatic conditions [30]. The interaction of these three directly shapes and controls the architecture of deep-water sedimentary systems. This study adopted the ternary classification system proposed by Gong et al. [3], with appropriate adjustments (Figure 3) to describe the differences in shelf-edge structures, conducting semi-quantitative analysis of the aforementioned three variables [31]. Multiple structural parameters (Table 1) were selected as inputs for the volumetric balance model to determine sediment supply (Qs), shelf accommodation space (δx), and climatic conditions. Using measured seismic profiles from both basins in previous studies and internal research data, we systematically measured the depositional characteristics of the shelf edge [32–34].

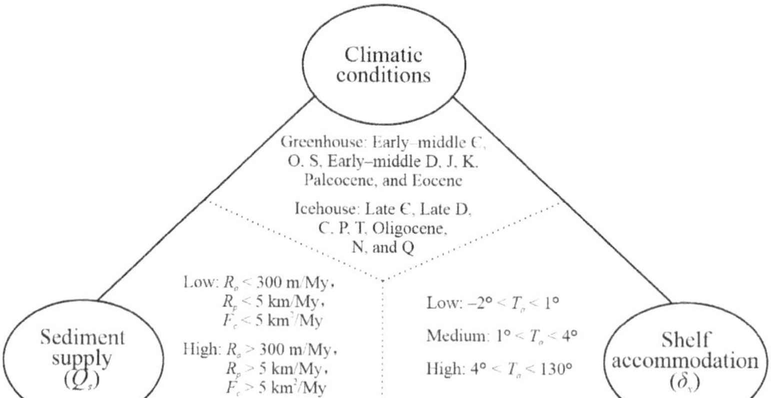

Figure 3. Ternary diagram classifying shelf-margin architectural styles using sediment supply, accommodation space, and climatic conditions as classification axes.

Table 1. Comparison of shelf-margin structural and morphological parameters for the two basins.

Basin Name	Age	Structural Parameters							Morphological Parameters	
		A^a (m)	P (km)	T (My)	R_a (m/My)	R_p (km/My)	F_c (km²/My)	T_a (°)	Curvature Index	E/C Ratio
Lower Congo Basin	Quaternary	629.0	15.3	2.5	251.6	6.1	3.8	2.4	Upstream channels dominate, decreasing seaward	0.2~0.6/0.5
	Pliocene	750.5	29.3	2.7	278.0	10.9	8.2	1.5	NA	NA
		569.6	−22.0 [b]	5.2	109.5	−4.2	2.4	−1.5		
	Miocene	515.4	−15.7	17.7	29.1	−0.9	0.5	−1.9	1.1~1.5/1.2	0.8~1.6/1.3
		1354.9	54.5	17.7	76.5	3.1	4.2	1.4	1.1~1.5/1.3	0.2~4.2/2.8
	Oligocene	705.3	28.9	10.8	65.3	2.7	1.9	1.4	1.4~1.6/1.5	0.5~0.6/0.6
Niger Delta Basin	Quaternary	525.9	9.2	1.3	404.6	7.1	3.7	3.3	1.0~1.4/1.2	
		542.3	11.1	1.3	417.2	8.5	4.6	2.8	1.4~2.5/1.8	0.8~3.1/1.9
		324.0	5.3	0.7	462.9	7.6	2.5	3.5	1.3~2.0/1.5	0.8~1.4/1.1
		458.8	12.4	1.4	327.7	8.8	4.1	2.1		
	Pliocene	492.9	4.4	1.2	410.8	3.7	1.4	6.3	NA	Co-occurrence of erosional braided channels and constructive lobe-channel complexes
		329.9	10.7	1.2	274.9	8.9	2.9	1.8		
		661.5	14.8	1.1	601.4	13.5	8.9	2.6		
		501.2	10.5	1.1	455.6	9.5	4.7	2.7		

Notes: Definition of structural parameters: A—Depositional thickness at shelf margin; P—Progradation distance; $R_a = A/T$—Sedimentation rate; $R_p = P/T$—Progradation rate; $F_c = R_p \times A$—Net sediment flux (cross-section); $T_a = arctan\ (A/P)$—Trajectory angle. Symbol annotations: [a] Average seismic velocity: 1600 m/s (sandstone-dominated strata); [b] Negative P and Rp values indicate retrogradation; Value range format: minimum–maximum/mean (e.g., 1.1~1.5/1.2)

By comparing deep-water system characteristics, sedimentary architectures, and shelf-margin configurations between the two basins, this study analyzes controlling factors-including regional tectonic activity, hydrodynamic mechanisms (global sea-level changes and paleo-drainage system stability), subcrop geomorphology (e.g., paleo-uplifts and diapirism), and shelf-slope types. We investigate their controls on lithofacies associations, morphological features, development scale, and distribution patterns of deep-water sediments during the late-stage continental drift.

4. Results

4.1. Deep-Water Sedimentary Systems Description

4.1.1. Lithofacies Types

In the Lower Congo Basin, the main vertical lithofacies include deep-water channels, natural levees, lobes, and deep-sea mud (Figure 4a). The lithology of deep-water channels is mainly medium- to fine-grained sandstone interbedded with thin-layered mudstone, with a maximum depth of 180 m, cumulative sandstone thickness of 165 m, and a sand-to-mud ratio exceeding 90%. Natural levees thin laterally in a wedge shape, characterized by thin sand bodies and sand-mud interlayers with low sand content. Lobes develop at the ends of deep-water distributary channels, exhibiting thin sand-mud interlayers and thick sandstones interbedded with thin-layered mudstones, with lobe sandstone thickness of 15–35 m and a sand-to-mud ratio of 40–70%.

In the Niger Delta Basin, the main vertical lithofacies include fan lobes, channels, and semi-deep-sea mud (Figure 4b). Fan lobes can be divided into single lobes and lobe complexes. Single lobes have thinner sand bodies, generally fine- to silt-grained sandstone interlayered with mudstone, with a cumulative sandstone thickness of 15–25 m and a sand-to-mud ratio of 30–50%. Lobe complexes have thicker sand bodies, mainly composed of lobe bodies, channelized lobes, and lobe margins, generally thick medium- to fine-grained sandstone interbedded with thin-layered mudstone, with a maximum cumulative sandstone thickness of 80 m and a sand-to-mud ratio of 50–70%. Sedimentary channels consist of thick medium- to fine-grained sandstone interbedded with thin-layered mudstone, with a cumulative sandstone thickness of 40–90 m and a sand-to-mud ratio of up to 80%.

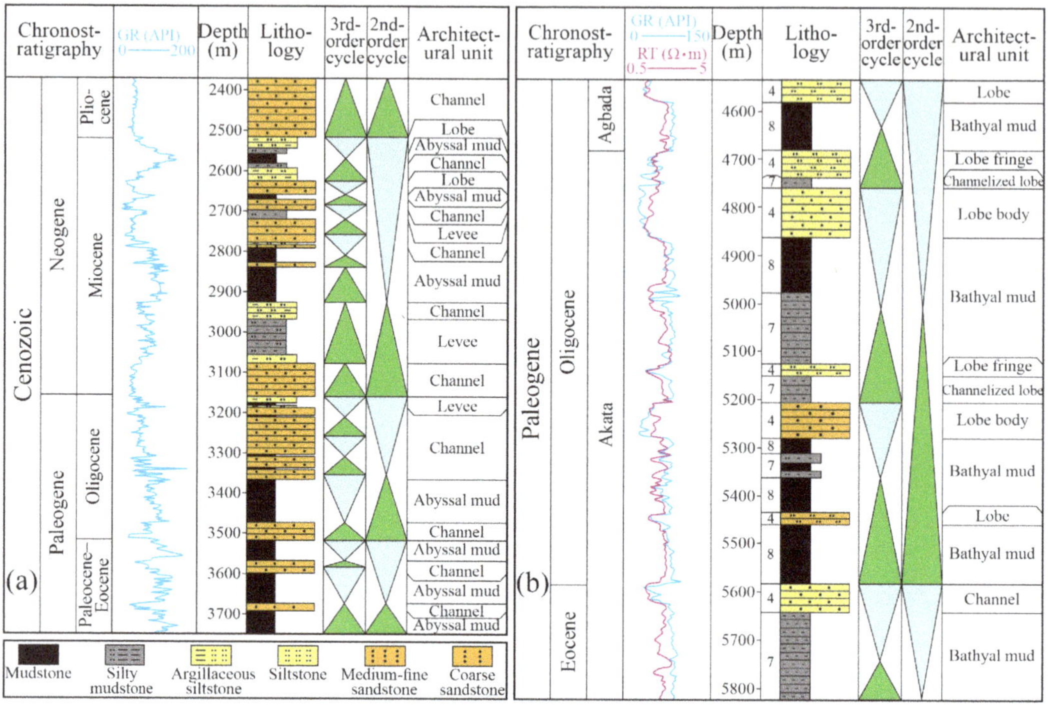

Figure 4. Sequence stratigraphic division and depositional facies interpretation from wireline logs in well Nzanza-1, the Lower Congo Basin (**a**) and well Ikebiri-1, the Niger Delta Basin (**b**).

4.1.2. Major Architectural Unit Characteristics

Deep-Water Channel-Levee Complexes

The main sedimentary system in the Lower Congo Basin is deep-water channels, followed by levees and lobes. According to channel location, deep-water channels can be divided into bypass channels, single-stage confined channels, and multi-stage (weakly) confined composite channels (Figure 5a). Turbidity currents often overflow the channel edges, leading to levee deposits on both sides of the channel, forming deep-water channel-levee complexes. Overbank deposits become progressively finer in grain size, thinner in thickness, and lower in flow energy with in-creasing distance outward from the channel margins. Channel-levee systems often exhibit a "U" or "V" shape on seismic profiles, with concaved complexes of 1.5–6 km width in RMS amplitude. They can be divided into inner levee deposits (channel terraces) and channel axis fill deposits, with an erosional boundary containing semi-confined channel deposits at the bottom and wedge-shaped outer levee deposits at the top (Figure 5a). The lower strata of the channel have seismic reflection characteristics of weak amplitude, poor continuity, and chaotic fill [35], with well logs often showing finger-shaped or serrated patterns. The middle strata of the channel exhibit strong amplitude, medium-to-poor continuity, and sub-parallel or chaotic fill seismic facies [36], with well logs showing box-shaped or serrated box-shaped characteristics. The upper strata of the channel are mostly medium-to-weak amplitude, medium-to-good continuity, and parallel-to-sub-parallel seismic reflection characteristics [37], with well logs showing serrated or serrated bell-shaped patterns. Seismic facies cutting across the levees show medium-to-strong amplitude, with amplitude weakening away from the channel edge, forming "gull-wing" seismic reflections with the channel facies.

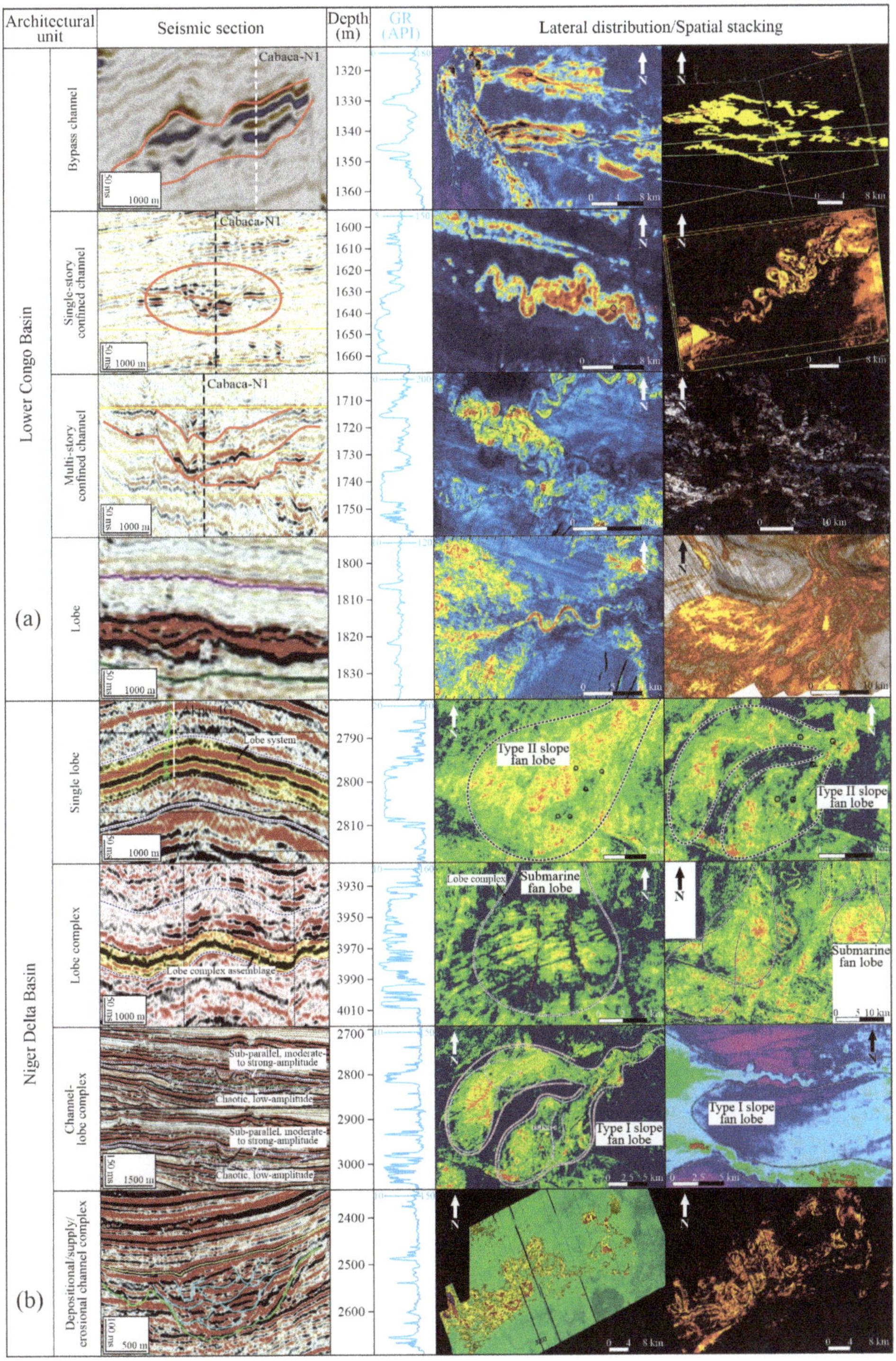

Figure 5. Seismic profiles, log responses, and spatial stacking patterns of sedimentary architecture units in the Lower Congo Basin (**a**) and the Niger Delta Basin (**b**).

Fan Lobes

The Niger Delta Basin primarily develops lobe deposits, with channels and levees as secondary features. Defined by their configuration, these lobes are classified into single lobes and lobe complexes, which may combine with channels to form channel-lobe complexes (Figure 5b). Seismically, lobes exhibit high-amplitude, continuous, parallel to sub-parallel reflections, occasionally showing low-angle shingled configurations [38]. They extend laterally over considerable distances with multiple reflection events, representing stacked composite lobe sand bodies. Well logs display high natural gamma ray amplitudes, showing variable patterns (box-shaped, bell-shaped, serrated) corresponding to architectural element combinations. Furthermore, depending on depositional environment, lobe systems divide into submarine fans and slope fans (Figure 5b). Submarine fan lobes demonstrate fan-shaped distributions on RMS amplitude maps. They feature large-scale, high-amplitude contiguous zones indicating sand-rich stacked deposits with minimal channelization [27]. In contrast, slope fan lobes are significantly smaller and comprise two subtypes: (1) Broom or tongue-shaped systems: Relatively larger with complex architectures. These composite lobes contain distinct channels, forming channel-lobe complexes that represent intermediate-stage channel progradation. (2) Elliptical or spindle-shaped systems: Smallest in scale, typically occurring at channel termini. These simpler units consist mainly of single lobes, indicating terminal-stage progradation.

4.2. Comparison of Shelf-Margin Structures and Sedimentary Architectures

During the late drift stage, the average *Ra* value of the Lower Congo Basin was 135.0 m/My, the average *Rp* value was 2.9 km/My, the average *Fc* value was 3.5 km^2/My, and the average *Ta* value was 0.6°, representing low sediment supply, medium-to-low accommodation space, and icehouse climate conditions (Table 1). In contrast, the average Ra value of the Niger Delta Basin during this period was 419.4 m/My, the average Rp value was 8.5 km/My, the average Fc value was 4.1 km^2/My, and the average Ta value was 3.1°, representing high sediment supply, medium accommodation space, and icehouse climate conditions (Table 1). In addition, the sinuosity index in the Lower Congo Basin has a mean of 1.3, compared to 1.6 in the Niger Delta Basin, indicating higher channel curvature and greater lateral migration in the latter. The *E/C* values are 1.3 and 1.5 in the Lower Congo and Niger Delta Basins, respectively, suggesting that due to the extensional zone of the two basins mainly being located on the middle-to-upper slope since the Miocene, the channel *E/C* values are larger, exhibiting overall erosional characteristics (Table 1).

In the early stage of gravity flow formation, the fluid transport capacity is strong, and the erosion of the lower strata is obvious, leading to the phenomenon of sedimentary bypass. In the middle stage, the fluid energy gradually decreases, forming confined channels rich in basal lag deposits and slump debris flow deposits. In the late stage of gravity flow development, the energy carried by the fluid decreases, and sand-rich sediments begin to accumulate in the incised valley, forming high sand-to-mud ratio weakly confined stacked channels. As the fluid energy further decreases, fine-grained sediments gradually deposit, forming low sand-to-mud ratio channels and basin floor channel-end lobes [36,39] (Figure 6a). In the Lower Congo Basin during the late drift stage, the main architectural units are deep-water channel-levee systems, with proximal lobes composed of distributary channels and sheet sands, and distal lobes composed of thin-layered sheet fine sandstones and mudstone interlayers. Composite channels are an important component of deep-water channel sedimentary architecture, including layers (layer sets), channel units, and single channels (Figure 6b). In the Niger Delta Basin, the sedimentary architecture is mainly fan lobe systems, including medium-to-small scale submarine fans and small-scale slope fans. The key architectural unit, the lobe complex, includes four main types: single swing, double swing, retrogradational, and progradational (Figure 6c). Submarine fan lobe systems are supplied by gravity flow channels on the slope, rapidly depositing in the relatively gentle slope zone (slope fans) and abyssal plain area (submarine fans), exhibiting large-area continuous distributions in plan view [4,22]. The lobe systems are mainly distributed within parasequence sets, composed of multi-stage lobe complexes, with a vertical thickness of up to 100 m. The lithology inside the lobe system is diverse, mainly thick and stable medium-to-fine grained and medium-to-coarse grained sandstones, with a single layer thickness of 5–15 m, with local development of thin sand–mud interlayers. The lobe complex is a high-order level of the fan lobe system, including layers (layer sets), lobe units, and single lobes (Figure 6d). Lobe units embedded in the lobe complex can be locally identified by high-resolution seismic tools (such as seafloor stratigraphic profilers), such as modern lobe complexes [40,41].

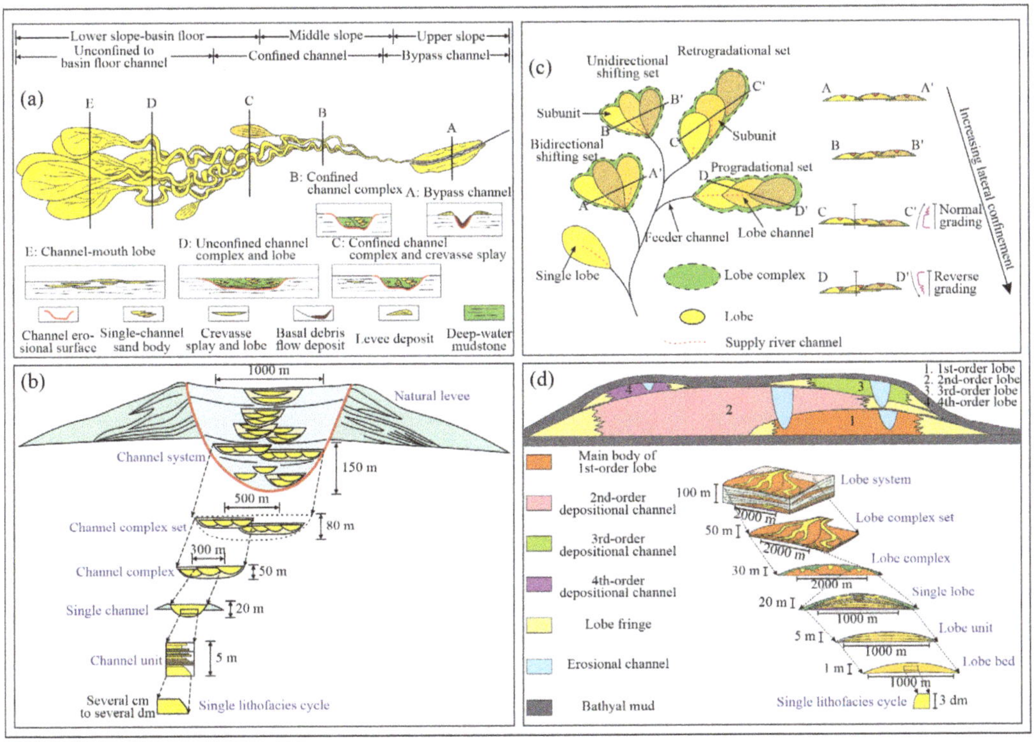

Figure 6. Stacking architecture of channel-lobe complexes in the Lower Congo Basin (**a**,**b**) and the Niger Delta Basin (**c**,**d**).

5. Discussion: Controlling Factors for Sedimentary Heterogeneity

5.1. Regional Tectonic Setting

Due to different plate tectonic locations, the two basins exhibit significant differences in tectonic background. The Lower Congo Basin is located in the central South Atlantic region, experiencing water evaporation and salinization after intracontinental rifting, and eventually developing marine sediments. The Niger Delta Basin is located at the intersection of the equatorial and South Atlantic, directly entering marine sedimentation after intracontinental rifting, without the formation of an evaporite environment. Due to the periodic activity of the Walvis Ridge, the initial ocean north of it was sometimes connected with the South Atlantic, resulting in the development of large areas of alternating terrestrial-marine evaporite deposits from the Santos Basin to the Douala Basin during the Aptian [42].

During the continental drift stage, both basins saw the emergence of gravity-driven structures related to plastic salt and shale layers. The main difference in tectonic activity lies in the differences in the tectonic processes and characteristics of salt tectonics and mud tectonics. On the central West African margin, the Aptian salt layer only developed one detachment, but it was activated in two main stages, resulting in four salt tectonic activities (Figure 7a). On the equatorial West African margin, three detachments of the intrabasinal shale layer led to extensional structures on the upper slope and thrust structures on the lower slope (Figure 7b). The evolution of sedimentation rate over time also shows that although the tectonic styles of the marginal basins in the two regions are different, they are closely related to gravity tectonics (Figure 7). On the central West African margin, the rapidly changing sedimentation rate is related to salt tectonic activity since the Late Cretaceous and the Oligocene [43]. In contrast, the equatorial West African margin is driven by an accelerating sedimentation rate, and long-term tectonic deformation activates the overlying detachment surface. The above all indicate that sedimentary loading plays a leading role in the activation of gravity tectonics. Another key parameter is the slope of the detachment, which is mainly controlled by the marginal tilting caused by thermal subsidence and sediment loading, and is also affected by regional tectonics [44] (Figure 7). In addition, since the Oligocene, the Lower Congo Basin has continued to deposit, while the onshore part has continued to uplift, and has experienced large-scale erosion during low water levels. The Niger Delta, on the other hand, was affected by the Cameroon volcanic belt on the east side in the Miocene, resulting in a large number of collapses, which also provided a large amount of volcanic debris source for the basin.

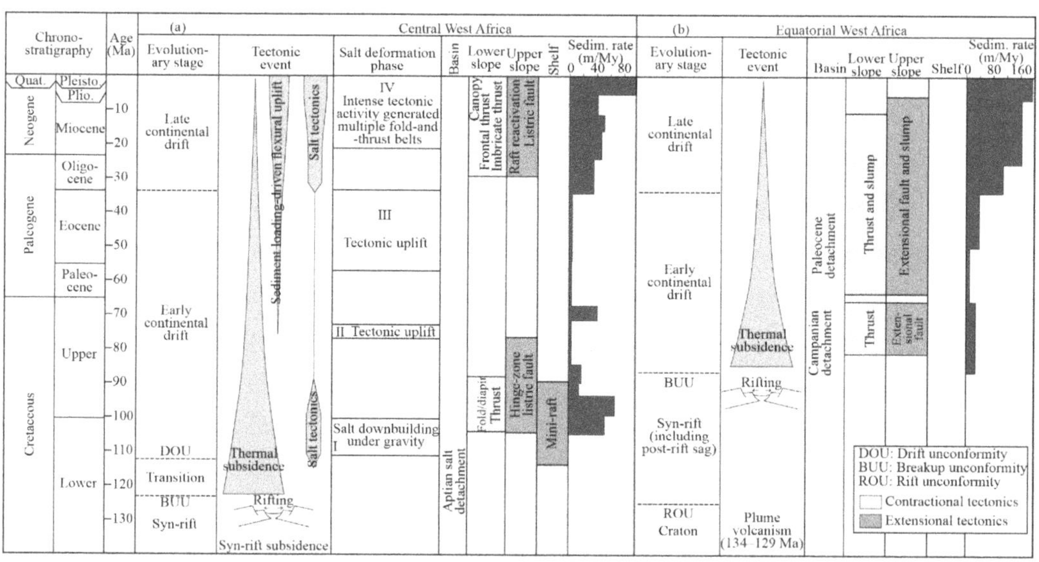

Figure 7. Tectonothermal evolution and gravity-driven deformation with sedimentation rate settings in the central (**a**) and equatorial (**b**) segments of West African passive margin basins.

The uplift of the southwestern African margin in the Miocene led to increased onshore erosion and deep-cut canyon development in the Lower Congo Basin, making the overall terrain of the basin steeper, enhancing the erosion capacity of gravity flow channels, and significantly increasing the E/C value (Table 1). A comparison of the sedimentary characteristics of the channel systems in different periods in the Lower Congo Basin shows that the Oligocene period was dominated by the development of high-migration channels, with an E/C value of only 0.6, representing depositional channels with low erosion capacity, and lacking uniform outer levee confinement between multiple single channels. The Miocene channel system is characterized by relatively low-migration channels, with an E/C value greater than 2.0, representing a high-erosion capacity channel system, with internal single channels confined (Figure 8a).

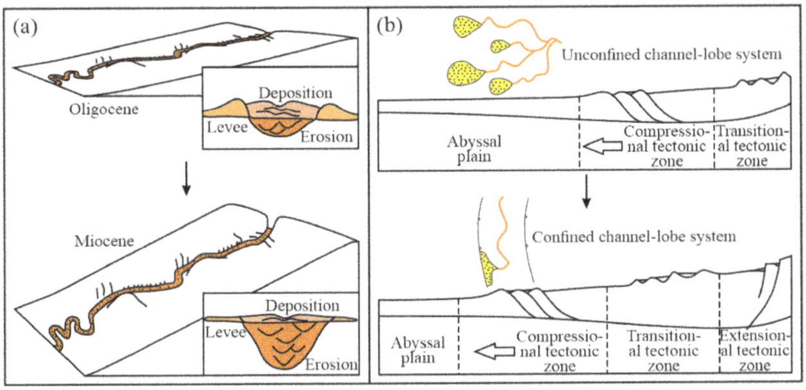

Figure 8. Evolutionary models of deep-water depositional systems during the late continental drift stage in the Lower Congo Basin (**a**) and the Niger Delta Basin (**b**).

Since the Eocene, the Niger Delta Basin has formed a progradational delta. With the continuous accumulation and thickening of detritus on the margin, the transitional crust and the basin as a whole have continued to subside [45]. The basin as a whole continues to prograde in a step-like manner, with continuous evolution of sedimentary belts. Although the overall tectonic pattern changes little, the extensional, transitional, and compressional tectonic zones have developed in each period, and the location is constantly migrating towards the sea. This leads to a large difference in the sedimentary characteristics between the early deep-water sedimentary systems and the late deep-water sedimentary systems in the same location. The early abyssal plain area develops unconfined sedimentary systems, while the late stages form thrust-fold structures as the compressional tectonic zone migrates seaward, limiting the development range of gravity flow channels (Figure 8b).

5.2. Oceanographic Controls

5.2.1. Global Sea Level Changes

Since the late continental drift (Eocene), the Lower Congo Basin began to form turbidite fans, while the Niger Delta Basin began to develop true delta progradational bodies. On the long-term cycle scale of global sea level, the overall trend is downward, so the Oligocene strata of both basins generally show a progradational development trend, and the progradation rate has significantly accelerated since the Miocene (Figure 9). In the Lower Congo Basin, tectonic uplift (with a maximum uplift height of 500 m) has caused the Oligocene and its overlying strata to be eroded on the landward side, and the change in the base level has caused the later channels to erode the bypass fan. During the low sea level period, submarine canyons developed in the basin, and the strata in a certain stage of the Miocene rapidly prograded seaward and then rapidly retrograded, which may be related to rapid tectonic uplift [7]. The sharp drop in sea level near the Niger Delta Basin has led to the formation of shelf erosion canyons, and changes in relative sea level in some areas have also changed the base level, thereby affecting the development style of deep-water sediments.

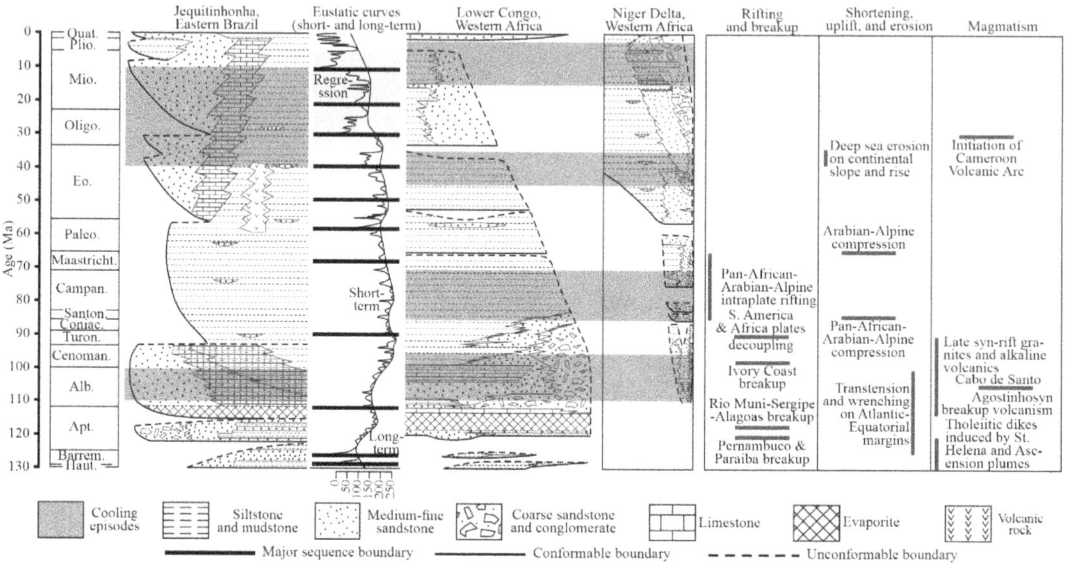

Figure 9. Eustatic curves along both South Atlantic margins and correlation of erosional phases in the two basins (the eustatic curves is referenced from [46]).

Since the Oligocene, the Niger Delta Basin has experienced multiple low sea level periods, forming specific sedimentary environments on the shelf edge. The Lower Congo Basin, on the other hand, has multiple unconformities, resulting in the erosion of strata in the terrestrial-marine transition area, making its paleogeomorphic pattern difficult to determine, and making it impossible to infer whether there are corresponding sedimentary systems at different sea level positions that change the position on the shelf (Figure 9). In addition, studies on the Oligocene of the Lower Congo Basin show that gravity flow sediments are fully developed, especially under the conditions of superimposed third-level and fourth-level sequence lowstand periods.

5.2.2. Paleodrainage Stability

The reason the seawater in the Niger Delta Basin has not been salinized and has a stable ancient water system is that the freshwater injection and tectonic location are relatively stable. In contrast, the ancient water system stability of the Lower Congo Basin is poor [47]. In the Niger Delta Basin, the uplift of the terrestrial shoulder allows the ancient water system to flow along the fault zone, while the ancient water system development of the Lower Congo Basin is controlled by strike-slip fault zones. The extensional zone of the Lower Congo Basin can be divided into small salt raft areas and giant salt raft areas (Figure 1d). The rolling fault blocks in the giant salt raft area have large lateral spans (>20 km), including the entire post-salt strata since the Early Cretaceous, with a thickness of 3–6 km. The faults in this area have large fault throws and are typical listric faults. The Cretaceous fault blocks are separated by Tertiary sediments. The section direction of the listric fault can be forward (inclined to the east) or reverse (inclined to the west), so local grabens are easily formed between these two major faults, thereby affecting the basin's water system distribution and sedimentary center position. During the Oligocene to Miocene period, the Lower Congo Basin successively saw two sedimentary centers:

one located in the southeastern part of the basin within Angola, roughly extending along the slope direction and parallel to the strike of the listric faults forming the graben; the other located in the northwestern part of the basin within Congo, radiating outwards from the center of the current canyon axis (Figure 10a). From the Oligocene to the mid-Miocene, turbidite sediments were mainly concentrated in the southeastern sedimentary center. It was not until the late Miocene that the sedimentary center began to migrate to the northwest, while the original southeastern sedimentary center had almost no sedimentary turbidite sand bodies during this period [19].

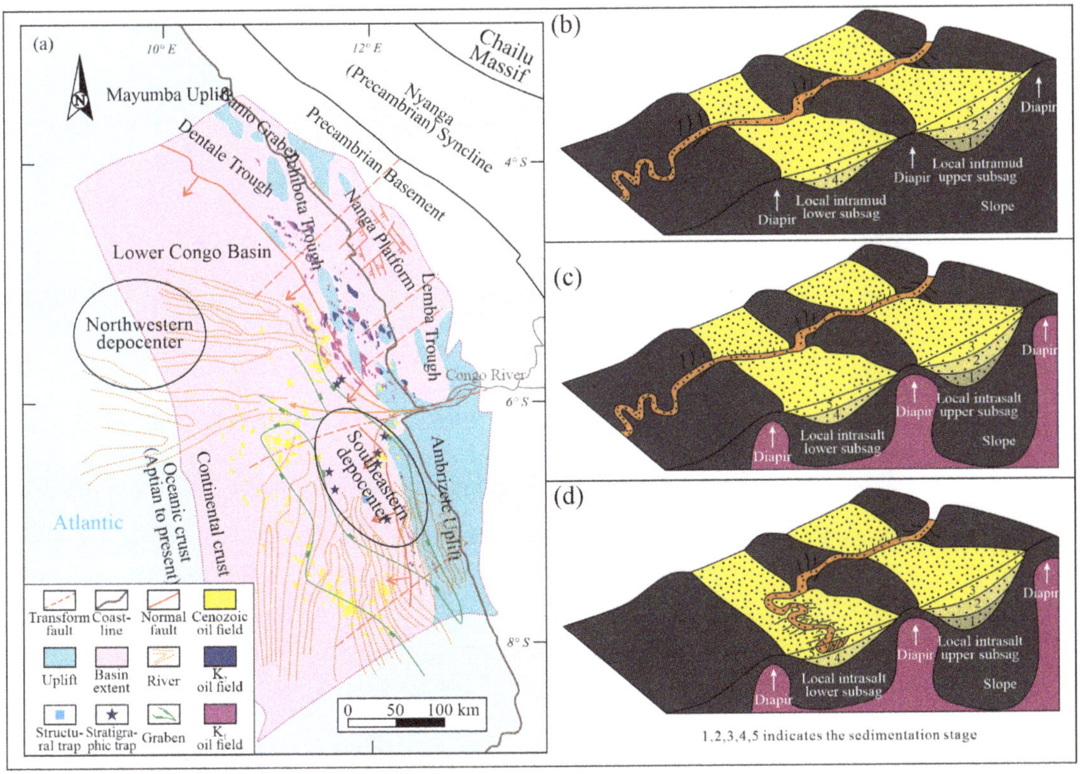

Figure 10. Oligocene–Miocene drainage distribution and basement paleouplift in the Lower Congo Basin (**a**), with mud/salt-diapir sub-basin fill patterns (**b**–**d**).

5.3. Pre-Depositional Substrate Geometry

5.3.1. Paleouplift

Compared with the overall tectonic pattern of the basin, the migration of the sedimentary center in the deep-water area is more affected by local geomorphic changes. In the Lower Congo Basin, local highs and paleohighs play an important role in deep-water sedimentary characteristics and hydrocarbon enrichment. In addition to promoting water system aggregation and outcrop erosion (often occurring in gentle slope zones), and being conducive to the development of carbonate reservoirs (providing growth conditions for bioherms), the up-dip direction of paleohighs also easily forms effective traps (conducive to the development of fractured and shallow-marine clastic rock reservoirs) and serves as an important channel for oil and gas migration [48]. For example, the oil and gas reservoirs of the Lower Cretaceous Bucomazi Formation source rocks in the Lower Congo Basin are mainly distributed in the Mayumba uplift, the Ambrizete uplift, and the high positions between the two (Figure 10a). Compared with the basement paleohighs before the drift period in the Lower Congo Basin, the Niger Delta Basin has a later development time, so it is mostly dominated by structural uplifts caused by mud diapirism.

5.3.2. Shale and Salt Diapirism

Shale usually only has plasticity and can move under overpressure conditions, so its development and evolution often show phased characteristics. Previous studies have used the "step-like" evolution model to explain this phased evolution, that is, the location of the diapir transition zone in a certain period is usually the formation location of the main fault zone in the next period [47]. The formation of salt structures still continues after salt deposition, and its formation and development are controlled by multiple factors, such as the source supply and dissolution rate of salt, the deposition and erosion rate of overlying strata, and the degree of extension and compression. The factors that cause salt to flow and deform mainly include differential loading, density difference with surrounding rocks, thermal convection, and stress environment, of which density difference is considered to be the most important triggering factor [49].

Salt structures and mud structures have differences in formation mechanism and evolution process, which promote the geometry, development and change in salt rocks to be more diverse, and not as regular as mud rocks. In the near-land area of salt-bearing basins, salt welding and salt raft structures under extensional conditions are generally developed, while the transition zone mainly develops salt stocks, salt walls, salt diapirs, and turtle back structures formed in the transition environment. In the compression zone at the foot of the slope, compressive salt structures are mainly developed, such as salt thrusts and compressive diapirs [47,50]. In contrast, the development amplitude and scale of mud structures are smaller, and they do not develop complex structures in multiple stages like salt rocks.

The diapirs in the transition tectonic zone of the Niger Delta Basin are mainly large mound-shaped mud diapirs, and the development density gradually increases from the extensional zone to the transition zone. The filling process of local depressions is shown in Figure 11b. Local uplifts are first filled, and then gravity flow enters the local downwarps after filling. As the base level drops, the diapir core and the lobes in the upper uplifts are eroded, and then the downwarps are also filled, and the channel enters the next-level downwarp. This process continues until the deep-sea plain. The salt diapirs developed in the Lower Congo Basin are slightly different from the mud diapirs. The geometry of salt diapirs is more diversified. Not only does the density increase from land to sea, but the diapir form also changes from isolated salt diapirs to laterally extending salt ridges or salt walls [47]. Multiple salt ridges enclose small depressions, and channels spread within them to form lobes or channelized lobes; trough-shaped terrains are easily formed between salt walls, forming channel confluence or restricting channel development. The development of local depressions in the basin salt structure can be divided into two situations: one is similar to mud structure sub-basins, where the channel directly erodes the diapir core (Figure 10c); the other is that the channel makes a turn and enters the next-level depression in other parts (Figure 10d).

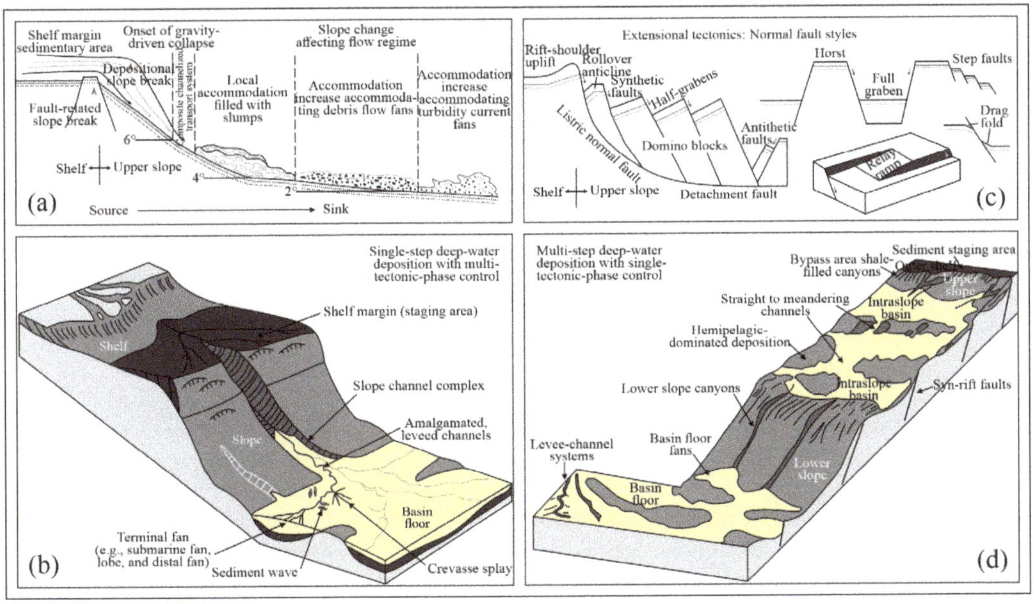

Figure 11. Depositional characteristics at shelf margins of the Lower Congo Basin (**a**,**b**) and the Niger Delta Basin (**c**,**d**).

5.4. Shelf-Slope Geometry

In terms of shelf and slope types, the Lower Congo Basin has a single-step narrow shelf-steep slope type, and the fan bodies are mainly formed in the lower slope and deep-water basin of the basin (Figure 11a,b); while the Niger Delta Basin has a multi-step wide shelf-gentle slope type, and the slope fan and submarine fan lobes are mostly developed on the upthrown side of the growth faults on the upper slope, the diapir depressions in the transition zone, or the basin floor area (Figure 11c,d).

The Congo Fan is mainly supplied by a single point source from the Congo River, and the source is continuously deposited at the fault slope break on the shelf edge. Because the structural pattern of the slope is simple, only a single-stage stepped slope is developed, but it has also experienced multiple periods of structural erosion (Figure 9). After the progradation thickening causes gravity collapse to start, the sediments develop into a composite channel transport system. Then, as the slope gradually decreases and the accommodation space continues to expand, the flow state of the slope sediments gradually changes from a slump body to a debris flow fan body and finally to a turbidite fan body (Figure 11a). The Lower Congo Basin has a large topographic drop and turbulent water flow due to tectonic uplift, and mud and sand are not easily deposited at the river mouth. Under the condition of continuous progradation, a large composite levee valley—the Congo Submarine Canyon—has been formed, with an average slope exceeding 40°, belonging to a high and steep canyon, so it is not easy to form a delta. The source detritus carried by the Congo River is directly transported to the basin floor plain through the Congo Canyon. The source erosion of the deep-cut canyon also narrows the shelf edge, making it difficult for deposition to occur on the shelf and upper slope. As the hydrodynamic force weakens, the channels begin to bifurcate on the lower slope and deep-sea plain, the scale gradually decreases and the number of tributaries increases, changing from erosional to aggradational, and forming submarine fans, frontal fans, and lobes at the end, and displaying a branching tree structure. The overall shape of the Congo Fan is a divergent tree-like shape, that is, a "single-step narrow shelf-steep slope rooted deep-water fan" (Figure 11b). The fan root area of this type of deep-water fan develops canyons and large channel complexes, the channel scale in the fan middle becomes smaller, and the fan end develops natural levees and channel end fans. The whole is mainly composed of channel deposits, followed by natural levees and fan lobes [22].

The Niger Fan is supplied by multiple points from the Niger River and the Niger Delta. Because the structural pattern of the slope is relatively complex, a multi-stage stepped slope is developed, but only a single-stage structural erosion has occurred (Figure 9). The sediments flow from the graben shoulder uplift of the shelf edge through the conversion slope to form a stepped upper slope, and finally extend to the deep sea or basin floor plain. The stepped upper slope is usually a domino-style fault block jointly composed of listric normal faults and detachment faults, including rolling anticlines, syn/antithetic faults, and half-grabens, or step-like faults formed by alternating horsts and grabens (Figure 11c). Since the Miocene, the Niger Delta Basin has continued to subside, the terrain is gentle, and the water flow is slow, leading to a lack of development of large canyons on the landward side of the Niger Fan, while mud and sand are easily deposited at the river mouth to form deltas. Under the gravity drive, terrestrial detritus is transported from the sediment accumulation area on the shelf edge to the deep-water area along the upper slope shale-filled canyons, straight-curved channels, and lower slope small canyon channels, forming a sheet-like deep-water submarine fan on the basin floor. Due to the gentle slope and the existence of synsedimentary stepped faults, some detrital materials are deposited in the intra-slope basins of the upper slope during the migration process to form slope fans, that is, "multi-step wide shelf-gentle slope unrooted deep-water fans" (Figure 11d). The fan root area of this type of deep-water fan does not develop large canyons and channels, and the fan middle and fan end mainly deposit lobe bodies, followed by deep-water channels [22].

5.5. Impact on Oil and Gas Exploration

This comparative study of deep-water sedimentary architectures in the Lower Congo and Niger Delta basins provides significant value for hydrocarbon exploration and development. It establishes a tectono-sedimentary framework linking salt/mud diapirism, shelf-slope morphology, and sediment pathways, enabling more accurate prediction of turbidite reservoir distribution (e.g., channel-levee complexes vs. fan lobes) in passive margins influenced by halokinesis. By quantifying how diapir-induced seafloor relief redirects sediment transport and amplifies facies heterogeneity, the research reduces exploration uncertainty in analogous basins, particularly regarding reservoir connectivity and compartmentalization. The identified contrast between the Lower Congo's narrow-shelf, erosional systems dominated by channels/terminal lobes and the Niger Delta's broad-shelf, progradational systems rich in delta-fed turbidites and lobes offers critical guidance for targeted exploration strategies: prioritizing terminal lobes in salt-influenced settings and delta-fed turbidites in high-sediment-supply regions. Furthermore, understanding the architectural controls (e.g., tectonic uplift, diapir evolution) optimizes well placement and development plans for complex deep-water reservoirs. These insights refine conventional "source-to-sink" models and directly support efficient resource exploitation in West Africa and global deep-water analogs.

6. Conclusions

(1) Since the Eocene, the Lower Congo Basin has mainly formed a narrow shelf-steep slope retrogradational deep-water turbidite sand system, showing that the basin fan bodies have changed from multi-source small to single-source large, single-step deposition, mainly developing deep-water turbidite/erosion channels, levees, large-

scale terminal lobes, and deep-sea mud, controlled by margin uplift tilting and the degree of salt tectonic activity, gradually transitioning from salt diapirs to salt walls from land to sea, resulting in channel flow changes or lobe formation.

(2) In the late stage of continental drift, the Niger Delta Basin is mainly a wide shelf-gentle slope large-scale progradational delta and its distal sand system, multi-point source supply, multi-step development, mainly depositing delta front underwater distributary channels, prodelta-semi-deep-sea mud, and medium-to-small-scale submarine fan lobes and small-scale slope fan lobes dominated by sedimentary channel-fan lobes, controlled by gravity-driven systems with mud structures as detachment surfaces, and mound-shaped mud diapirs form local uplifts.

(3) A comparison of the shelf-edge structure styles and sedimentary architectural characteristics in the late stage of continental drift shows that the Lower Congo Basin is in a condition of low sediment supply, medium-to-low equal accommodation space, and icehouse climate, mainly developing by-pass type and single (multi)-period confined deep-water channel-natural levee systems; while the Niger Delta Basin belongs to a high sediment supply, medium accommodation space, and icehouse environment, mainly composed of various lobe systems such as retro(pro)gradational and single (double)-directional swing types.

(4) Deltaic sedimentation can be regarded as a turbidite sedimentary system less affected by upwelling. Due to the constraints of various factors such as tectonic background, sea level changes, paleo-geomorphology, mud/salt diapirism, and shelf-slope types, there are differences in shelf-edge structure styles and sedimentary configurations between the two types of passive margin basins. Among them, regional or local tectonic activity is the main factor controlling deep-water turbidite sedimentation in the late stage of continental drift in the two basins, and sea level changes also affect the development of deep-water systems to a certain extent, while the differences in shelf-slope types and basin floor paleohighs are specific manifestations of the former two in the deep-water sedimentation process.

Author Contributions: Conceptualization, X.G. and J.Y.; methodology, F.Q.; software, F.Q.; validation, L.G. and X.G.; investigation, X.G.; resources, X.G. and J.Y.; writing—original draft preparation, F.Q.; writing—review and editing, L.G.; supervision, X.G.; project administration, X.G.; funding acquisition, X.G. and L.G. All authors have read and agreed to the published version of the manuscript.

Funding: This work was jointly funded by the National Science and Technology Major Projects of China (Grant No. 2016ZX05033-001) and the Special Scientific Research Fund of Yunnan Communications Vocational and Technical College (Grant No. 202501JTBS001).

Data Availability Statement: The original contributions presented in this study are included in the article material. Further inquiries can be directed to the corresponding author.

Acknowledgments: We thank the Petroleum Exploration and Production Research Institute, SINOPEC Group for providing core samples, drilling/logging records, seismic data, and publication permission. We acknowledge Zequan Hu, Yubin Hu, and Weiting Wang for their preliminary work on core analysis, data processing, and seismic interpretation. We are grateful to the editors and reviewers for their insightful comments, which significantly improved this manuscript.

Conflicts of Interest: The authors declare no conflicts of interest. The funders had no role in the design of the study; in the collection, analyses, or interpretation of data; in the writing of the manuscript; or in the decision to publish the results.

References

1. Reading, H.G.; Richards, M. Turbidite systems in deep-water basin margins classified by grain size and feeder system. *AAPG Bull.* **1994**, *78*, 792–822. [CrossRef]
2. Marcano, G.; Anka, Z.; di Primio, R. Major controlling factors on hydrocarbon generation and leakage in South Atlantic conjugate margins: A comparative study of Colorado, Orange, Campos and Lower Congo basins. *Tectonophysics* **2013**, *60*, 172–190. [CrossRef]
3. Gong, C.; Steel, R.J.; Wang, Y.; Lin, C.; Olariu, C. Shelf-margin architecture variability and its role in sediment-budget partitioning into deep-water areas. *Earth-Sci. Rev.* **2016**, *154*, 72–101. [CrossRef]
4. Zhang, J.J.; Wu, S.; Hu, G.; Fan, T.-E.; Yu, B.; Lin, P.; Jiang, S. Sea-level control on the submarine fan architecture in a deepwater sequence of the Niger Delta Basin. *Mar. Pet. Geol.* **2018**, *94*, 179–197. [CrossRef]
5. Wang, L.L.; Wang, Z.Q.; Xiao, P. Characterization of deep water sedimentary system in the Miocene of Block A in Lower Congo Basin. *Oil Gas Geol.* **2015**, *36*, 963–974.
6. Zhao, P.F.; Zhang, H.; Bu, F.; Wu, J.; Yang, X. The sedimentary architecture characteristics and fluid system of the deep sea turbidite-lobe complex sandbodies: A case study of the deep-water region in the Niger Delta Front. *Earth Sci.* **2017**, *42*, 1972–1983.
7. Li, Q.; Wu, W.; Kang, H.Q.; Ren, S.J.; Pang, L.A.; Yang, T.; Cai, L.L.; Liu, X.L. Characteristics and controlling factors of deep-water channel sedimentation in Lower Congo Basin, West Africa. *Oil Gas Geol.* **2019**, *40*, 917–929.
8. Weimer, P.; Slatt, R.M. *Introduction to the Petroleum Geology of Deepwater Settings*; AAPG/Datapages: Tulsa, OK, USA, 2007.
9. Qin, Y.Q.; Ba, D.; Xu, H.L.; Liang, Y.B.; Yang, Z.; Liang, X.; Wang, X.L. Sedimentation and hydrocarbon accumulation of deep-water fan fed by large canyon in passive continental margin: A case of Congo fan in West Africa. *China Pet. Explor.* **2018**, *23*, 59–68.
10. Posamentier, H.; Walker, R. *Deep-Water Turbidites and Submarine Fans*; SEPM Special Publication: Tulsa, OK, USA, 2006; pp. 399–520.
11. Hüneke, H.; Mulder, T. *Deep-Sea Sediments*; Elsevier: Amsterdam, The Netherlands, 2011.
12. Liu, X.Y. Depositional characteristics and evolution of the Tertiary deep-water fan in West Africa. *J. Northeast. Pet. Univ.* **2013**, *37*, 24–31.
13. Wang, W.G.; Tong, X.G.; Zhang, Y.X.; Cheng, K.N. Features of major passive continental margin basins, South Atlantic Ocean. *China Pet. Explor.* **2012**, *17*, 62–69.
14. Estrella, G. Breaking paradigms giant and super-giant discoveries in Brazil. In Proceedings of the AAPG Annual Conventional and Exhibition, Denver, CO, USA, 7–10 June 2009.
15. Qin, Y.Q.; Zhang, G.Y.; Liang, Y.B.; Wen, Z.X.; Wang, Z.M.; Zhang, L.; Ba, D.; Wang, Y.H. Distribution characteristics, accumulation rules and exploration directions of deep water hydrocarbon in South Atlantic. *Nat. Gas Geosci.* **2016**, *27*, 229–240.
16. IHS. *International Energy Oil & Gas Industry Solutions*; IHS: Rockville, MD, USA, 2014.
17. Feng, Y.W.; Qu, H.J.; Zhang, G.C.; Mi, L.J.; Fan, Y.H.; Guan, L.Q. Toctonic-Sedimentary Evolution and its Control on Source-Reservior-Cap Rocks in Passive Continental Margin, West Africa. *Mar. Orig. Pet. Geol.* **2010**, *15*, 45–51.
18. Luo, Y.; Li, J.H.; Yang, M.L. A comparative study of tectonic and sedimentary evolution of West African coastal basins in the south Atlantic Ocean region. *Geol. China* **2021**, *48*, 120–128.
19. Anka, Z.; Séranne, M.; Lopez, M.; Scheck-Wenderoth, M.; Savoye, B. The long-term evolution of the Congo deep-sea fan: A basin-wide view of the interaction between a giant submarine fan and a mature passive margin (ZaiAngo project). *Tectonophysics* **2009**, *470*, 42–56. [CrossRef]
20. Deng, R.J.; Deng, Y.H.; Yu, S.; Hou, D.J. Hydrocarbon geology and reservoir formation characteristics of Niger Delta Basin. *Pet. Explor. Dev.* **2008**, *35*, 755–762.
21. Navarre, J.-C.; Claude, D.; Liberelle, E.; Safa, P.; Vallon, G.; Keskes, N. Deepwater turbidite system analysis, West Africa: Sedimentary model and implications for reservoir model construction. *Lead. Edge* **2002**, *21*, 1132–1139. [CrossRef]
22. Cai, L.L.; Wang, Y.; Wang, Y.; Dong, S.; Zhu, S.; Liao, J.; Zhao, Z.; Xue, D. Type features and hydrocarbon exploration significance of deepwater sedimentary in West Africa. *Acta Pet. Sin.* **2016**, *37*, 131–142.
23. Liu, X.Y.; Yu, S.; Tao, W.X.; Hu, X.L.; Hao, L.H. Filling architecture and evolution of Upper Miocene deep-water channel in Congo Fan Basin. *Earth Sci.* **2012**, *37*, 105–112.
24. Hooper, R.J.; Fitzsimmons, R.J.; Grant, N.; Vendeville, B.C. The role of deformation in controlling depositional patterns in the south-central Niger Delta, West Africa. *J. Struct. Geol.* **2002**, *24*, 847–859. [CrossRef]
25. Shi, S.Y.; Yu, Y.X.; Yin, J.Y.; Wu, C.W.; Liu, J.J.; Liu, Y.L.; Wang, B. Deformation characteristics and formation mechanisms of salt structures in the Lower Congo Basin. *Oil Gas Geol.* **2020**, *41*, 1092–1099.
26. Feng, Z.Q.; Guo, F.T.; Zhang, Z.M.; Yin, J.Y.; Tian, N.X.; Su, Y.S.; Guo, R.T.; Wu, G.K. A transform margin: Review of genetic types of the Niger Delta Basin. *Acta Petrol. Sin.* **2022**, *38*, 2565–2580.
27. Cai, L.L.; Xie, X.J.; Li, J.P.; Liao, J.H. Influence of different modes of deep-water sedimentation on oil and gas distribution: A case study of deep-water fans in eastern and western Niger Delta Basin. *Acta Sedimentol. Sin.* **2022**, *40*, 229–243.

28. Partyka, G.; Gridley, J.; Lopez, J. Interpretational applications of spectral decomposition in reservoir characterization. *Lead. Edge* **1999**, *18*, 353–360. [CrossRef]
29. Pigott, J.D.; Abdel-Fattah, M. Seismic stratigraphy and geomorphology of offshore Tanzania. *Interpretation* **2014**, *2*, 1–13.
30. Catuneanu, O.; Galloway, W.E.; Kendall, C.G.; Miall, A.D.; Posamentier, H.W.; Strasser, A.; Tucker, M.E. Sequence stratigraphy: Methodology and nomenclature. *Newsl. Stratigr.* **2011**, *44*, 173–245. [CrossRef]
31. Allen, P.A.; Armitage, J.J.; Carter, A.; Duller, R.A.; Michael, N.A.; Sinclair, H.D.; Whitchurch, A.L.; Whittaker, A.C. The Qs problem: Sediment volumetric balance of proximal foreland basin systems. *Sedimentology* **2013**, *60*, 102–130. [CrossRef]
32. Jermannaud, P.; Rouby, D.; Robin, C.; Nalpas, T.; Guillocheau, F.; Raillard, S. Plio-Pleistocene sequence stratigraphic architecture of the eastern Niger Delta: A record of eustasy and aridification of Africa. *Mar. Pet. Geol.* **2010**, *27*, 810–821. [CrossRef]
33. Armitage, D.A.; McHargue, T.; Fildani, A.; Graham, S.A. Postavulsion channel evolution: Niger Delta continental slope. *AAPG Bull.* **2012**, *96*, 823–843. [CrossRef]
34. Wang, L.L.; Wang, Z.Q.; Yu, S.N.N.R. Seismic responses and controlling factors of Miocene deepwater gravity-flow deposits in Block A, Lower Congo Basin. *J. Afr. Earth Sci.* **2016**, *120*, 31–43. [CrossRef]
35. Schwab, A.M.; Cronin, B.T.; Ferreira, H. Seismic expression of channel outcrops: Offset stacked versus amalgamated channel systems. *Mar. Pet. Geol.* **2007**, *24*, 504–514. [CrossRef]
36. Mayall, M.; Jones, E.; Casey, M. Turbidite channel reservoirs—Key elements in facies prediction and effective development. *Mar. Pet. Geol.* **2006**, *23*, 821–841. [CrossRef]
37. Posamentier, H.W.; Kolla, V. Seismic geomorphology and stratigraphy of depositional elements in deep-water settings. *J. Sediment. Res.* **2003**, *73*, 367–388. [CrossRef]
38. Wang, W.W.; Wang, D.; Sun, J.; Shao, D.; Lu, Y.; Chen, Y.; Wu, S. Evolution of deepwater turbidite bedforms in the Huaguang channel–lobe transition zone revealed by 3D seismic data in the Qiongdongnan Basin, South China Sea. *Geomorphology* **2020**, *370*, 107412. [CrossRef]
39. Guo, X.; Liu, X.; Zheng, T.; Zhang, H.; Lu, Y.; Li, T. A mass transfer-based LES modelling methodology for analyzing the movement of submarine sediment flows with extensive shear behavior. *Coast. Eng.* **2024**, *191*, 104531. [CrossRef]
40. Prélat, A.; Hodgson, D.M.; Flint, S.S. Evolution, architecture and hierarchy of distributary deep-water deposits: A high-resolution outcrop investigation from the Permian Karoo Basin, South Africa. *Sedimentology* **2009**, *56*, 2132–2154. [CrossRef]
41. Picot, M.; Droz, L.; Marsset, T.; Dennielou, B.; Bez, M. Controls on turbidite sedimentation: Insights from a quantitative approach of submarine channel and lobe architecture (Late Quaternary Congo Fan). *Mar. Pet. Geol.* **2016**, *72*, 423–446. [CrossRef]
42. Yu, X.; Hou, G.T.; Dai, S.H.; Han, Y.C.; Xie, J.L. Comparison of petroleum geology characteristics and hydrocarbon accumulation regularity of deep water basins in eastern part of Brazil. *Mar. Orig. Pet. Geol.* **2016**, *21*, 61–72.
43. Leturmy, P.; Lucazeau, F.; Brigaud, F. Dynamic interactions between the Gulf of Guinea passive margin and the Congo River drainage basin: Morphology and mass balance. *J. Geophys. Res.* **2003**, *108*, 1–19. [CrossRef]
44. Séranne, M.; Anka, Z. South Atlantic continental margins of Africa: A comparison of the tectonic vs climate interplay on the evolution of equatorial west Africa and SW Africa margins. *J. Afr. Earth Sci.* **2005**, *43*, 283–300. [CrossRef]
45. Reijers, T.J.A. Stratigraphy and sedimentology of the Niger Delta. *Geologos* **2011**, *17*, 133–162. [CrossRef]
46. Vail, P.R.; Mitchum, R.M., Jr. Global Cycles of Relative Changes of Sea Level from Seismic Stratigraphy1. In *Geological and Geophysical Investigations of Continental Margins*; Watkins, J.S., Montadert, L., Dickerson, P.W., Eds.; American Association of Petroleum Geologists: Tulsa, OK, USA, 1979.
47. Hong, F.H. *Comparison on Deep Water Deposition of West Africa Passive Margin Basins: Case Studies of Niger Delta Basin and Lower Congo Basin*; China University of Petroleum (Beijing): Beijing, China, 2016.
48. Gong, L.; Gao, X.Z.; Qu, F.T.; Zhang, Y.S.; Zhang, G.Y.; Zhu, J. Reservoir quality and controlling mechanism of the Upper Paleogene fine-grained sandstones in lacustrine basin in the hinterlands of northern Qaidam Basin, NW China. *J. Earth Sci.* **2023**, *34*, 806–823. [CrossRef]
49. Jackson, M.P.A.; Hudec, M.R. *Salt Tectonics: Principles and Practice*; Cambridge University Press: Cambridge, UK, 2017.
50. Wang, D.J.; Li, J.H.; Cheng, P.; Li, Y.H.; Liu, Z.Q.; Wang, Y. Controlling factors of salt tectonic formation in the South Atlantic passive margin basins. *Geotecton. Metallog.* **2019**, *43*, 924–933.

Journal of
Marine Science and Engineering

MDPI

Article

Large-Eddy Simulation of Flows Past an Isolated Lateral Semi-Circular Cavity

Yiqing Gong [1], Yun Xu [2], Jingqiao Mao [3], Jie Dai [3,*], Lei He [4,*], Hao Zhang [3] and Qianshun Xu [5]

1 Institute of Water Science and Technology, Hohai University, Nanjing 211106, China; yqgong@hhu.edu.cn
2 Zhejiang Institute of Communications Co., Ltd., Hangzhou 310012, China; xuyuncz@163.com
3 College of Water Conservancy and Hydropower Engineering, Hohai University, Nanjing 210098, China; maojq@hhu.edu.cn (J.M.); haozh2597@163.com (H.Z.)
4 College of Architectural Engineering, Zhejiang Tongji Vocational College of Science and Technology, Hangzhou 311231, China
5 Yellow River Engineering Consulting Co., Ltd., Zhengzhou 450003, China; xuqianshun@yrec.cn
* Correspondence: jiedai@hhu.edu.cn (J.D.); leidake@foxmail.com (L.H.)

Abstract: Lateral cavities along coastlines strongly influence sedimentary morphology and ecological processes by modifying local flow dynamics. This study employed high-resolution large-eddy simulation to investigate flow structures and momentum exchange mechanisms in a semi-circular lateral cavity driven by longshore currents. Model validation against experimental data confirmed the LES's capability to capture both recirculating flow and turbulent structures accurately. The impact of Reynolds number was examined across three cases (Re = 12,000, 17,000, and 22,000). From Re = 12,000 to 17,000, a significant upstream shift of the primary vortex core occurred, accompanied by stronger shear layer turbulence and intensified secondary vortices. Between Re = 17,000 and 22,000, the flow features stabilized, indicating a transition toward quasi-equilibrium. These changes enhanced vertical momentum transfer and turbulence production within the cavity. Spectral analysis revealed dominant KH frequencies governing periodic momentum exchange and indicating a transition from viscosity-damped upstream turbulence to fully developed shedding downstream.

Keywords: large-eddy simulation; lateral semi-circular cavity; shear layer; Kelvin–Helmholtz instability

1. Introduction

Coastal morphological evolution is governed by the interplay between hydrodynamic forces and sediment transport processes [1–4]. Among these, alongshore currents play a key role and are strongly influenced by shoreline geometry. Arc-shaped embayments and recessed features along the coast, which often result from morphological evolution, can significantly modify local hydrodynamics. These processes shape meandering shorelines and recessed morphological features that serve as critical zones for sediment retention and biodiversity enhancement [5,6]. Semi-enclosed natural cavities, such as bays and lagoons, facilitate vegetation colonization while providing essential fish nurseries [7,8]. Similarly, cuspate forelands and tombolos create localized embayments where flow attenuation promotes the selective deposition of fine sediments, reshaping coastal geomorphology and fostering niche habitats [9–11]. In estuarine systems, sinuous channel margins are often lined with lateral cavities that serve as brackish mixing zones and critical ecological transition corridors [12,13]. Lateral cavities enhance ecological resilience and support the long-term stability of coastal landscapes by promoting sedimentary self-organization and enabling hydrodynamic–biotic feedback. Artificial lateral cavities have also been adopted in nature-based coastal restoration strategies. In Sydney Harbour, localized cavity structures integrated into seawalls by Australian engineers have been shown to increase both the abundance and diversity of sessile species [14,15].

The ecological performance of natural and artificial cavities is tightly coupled to by their influence on local flow dynamics [16–18]. Mass transport and exchange across the cavity interface are governed by recirculating flow patterns, vortex shedding dynamics, and shear layer instabilities that develop within and around the cavities. Recent advances combining field observations, laboratory experiments, and numerical simulations have progressively clarified the fundamental physics of cavity flows, particularly the interdependent effects of Reynolds number (Re), aspect ratio, and morphological complexity on flow properties. An increase in Re typically results in a shift of the main vortex (MV) core, the emergence of additional counter-rotating corner vortices, and enhanced mass and momentum exchange across the cavity mouth [19–21]. The aspect ratio determines the spatial extent of the main recirculation zone and the threshold for secondary vortex formation [22–24]. Furthermore, topographic undulations and curved geometries further complicate the flow field by introducing an additional secondary corner and bottom vortices [25,26].

Despite significant progress in understanding cavity-induced hydrodynamics, most existing insights are derived from flow characteristics observed in fixed-spacing rectangular cavity arrays, such as groyne fields [27,28]. In these engineered multi-cavity systems, the shear layer originates at the leading edge of the first cavity, gradually develops downstream along the cavity–channel interface, and typically reaches a fully developed state by the fourth to sixth cavity [29,30]. Coherent vortical structures span the entire interface, facilitating a continuous momentum exchange across the cavity mouths. In contrast, natural coastal cavities differ fundamentally from such engineered configurations. They are discretely distributed with a significantly larger spacing, resulting in minimal upstream influence on downstream cavities under longshore currents. The upstream incoming boundary layer is different between the two cases: it is fully turbulent for continuous cavity arrays but remains laminar for an isolated cavity. Grace et al. [31] showed that a laminar incoming boundary layer produces a shear layer that remains stable near the cavity's leading edge and gradually transitions to turbulence downstream, while a turbulent incoming boundary layer yields a shear layer that becomes fully turbulent immediately over the cavity opening.

A critical gap persists in the geometric representation of cavities. While the majority of existing studies have focused on orthogonal geometries (e.g., square or rectangular cavities) [32,33], natural lateral cavities predominantly exhibit curvilinear wetted perimeters [25,34]. However, the hydrodynamics of non-rectangular lateral cavities have received relatively limited attention. In particular, curved cavity geometries that arise from shoreline evolution and sedimentary processes—such as arc-shaped embayments, gullies, and wetland bays—are rarely isolated for detailed hydrodynamic investigations. Addressing this gap is essential for advancing our understanding of cavity–flow interactions and improving the predictive modeling of natural and engineered coastal systems. Some efforts have been made to examine shape effects. Ozalp et al. [35] experimentally investigated the influence of cavity shape on vortex formation, turbulence intensity, and flow entrainment. Their results showed that rectangular and triangular cavities generated stronger turbulence and more pronounced vortex structures compared to semi-circular cavities. Jackson et al. [36] further examined the effect of semi-circular cavity deformation on mean residence time, reporting that forward-conic cavities exhibited the highest exchange rates. While experimental measurements provide valuable insights at selected cross-sections and gauging points, numerical simulations offer a more comprehensive view of the full flow field. To our knowledge, limited 3D LES studies have been conducted on semi-circular cavity flows that resemble naturally evolved lateral indentations. Most existing numerical studies on semi-circular cavities are limited to idealized two-dimensional lid-driven flows, which do not realistically capture three-dimensional flow dynamics in open cavity systems. Indeed, Koseff and Street's [37] experimental measurements demonstrated that 3D structures at the mid-plane differ from those predicted by 2D simulations.

To fully capture the physics of three-dimensional cavity flows, researchers have increasingly adopted high-fidelity numerical approaches. Among these, detached-eddy simulation (DES) and large-eddy simulation (LES) are widely utilized to resolve multi-scale turbulent structures in flows involving groynes, embayments, and lateral cavities [38,39]. While DES offers a compromise between accuracy and computational cost, it has been shown to be less accurate than LES in capturing the complex unsteady features of cavity flows [40]. In a recent benchmark study, Ouro et al. [41] systematically evaluated various computational models for simulating flows past lateral cavities, and they demonstrated the superior predictive performance of LES. While two-dimensional shallow-water models are capable of reproducing large-scale flow patterns, they fall short in capturing the fine-scale features critical to cavity–channel interactions. In contrast, LES offers a significantly improved accuracy in resolving velocity distributions, shear-layer instabilities, and unsteady vortex dynamics [42,43].

The present study employs LES to numerically investigate the flow patterns and momentum exchange mechanisms associated with a lateral semi-circular cavity. A no-slip boundary condition is imposed using the immersed boundary method (IBM) [44], which also provides flexibility for future investigations involving cavity shape deformation. To validate the LES results, time-averaged velocities and turbulence intensities are compared against laboratory measurements at selected cross-sectional profiles. The analysis focuses on the mean flow structures, turbulence statistics, and coherent vortical features across a range of Reynolds numbers. Furthermore, the generation and transport of the unsteady shear layer are quantitatively characterized through the spectral analysis of spanwise fluctuating velocity at four gauging points along the cavity interface.

2. Computational Method and Setup

2.1. Numerical Framework

In this study, high-resolution large-eddy simulations (LESs) were conducted using the open-source code Hydro3D, which has been extensively validated and applied to a wide range of complex hydrodynamic problems [45–50]. The governing equations for turbulent, incompressible, viscous flow are solved on a uniform Cartesian grid using the spatially filtered Navier–Stokes equations.

$$\frac{\partial u_i}{\partial x_i} = 0 \tag{1}$$

$$\frac{\partial u_i}{\partial t} + \frac{\partial u_i u_j}{\partial x_j} = -\frac{\partial p}{\partial x_i} + \nu \frac{\partial^2 u_i}{\partial x_i x_j} - \frac{\partial \tau_{ij}}{\partial x_j} + f_i \tag{2}$$

where u_i and u_j (i or j = 1, 2, 3) are the filtered fluid velocity vectors, p is the filtered pressure, and ν is the kinematic viscosity. The term τ_{ij} is the subgrid scale stresses tensor, and f_i is the external forcing term due to the immersed boundary (IB) method, which follows the discrete IB approach known as the direct forcing method introduced by Fadlun et al. [51] and later improved by Uhlmann [44].

The effects of the unresolved subgrid scales are modeled by the wall-adapting local eddy-viscosity (WALE) model [52], which has proven effective in capturing the influence of small-scale turbulence on the resolved large eddies.The diffusive terms in the momentum equations are discretized using fourth-order central differences, ensuring high accuracy and minimal numerical dissipation. Similarly, the convective fluxes are also approximated using a fourth-order central difference scheme, which provides a robust balance between computational accuracy and stability, particularly important in regions with complex flow features. The force exerted by immersed solid bodies is incorporated via a direct forcing immersed boundary method. In this approach, the no-slip boundary of the cavity is enforced by transferring the predicted Eulerian velocities onto a set of Lagrangian markers through delta function interpolation, computing the necessary force corrections, and then distributing these forces back onto the Eulerian grid.

A three-step Runge–Kutta scheme is employed to integrate the governing equations in time, ensuring temporal accuracy and stability during the predictor stage [38,53,54]. Following the prediction, a fractional-step method is applied [55], where a pressure-correction equation is formulated to enforce the incompressibility constraint. This Poisson equation is solved efficiently using a multigrid method, thereby yielding a divergence-free velocity field at the end of each time step. A grid configuration employing the power-of-two discretization in three spatial directions is commonly adopted to improve the performance of the multigrid method [56]. The overall computational framework is parallelized through domain decomposition, with inter-domain communication managed via the Message Passing Interface (MPI) [57]. This hybrid approach enables Hydro3D to leverage high-performance computing resources effectively, making it a powerful tool for simulating complex turbulent flows with immersed boundaries.

2.2. Computational Setup

The LES method was validated against laboratory experiments conducted by Ozalp et al. [35]. The experiments utilized a recirculating open-channel flume measuring 8.0 m long, 1.0 m wide, and 0.75 m high. A radial flow pump equipped with a variable-speed control unit regulated the water flow rate to maintain steady flow conditions. Prior to entering the test section, water passed through a settling chamber containing a honeycomb structure and a 2:1 contraction zone to reduce flow heterogeneity. The bed of the flume and the cavity models were made of PVC. Experiments were conducted to compare flow properties within rectangular, triangular, and semi-circular cavities, all with a uniform length-to-width ratio of 2:1. For the semi-circular cavity, the width (i.e., the radius) was 0.05 m. Velocity measurements were obtained using a two-dimensional Particle Image Velocimetry (PIV) system, enabling instantaneous and time-averaged analyses of the velocity field within and near the semi-circular cavity. A horizontal laser sheet (<1.5 mm thick) was generated using a cylindrical lens followed by a spherical lens to illuminate seeded tracer particles at the mid depth $z/h = 0.5$. Time-averaged first- and second-order hydrodynamic statistics were derived from the PIV data.

The computational domain, illustrated in Figure 1, is based on the flume experiment by Ozalp et al. [35] and represents a localized section containing a lateral semi-circular cavity. A similar modeling configuration has also been employed in recent numerical studies, such as Zhang et al. [58]. The cavity is subjected to longshore current forcing to replicate nearshore flow conditions. The cavity length $L = 0.1$ m is adopted as the characteristic length scale due to the absence of a definitive water depth in the physical experiments. The main channel extends $5L$ in the streamwise (x) direction, $2L$ in the spanwise (y) direction, and $1.2L$ in the vertical (z) direction based on the dimensions of the experimental flume. The leading edge of the semi-circular cavity (radius $R = 0.5L$) is located at $x = 1.0L$, downstream from the inlet. The synthetic eddy method (SEM) is employed at the inlet to produce a fully-developed channel flow [59]. This method reliably generates synthetic turbulence fields that match prescribed first- and second-order statistics, as well as characteristic length and time scales, and it has been validated in turbulent channel flow simulations. At the outlet, a convective boundary condition is applied to allow large-scale coherent structures to exit the domain smoothly without introducing numerical instabilities [60]. No-slip boundary conditions are imposed at the channel bed and at the cavity-adjacent sidewall, enforced using near-wall grids that resolve the viscous sublayer. The semi-circular arc of the cavity is accurately represented using immersed boundary points, ensuring a faithful application of the no-slip condition at the fluid–solid interface, as validated in prior studies [45,49]. The use of IBM also enables future deformation of the cavity into arbitrary geometries, including prescribed shapes or naturally indented coastlines. A rigid-lid approximation is applied at the free surface, a common simplification in CFD for Froude numbers below 0.35 [27]. This approach reduces computational cost to approximately one-fifth of that required by the level-set method while maintaining negligible errors, as demonstrated in previous simulations of lateral cavities in open-channel flows under similar flow conditions. On the wall opposite the cavity, a symmetry boundary condition is imposed, given that the far-field flow in the main channel is nearly streamwise-parallel and spanwise-uniform. This assumption maintains physical realism while reducing the overall computational domain size [58].

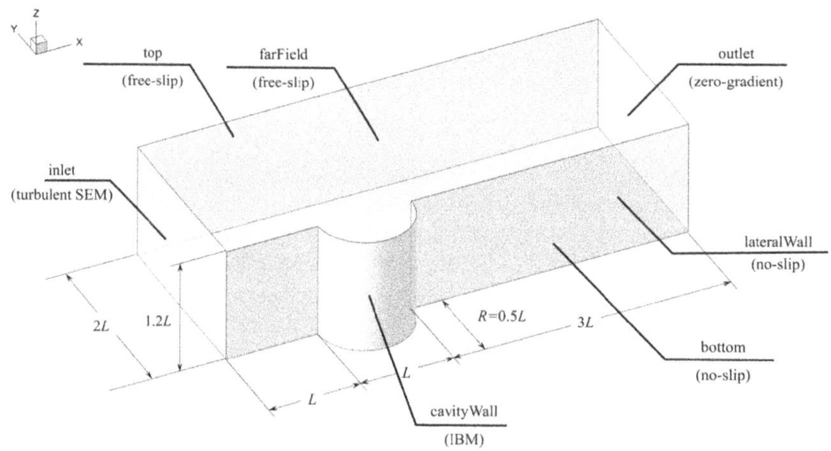

Figure 1. Geometric configuration of the computational domain, which corresponds to the semi-circular cavity in the laboratory work of Ozalp et al. [35].

The mean streamwise velocity in the main channel, U_0, is used to compute the Reynolds and Froude numbers, defined as $Re = U_0 L/\nu$ and $Fr = U_0/\sqrt{gh}$, respectively, and listed in Table 1. The simulations used water as the working fluid rather than a mixture based on the experimental setup, with the corresponding density and kinematic viscosity values adopted from standard water parameters. The simulation domain spans the entire water column from the channel bed to the free surface. Case 1 is configured based on the flume experiment using the semi-circular cavity, while two additional cases are introduced to investigate the effect of increasing Reynolds number on flow characteristics. A uniform horizontal grid spacing of 0.625 mm is adopted across all cases. The vertical grid spacing is refined from 1 mm in Cases 1 and 2 to 0.5 mm in Case 3 to resolve near-wall gradients at higher Reynolds numbers. This vertical refinement ensures that the first off-wall grid points reside within the viscous sublayer, resulting in total mesh sizes of approximately 33.79, 33.79, and 67.58 million elements for Cases 1–3, respectively. The mesh resolution in wall units is computed along the cavity surface, including the upstream and downstream bottom and lateral walls, as illustrated in Figure 2. It should be noted that the color bar range is slightly adjusted between cases, and the layout is mirrored for enhanced visual clarity. The wall-adjacent cell values are computed by

$$\Delta\eta^+ = \Delta\eta \cdot u^*/\nu \tag{3}$$

in which the friction velocity, u^*, is given by

$$u^* = (\nu \cdot u_\eta/\Delta\eta)^{1/2} \tag{4}$$

Here, u_η denotes the tangential velocity component along the local wall direction, obtained by projecting the time-averaged Cartesian velocities onto the no-slip surface. On the curved cavity surface, u_η is synthesized from all three velocity components, whereas on flat surfaces, it is derived from two components. Due to the relatively low flow rate inside the semi-circular cavity, the maximum values of $\Delta\eta^+$ appear near the bottom wall of the main channel. While $\Delta\eta^+$ along the downstream lateral wall of the cavity increases slightly with the Reynolds number, the maximum value remains below 5.0 across all cases. The contour plots reveal that both horizontal and vertical surfaces are fully recovered by the boundary layer and also indicate localized backflow near the cavity walls, particularly within the semi-circular region. These results confirm that the no-slip condition is accurately enforced via the immersed boundary method and that the near-wall grid resolution remains within the viscous sublayer, ensuring that the simulations are wall-resolved.

Table 1. Details of grid resolution and the hydrodynamic conditions for the cases studied.

Case	$\Delta x = \Delta y$ (mm)	Δz (mm)	U_0 (m/s)	Re	Fr
1	0.625	1.0	0.12	12,000	0.11
2	0.625	1.0	0.17	17,000	0.15
3	0.625	0.5	0.22	22,000	0.20

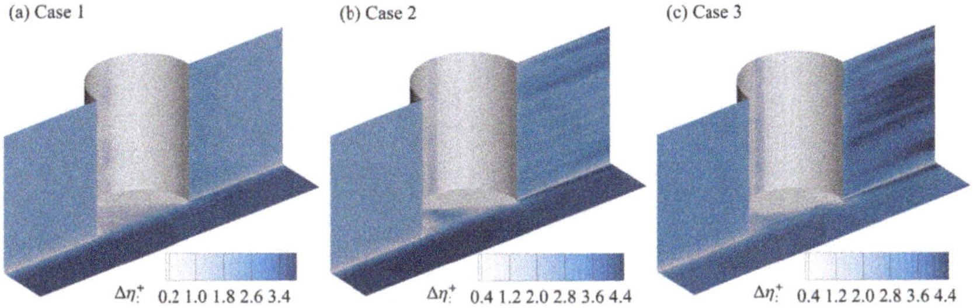

Figure 2. Contours of the dimensionless wall distance around the cavity section at different Reynolds numbers.

In this study, a variable time-stepping scheme is employed with a constant Courant–Friedrichs–Lewy (CFL) number of 0.4 to balance numerical stability and computational efficiency [38]. The residual criterion was restricted to a maximum value of 10^{-5} to preserve mass conservation. Given different Reynolds numbers across cases, each simulation was run for approximately 100 cavity flow-through times ($t_c = L/U_0$) to ensure full flow development within the cavity, followed by an additional 200 t_c for adequate statistical convergence. The computations were performed on 128 cores of AMD EPYC 7452. Cases 1 and 2 consumed approximately 4600 CPU hours, and Case 3 consumed approximately 7800 CPU hours.

3. Results and Discussion

3.1. Verification

Time-averaged statistics from Case 1 are compared with experimental measurements in Figure 3, including profiles of $\overline{u}/U_0$, u'/U_0, and v'/U_0 along the center radial (spanwise) line. The solid lines represent the results obtained using the fine grid defined in the previous Computational Setup Section, whereas the dashed lines correspond to those from a coarse grid with doubled grid spacings. The coarse grid slightly underpredicts the mean velocity compared to the fine grid. In terms of turbulence intensity, the coarse grid yields higher values near the bed but lower values within the shear layer. However, the impact of grid refinement on the simulation accuracy is not substantial, as the fine grid is already adequately resolved. Accordingly, all subsequent plots and analyses in this paper are based on the results of the fine-grid simulation.

Due to the recirculation inside the cavity, negative streamwise velocities are observed for $y/L < 0.35$ in both the experimental and LES results. However, LES slightly overpredicts the backflow strength near the cavity bottom. The maximum discrepancy occurs around $y/L = 0.1$, with a relative error of up to 35%, whereas LES matches the PIV measurements well in the upper region of the cavity. This deviation is primarily attributed to inherent limitations of 2D-PIV in resolving near-wall flow, where strong velocity gradients and sparse seeding particles can bias measurements. In particular, the measured streamwise velocity increases almost linearly from the cavity bottom, while the LES profile exhibits a more physically consistent curvature near the wall. Both streamwise and spanwise turbulence intensities remain relatively constant within the recirculation zone but show a marked increase in the upper region of the cavity. In the transitional region ($0.3 < y/L < 0.4$), LES predictions of turbulence intensity exceed the experimental values by 20–30%, which may be partially attributed to the relatively low flow rate and spatial resolution of the experimental setup. Overall, the favorable agreement between LES and experimental data supports the capability of LES in accurately resolving the mean and turbulent flow features within a semi-circular cavity.

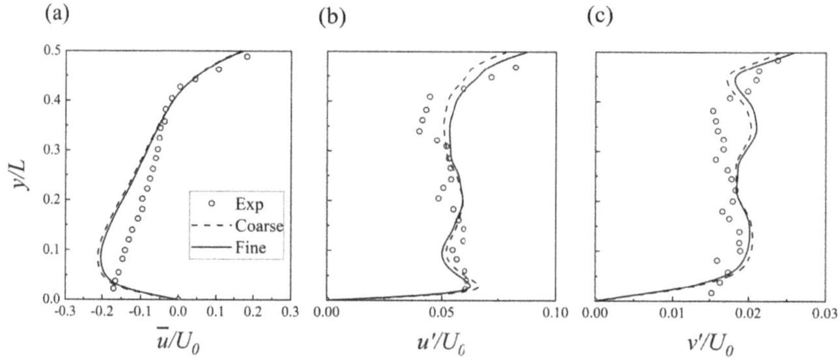

Figure 3. Transverse profiles of time-average streamwise velocity $\overline{u}/U_0$ (**a**) and streamwise (**b**) and spanwise turbulence intensities (**c**) at x/L =1.5. Comparison between experiments and the present Case 1 LES results. Experimental data (circles), coarse-mesh LES (dashed line), and fine-mesh LES (solid line).

3.2. Time-Averaged Flow Structures

The time-averaged flow field in the semi-circular cavity and adjacent main channel region is illustrated in Figure 4 at the mid-depth horizontal plane (z/h = 0.5). The left column presents the normalized streamwise velocity overlaid with streamlines, while the right column shows the normalized spanwise velocity. Streamlines in the main channel remain nearly parallel, except for localized distortions near the trailing edge of the cavity. The incoming boundary-layer flow separates at the leading edge of the cavity, develops into a shear layer along the channel–cavity interface, and impinges upon the trailing edge, where it is deflected into the cavity along the wetted perimeter. This deflection results in the formation of a prominent negative streamwise velocity region. After reaching the bottom point, the reversed flow turns upward and merges with the incoming shear layer, forming a closed-loop clockwise circulation, referred to as the MV, which occupies the majority of the cavity. The corner vortex (CV) forms near the leading edge of the semi-circular cavity and is the only secondary structure caused by the absence of orthogonal bottom corners unlike in rectangular cavities [27,61]. The overall flow pattern inside the cavity remains qualitatively similar across all three cases, aligning with characteristics of the skimming flow regime [62]. However, the velocity distributions and vortical structures exhibit significant variations with respect to the Reynolds number.

As the Reynolds number increases from 12,000 to 17,000, the most prominent change is the displacement of the MV core from the downstream cavity corner ($x/L = 1.85$) to a location slightly upstream of the cavity's bottom center ($x/L = 1.45$), as seen by comparing Figure 4a,c. A similar vortex core migration was reported in previous studies investigating the effects of Reynolds number on rectangular cavity flows [19]. At higher *Re* values, inertial forces increasingly outweigh viscous effects and promote a stronger momentum transfer from the shear layer into the cavity. This leads to a steeper velocity gradient along the wetted perimeter, as shown in Figure 4c, and expands the region of negative spanwise velocity, as illustrated in Figure 4e. To maintain momentum and pressure balance, the MV core shifts toward the low-pressure zone near the cavity bottom. The CV at the upstream corner also grows significantly. Its spatial extent expands from a small core (diameter $\approx 0.01L$) in Figure 4a to a broader zone spanning $0.4 < y/L < 0.5$ in Figure 4c. Although Cases 1 and 2 show marked differences in their time-averaged flow fields, Cases 2 and 3 appear remarkably similar. In Case 3, the MV core moves only slightly upstream (about $0.1L$), while the size of the CV remains nearly constant. The streamwise and spanwise velocities within the recirculating region show mild enhancements. The similarity between Cases 2 and 3 indicates that a dynamic equilibrium has formed near the MV core, where pressure gradients, velocity shear, and wall damping collectively prevent further displacement.

The time-averaged velocity distributions at the cavity–main channel interface for the three cases are shown in Figure 5. Under the lower Reynolds number in Case 1, the flow field exhibits near symmetry about the mid-height plane ($z/L = 0.6$), as seen in Figure 5a. The highest streamwise velocity appears near the upstream and downstream corners, while peak spanwise velocity occurs along the mid-height line. Blue contours of spanwise velocity highlight fluid entrainment into the cavity, with a typical entrainment width of approximately one-tenth the cavity mouth width. However, this width increases significantly near the top and bottom boundaries, reaching up to four-fifths of the cavity width. In Cases 2 and 3, corresponding to higher Reynolds numbers, the velocity fields are nearly identical but deviate notably from Case 1. The region of maximum streamwise velocity shifts upward, peaking around $z/L = 0.7$ in Figure 5b,c, with a secondary maximum symmetrically located near $z/L = 0.5$. The maximum spanwise velocity is concentrated along the downstream sidewall in the range $0.2 < y/L < 0.4$, as shown in Figure 5e,f. The vertical variations in streamwise and spanwise velocities confirm the three-dimensionality of the cavity flow [63,64]. The structure of the mean flow and vortex distribution changes considerably across vertical layers, and these Reynolds number-dependent vertical variations further contribute to the observed shift in the vortex's core position.

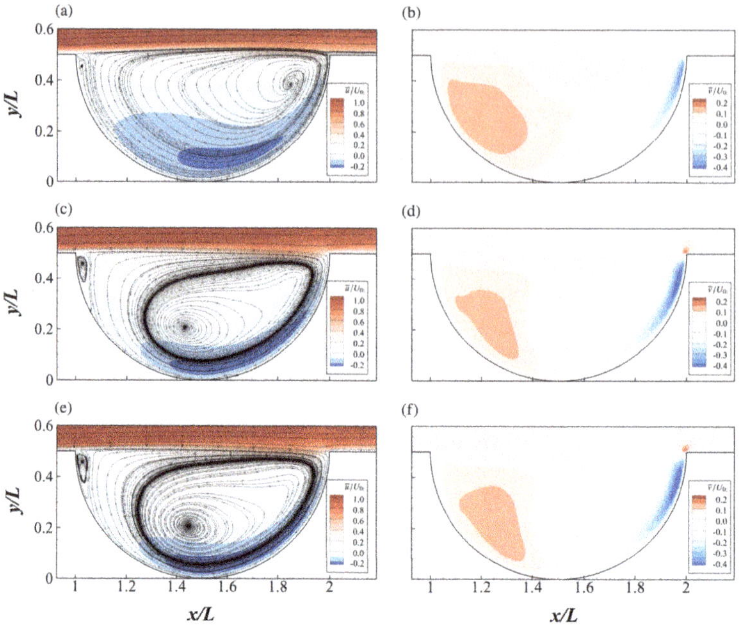

Figure 4. Distribution of normalized time-averaged streamwise ($\overline{u}/U_0$) and spanwise velocities $\overline{v}/U_0$ at the mid−depth plane for the three cases. Streamlines are only plotted on the contours of streamwise velocity for simplicity. (**a**,**b**) for Case 1; (**c**,**d**) for Case 2; (**e**,**f**) for Case 3.

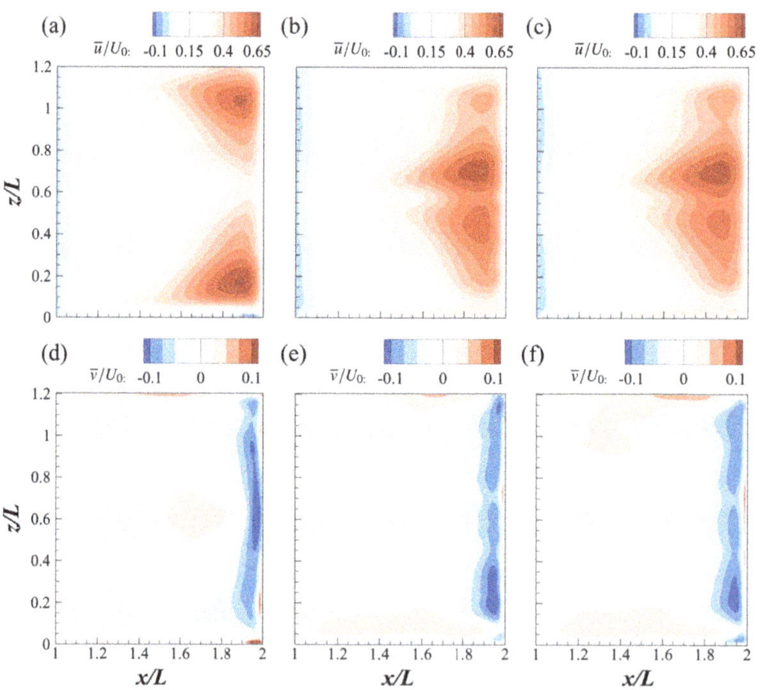

Figure 5. Distribution of normalized time-averaged streamwise $\overline{u}/U_0$ (the upper row) and spanwise velocities $\overline{v}/U_0$ (the lower row) through the mouth of the cavity at y/L = 0.5 for the three cases. (**a**,**d**) for Case 1; (**b**,**e**) for Case 2; (**c**,**f**) for Case 3.

3.3. Second-Order Turbulence Statistics

Figure 6 compares the streamwise turbulence intensity (u'/U_0), spanwise turbulence intensity (v'/U_0), and Reynolds shear stress ($\overline{uv}/U_0^2$) between Case 1 and Case 2 at the mid-depth horizontal plane. Case 3 is omitted due to its close similarity to Case 2. The turbulence statistics reveal two primary sources of turbulence: (1) the shear layer at the interface between the main channel and the cavity, and (2) a jet-like inflow penetrating the cavity along the wetted perimeter [36]. As shown in Figure 6a–d, turbulence intensities are concentrated along the shear layer. Streamwise flow instabilities emerge in three distinct regions: (1) the central part of the cavity, where recirculating flow from the MV reattaches to the shear layer, inducing unsteady acceleration; (2) the trailing corner, where flow impingement causes local deceleration and instability; and (3) the outer turbulent boundary layer near the downstream wall. Figure 6b shows that the increasing Reynolds number amplifies the streamwise turbulence intensity within the shear layer. Additionally, the upstream shift of the MV core displaces the high-turbulence region further upstream. For example, the $u'/U_0 = 0.1$ contour boundary moves from $x/L = 1.35$ in Figure 6a to $x/L = 1.15$ in Figure 6b.

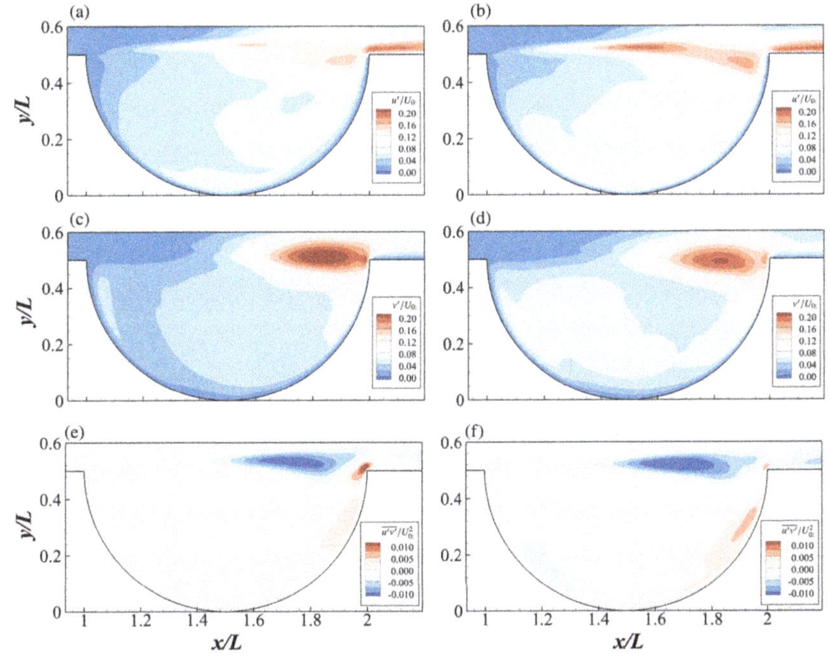

Figure 6. Distribution of normalized streamwise u'/U_0 and spanwise v'/U_0 turbulence intensities and Reynolds shear stress $\overline{u'v'}/U_0^2$ at the mid-depth plane. The left column is for Case 1 and the right one is for Case 2. (**a**,**c**,**e**) for Case 1; (**b**,**d**,**f**) for Case 2.

Positive contours of Reynolds shear stress ($\overline{uv}$) are distributed along the wetted perimeter of the downstream cavity, reflecting the contribution of turbulence to wall shear stress [38]. In contrast, negative contours are primarily concentrated within the downstream half of the cavity mouth, indicating momentum transfer from the outer high-momentum region to the inner low-momentum region via the shear layer. Notably, in Case 2, negative $\overline{uv}$ values appear along the upstream curved wall from $x/L = 1.1$ to 1.3, as shown in Figure 6f. This region corresponds to the boundary layer separation of the recirculating backflow illustrated in Figure 4c. In contrast, this feature is absent in the low-Reynolds-number

Case 1 (Figure 6e), where the recirculating streamlines remain attached to the wall up to the leading-edge CV. Although the CV contributes minimally to the streamwise and spanwise turbulence intensities, its presence is evident through localized positive $\overline{uv}$ contours near the cavity corner, highlighting its role in local Reynolds shear stress generation.

Figure 7 compares turbulence statistics across the transition plane between the main channel and the lateral cavity for Cases 1 and 2. The distributions of both turbulence intensity components and turbulent kinetic energy (TKE) are nearly two-dimensional, indicating lower turbulence levels in the upstream section of the cavity. However, downstream of $x/L = 1.6$, turbulence activity increases rapidly. The influence of Reynolds number is more pronounced for streamwise turbulence intensity (u'/U_0) than for spanwise (v'/U_0), consistent with experimental measurements [35]. While the spatial extent of $v'/U_0 > 0.15$ remains similar between the two cases, both the coverage and magnitude of $u'/U_0 > 0.1$ in Figure 7e exceed those in Figure 7a, contributing to the higher TKE observed in Figure 7h, where $tke = u_i'u_i'/2$. In Case 1, the Reynolds shear stress contours along the cavity mouth are predominantly negative and exhibit vertically uniform distributions. In Case 2, however, four distinct high-value regions ($\overline{u'v'}/U_0^2 \approx 0.1$) emerge near the upper, middle, and bottom layers ($z/L = 0.05, 0.50, 0.70$, and 0.95), indicating enhanced three-dimensionality of turbulence at higher Reynolds numbers. Additionally, localized positive Reynolds shear stress appears near the downstream edge at $x/L = 2$ in Figure 7e,g. This feature is attributed to cavity-corner-induced effects and is more clearly illustrated in Figure 6e,f.

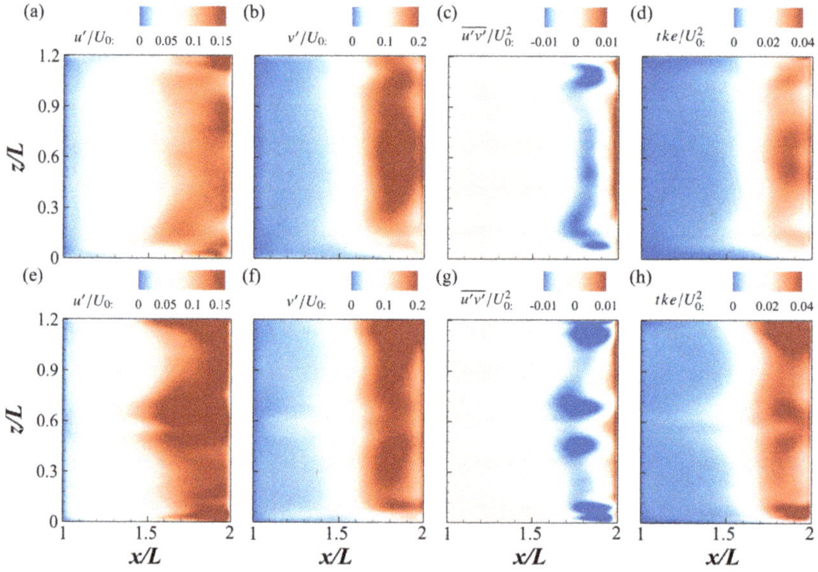

Figure 7. Distribution at a plane through the mouth of the cavity at y/L = 0.5 for Case 1 (**upper row**) and Case 2 (**lower row**); (**a**,**e**) are the streamwise turbulence intensity u'/U_0; (**b**,**f**) are the spanwise turbulence intensity u'/U_0; (**c**,**g**) are Reynolds shear stress $\overline{u'v'}/U_0^2$; (**d**,**h**) are total kinetic energy tke/U_0^2.

3.4. Instantaneous Turbulent Flow Structures

Figure 8 presents instantaneous pressure (p) contours overlaid with velocity vectors of the horizontal components (u) and (v) at the mid-depth plane for Cases 1 and 2, with snapshots taken at 0.20 s intervals. In both cases, visible unsteady pressure bubbles emerge in the mid-section of the cavity mouth while upstream velocity vectors remain nearly parallel, indicating a steady shear layer. Spanwise perturbations from the MV backflow gradually destabilize the shear layer and trigger the onset of Kelvin–Helmholtz (KH) instability. Initial vortex structures are subtle but grow rapidly in size and energy through interactions with the surrounding flow, propagating downstream along the cavity mouth. The KH vortices play a dominant role in driving mass and momentum exchange between the main stream and the cavity: negative-pressure cores entrain high-momentum channel flow into the cavity, while positive-pressure regions eject cavity fluid to the channel. At $t + 0.6$ s, KH vortices reach $x/L = 1.9$ with a maximum diameter of approximately $0.14L$. As approaching the trailing corner, vortices shrink and core negative pressure descends. Upon impinging on the downstream wall, part of the vortex structure is entrained into the cavity as small-scale secondary vortices (SVs) while the remaining portion merges into the near-wall flow of the main channel. This process governs turbulent scalar transport resembling other types of skimming flow but exhibiting a slight difference from the early stage in flows past a series of cavities or groynes. In such configurations, KH vortices originate at the upstream corner because the near-wall inflow is fully perturbed after passing multiple upstream structures. In contrast, the incoming near-wall flow for the isolated cavity herein remains a steady boundary layer of sufficient thickness. As the boundary layer transitions into a shear layer, its initial stability is preserved across the upstream section of the cavity mouth, resulting in lower turbulence levels as previously shown in Figure 6.

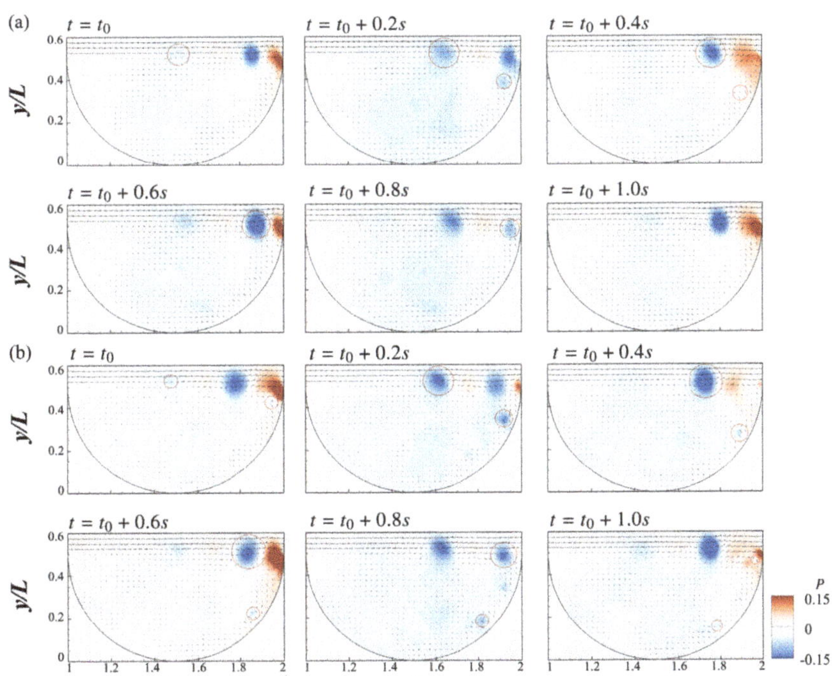

Figure 8. Contours of instantaneous pressure, p, with velocity vectors of instantaneous streamwise (u) and spanwise (v) components for Case 1 (**a**) and Case 2 (**b**). The snapshots have a time lapse of 0.20 s among them and the the red circles indicate the outlines of counter-clockwise vortices.

The evolution of Kelvin–Helmholtz (KH) vortices follows a similar pattern in Cases 1 and 2. However, the vortex core pressure magnitude in Case 2 is elevated due to stronger inertial forces at equivalent stages. Secondary vortices (SVs) propagate from the impingement location into the cavity along the wetted perimeter. In Case 1, SVs appear only in two consecutive snapshots, initially forming at $y/L = 0.4$ and dissipating rapidly before reaching $y/L = 0.3$. This rapid dissipation is attributed to insufficient energy transfer from KH vortices to SVs during impingement, limiting travel length to $<0.1L$. In contrast, Case 2 exhibits persistent SVs in all snapshots, propagating from $y/L = 0.42$ to $y/L = 0.16$. The intensified jet-like flow in Case 2 drives SVs deeper into the cavity, thereby inducing migration of the MV core from the trailing corner vicinity in Figure 4a to the cavity center in Figure 4c.

To better visualize the vortex structures around and inside the cavity, Figure 9 displays iso-surfaces of the instantaneous Q-criterion colored by a streamwise velocity for Case 1. The KH vortices originate from the mid-section of the cavity mouth as a result of spanwise perturbations in the shear layer. Due to their inherently three-dimensional nature, KH vortices maintain a relatively coherent vertical structure across the water depth during downstream convection as observed in the side view. The top view reveals a distinct velocity contrast between low-velocity cavity flow and high-velocity main-channel flow through color contours on KH vortex rollers. At the trailing corner, KH vortices are divided into high-momentum attached vortices and low-momentum SVs. Attached vortices are advected downstream along the sidewall, partially evolving into hairpin vortices. The turbulence level near the downstream sidewall is significantly elevated due to the uniform distribution of coherent structures in contrast to the steady boundary layer on the upstream sidewall. If a series of consecutive cavities were present downstream, KH vortices could potentially be triggered at the leading edge of each cavity. SVs characterized by negative (blue) streamwise velocity and small-scale roller are mainly concentrated near the separation point. Few SVs propagate toward the cavity crest along the wetted perimeter. In addition, coherent structures near the leading corner are absent for CVs as their size is negligibly small in Case 1.

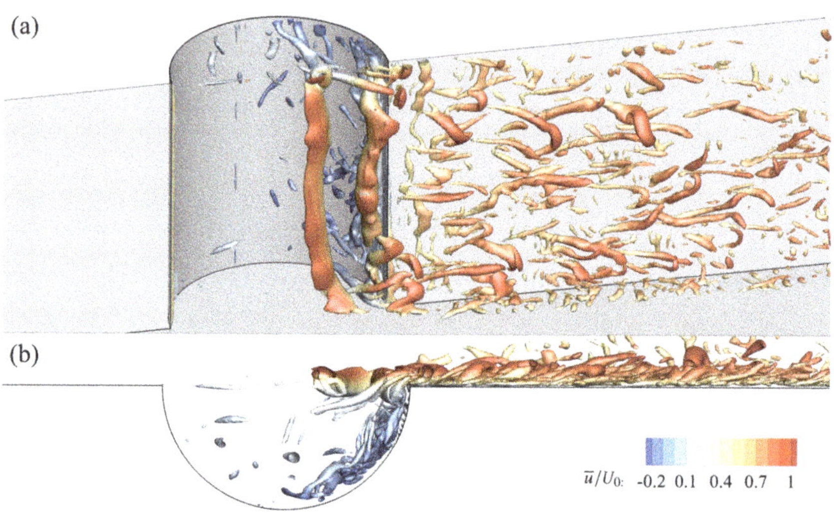

Figure 9. Instantaneous flow structures represented by iso-surfaces of the Q criterion with normalized streamwise velocity (**a**) in the side view and (**b**) in the top view for Case 1.

The power spectral density computed from the time series of the spanwise fluctuating velocity at four locations along the cavity mouth (at $y/L = 0.5$) at mid-depth is presented in Figure 10 for Case 1. The other two cases are not shown as they exhibit qualitatively similar trends. Transverse velocity spectra at all four locations reveal periodic shear layer oscillations at a dominant frequency $f = 1.90$. In Case 1 and Case 3, the corresponding peak frequencies are $f = 1.80$ and $f = 1.92$, respectively. However, the spectral energy at the peak frequency increases by 42 times from the second ($x/L = 1.4$) to the fourth point ($x/L = 1.8$). In the upstream half of the cavity mouth, turbulence in the shear layer remains underdeveloped and energy dissipation is primarily governed by viscosity. Insufficient scale separation prevents the formation of a Kolmogorov inertial subrange, bypassing intermediate scales to directly feed energy into dissipation. Conversely, the shear layer undergoes progressive energy cascade formation in the downstream half of the cavity mouth, eventually exhibiting a localized $-5/3$ spectral slope beyond the peak frequency. These observations are consistent with the experimental findings of Grace et al. [31], who reported a gradual transition to turbulence and increased Reynolds shear stress downstream in cases with a laminar incoming boundary layer. In addition to the primary energy peak associated with the large-scale coherent structures, secondary peaks are also present due to the interaction between the undulant shear layer and the fluctuating MV in the cavity.

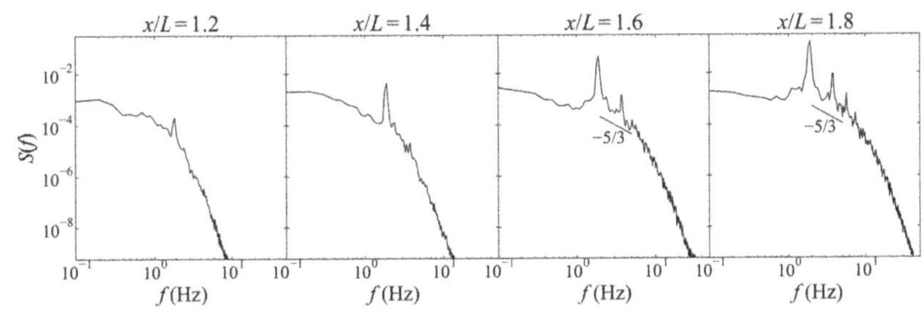

Figure 10. Power spectral density of the spanwise fluctuating velocity time-series at four spatial samples located at different transverse locations along the cavity mouth ($y/L = 0.5$) at half depth for Case 2.

4. Conclusions

This study employed high-resolution large-eddy simulation (LES) to investigate the flow dynamics and momentum exchange mechanisms within a lateral semi-circular cavity subjected to longshore current. The simulation results are validated against laboratory data, confirming the model's capability to capture key turbulence statistics and cavity-induced recirculation. Key findings are as follows:

(1) The time-averaged flow field features a dominant MV and a small upstream CV. With increasing Reynolds number, the MV core shifts upstream and shear layer interactions intensify, modifying the internal recirculation dynamics.

(2) As Re increases from 12,000 to 17,000, turbulence intensities and Reynolds shear stresses increase significantly near the cavity boundaries and shear interface, enhancing momentum exchange and mixing. However, a further increase to $Re = 22{,}000$ leads to a minimal change, indicating a transition toward a quasi-equilibrium regime.

(3) Instantaneous flow structures visualized by the Q-criterion reveal coherent KH vortices forming at the cavity mouth, with intensified secondary vortex (SV) generation and penetration at higher Reynolds numbers.

(4) The spectral analysis of spanwise velocity fluctuations reveals dominant KH instability frequencies governing the periodic transport of momentum and vorticity into the cavity. Fully developed turbulence emerges downstream of the cavity mouth, where an inertial subrange with a localized $-5/3$ slope appears. In contrast, the upstream region remains viscosity-dominated with underdeveloped shear layer turbulence.

The semi-circular cavity investigated in this study serves as an idealized representation of natural coastal lateral cavities under longshore current. The use of IBM in the present simulations was intended as a preparatory step toward modeling more complex, irregularly shaped cavities found in real-world settings. Future work will focus on extending the current approach to cavities with realistic geometries and bathymetries.

Author Contributions: Conceptualization, Y.G., Y.X. and J.M.; methodology, Y.G., Y.X. and J.D.; formal analysis, Y.X., H.Z. and Q.X.; writing—original draft preparation, Y.G., J.D. and L.H.; writing—review and editing, L.H. and J.M.; visualization, Y.G., Y.X., H.Z. and Q.X.; supervision, J.D. and L.H.; project administration, J.M. and L.H. All authors have read and agreed to the published version of the manuscript.

Funding: This work was supported by the National Natural Science Foundation of China (52279013) and National Key Research and Development Program of China (2023YFC3209200).

Data Availability Statement: Data can be obtained from the first author upon request.

Acknowledgments: The authors gratefully acknowledge the support from the funding listed above. We also appreciate the anonymous reviewers who gave comments to revise the paper.

Conflicts of Interest: Author Yun Xu was employed by the company Zhejiang Institute of Communications Co., Ltd., and author Qianshun Xu was employed by the company Yellow River Engineering Consulting Co., Ltd. The remaining authors declare that the research was conducted in the absence of any commercial or financial relationships that could be construed as potential conflicts of interest.

References

1. MacMahan, J.H.; Thornton, E.B.; Reniers, A.J. Deterministic Coastal Morphological and Sediment Transport Modeling: A Review and Discussion. *Rev. Geophys.* **2010**, *48*, RG3001. [CrossRef]
2. Falqués, A.; Calvete, D.; Dodd, N.; de Swart, H.E. Understanding Coastal Morphodynamic Patterns from Depth-Averaged Sediment Concentration. *Rev. Geophys.* **2014**, *52*, 201–237. [CrossRef]
3. Guo, X.; Fan, N.; Liu, Y.; Liu, X.; Wang, Z.; Xie, X.; Jia, Y. Deep seabed mining: Frontiers in engineering geology and environment. *Int. J. Coal Sci. Technol.* **2023**, *10*, 23. [CrossRef]
4. Guo, X.; Fan, N.; Zheng, D.; Fu, C.; Wu, H.; Zhang, Y.; Song, X.; Nian, T. Predicting impact forces on pipelines from deep-sea fluidized slides: A comprehensive review of key factors. *Int. J. Min. Sci. Technol.* **2024**, *34*, 211–225. [CrossRef]
5. Eriksson, B.K.; van der Heide, K.; van de Koppel, M.; Piersma, T.; van der Veer, H.W.; Olff, H. Ecosystem Engineering and Biodiversity in Coastal Sediments. *Helgol. Mar. Res.* **2009**, *63*, 23–32. [CrossRef]
6. Hall, A.E.; Herbert, R.J.; Britton, J.R.; Hull, S.L. Ecological Enhancement Techniques to Improve Habitat Heterogeneity on Coastal Defence Structures. *Estuar. Coast. Shelf Sci.* **2018**, *210*, 68–78. [CrossRef]
7. Boyd, L.B.; Janus, L.L.; Casselman, J.M. Submerged Aquatic Vegetation in Cook's Bay, Lake Simcoe. *J. Gt. Lakes Res.* **2011**, *37*, 93–100. [CrossRef]
8. Castellanos-Galindo, G.A.; Wolff, M.; Saint-Paul, U. Multilevel Decomposition of Spatial and Environmental Effects on Nearshore Fish Assemblages in Tropical Semi-Enclosed Ecosystems. *Estuar. Coast. Shelf Sci.* **2020**, *237*, 106691. [CrossRef]
9. Bird, E.C.F. *Coastal Geomorphology: An Introduction*, 2nd ed.; John Wiley & Sons: Hoboken, NJ, USA, 2008.
10. Symonds, J.P.; Jones, R.M. The Influence of Geomorphology and Sedimentary Processes on Benthic Habitat Distribution and Littoral Sediment Dynamics, Geraldton, Western Australia. *Estuar. Coast. Shelf Sci.* **2009**, *84*, 527–537. [CrossRef]
11. Eliot, M.; Eliot, G. Shoreline Salients, Cuspate Forelands and Tombolos on the Coast of Western Australia. *J. Coast. Res.* **2010**, *26*, 812–822.
12. Bo, T.; Ralston, D.K. Frontogenesis, Mixing, and Stratification in Estuarine Channels with Curvature. *J. Phys. Oceanogr.* **2022**, *52*, 1333–1352. [CrossRef]
13. Pasquaud, P.; Schinegger, J.; Blanchet, S. Estuarine Lateral Ecotones Shape Taxonomic and Functional Structure of Fish Assemblages. *Estuar. Coast. Shelf Sci.* **2024**, *313* , 109066. [CrossRef]
14. Chapman, M.G.; Underwood, A.J. Scales of variation of gastropod densities over multiple spatial scales: Comparison of common and rare species. *J. Exp. Mar. Biol. Ecol.* **2007**, *348*, 91–102. [CrossRef]
15. Browne, M.A.; Thompson, R.J.A.; Chapman, L.A. Designing for sedimentation: The role of urban structures in the accumulation and distribution of estuarine sediment. *Mar. Ecol. Prog. Ser.* **2014**, *507*, 15–30. [CrossRef]

16. Uijttewaal, W.S.J.; Lehmann, D.; van Mazijk, A. Exchange processes between a river and its groyne fields: Model experiments. *J. Hydraul. Eng.* **2001**, *127*, 928–936. [CrossRef]
17. Lesack, L.F.W.; Marsh, P. River-to-lake connectivities, water renewal, and aquatic habitat diversity in the Mackenzie River Delta. *Water Resour. Res.* **2010**, *46*, W12504. [CrossRef]
18. Mignot, E.; Barthelemy, E.; Hurther, A. Coherent turbulent structures at the mixing-interface of a square open-channel lateral cavity. *Phys. Fluids* **2016**, *28*, 045102. [CrossRef]
19. Chiang, T.P.; Sheu, W.H.; Hwang, R.R. Effect of Reynolds number on the eddy structure in a lid-driven cavity. *Int. J. Numer. Methods Fluids* **1998**, *26*, 557–579. [CrossRef]
20. Chang, K.S.; Constantinescu, S.G.; Park, S. Analysis of the flow and mass transfer processes for the incompressible flow past an open cavity with a laminar and a fully turbulent incoming boundary layer. *J. Fluid Mech.* **2006**, *561*, 113–145. [CrossRef]
21. Faure, T.M.; Adrianos, P.; Lusseyran, F.; Pastur, L. Visualizations of the flow inside an open cavity at medium range Reynolds numbers. *Exp. Fluids* **2007**, *42*, 169–184. [CrossRef]
22. Drost, K.; Jackson, T.; Haggerty, R.; Apte, S.V. RANS Predictions of Turbulent Scalar Transport in Dead Zones of Natural Streams. In Proceedings of the ASME 2012 Fluids Engineering Division Summer Meeting, Rio Grande, PR, USA, 8–12 July 2012; pp. 1167–1177. [CrossRef]
23. Drost, K.J.; Apte, S.V.; Haggerty, R.; Jackson, T.R. Parameterization of Mean Residence Times in Idealized Rectangular Dead Zones Representative of Natural Streams. *J. Hydraul. Eng.* **2014**, *140*, 04014035. [CrossRef]
24. Brevis, W.; García-Villalba, M.; Niño, Y.N. Experimental and large eddy simulation study of the flow developed by a sequence of lateral obstacles. *Environ. Fluid Mech.* **2014**, *14*, 873–893. [CrossRef]
25. Jackson, T.R.; Haggerty, R.; Apte, S.V.; Coleman, A.; Drost, K.J. Defining and measuring the mean residence time of lateral surface transient storage zones in small streams. *Water Resour. Res.* **2012**, *48*, W10501. [CrossRef]
26. Jackson, T.R.; Haggerty, R.; Apte, S.V.; O'Connor, B.L. A mean residence time relationship for lateral cavities in gravel-bed rivers and streams: Incorporating streambed roughness and cavity shape. *Water Resour. Res.* **2013**, *49*, 3642–3650. [CrossRef]
27. Constantinescu, G.; Sukhodolov, A.; McCoy, A. Mass exchange in a shallow channel flow with a series of groynes: LES study and comparison with laboratory and field experiments. *Environ. Fluid Mech.* **2009**, *9*, 587–615. [CrossRef]
28. Juez, C.; Bühlmann, I.; Maechler, G.; Schleiss, A.J.; Franca, M.J. Transport of suspended sediments under the influence of bank macro-roughness. *Earth Surf. Process. Landf.* **2018**, *43*, 271–284. [CrossRef]
29. McCoy, A.; Constantinescu, G.; Weber, L. A numerical investigation of coherent structures and mass exchange processes in channel flow with two lateral submerged groynes. *Water Resour. Res.* **2007**, *43*, W05445. [CrossRef]
30. Weitbrecht, V.; Socofolosky, S.A.; Jirka, G.H. Experiments on mass exchange between groin fields and main stream in rivers. *J. Hydraul. Eng.* **2008**, *134*, 173–183. [CrossRef]
31. Grace, S.M.; Dewar, W.G.; Wroblewski, D.E. Experimental investigation of the flow characteristics within a shallow wall cavity for both laminar and turbulent upstream boundary layers. *Exp. Fluids* **2004**, *36*, 791–804. [CrossRef]
32. Crook, S.D.; Lau, T.C.W.; Kelso, R.M. Three-dimensional flow within shallow, narrow cavities. *J. Fluid Mech.* **2013**, *735*, 587–612. [CrossRef]
33. Engelen, L.; Perrot-Minot, C.; Mignot, E.; Rivière, N.; De Mulder, T. Lagrangian study of the particle transport past a lateral, open-channel cavity. *Phys. Fluids* **2021**, *33*, 013303. [CrossRef]
34. Engelhardt, C.; Krüger, A.; Wüst, A.; Mohrholz, A. A study of phytoplankton spatial distributions, flow structure and characteristics of mixing in a river reach with groynes. *J. Plankton Res.* **2004**, *26*, 1351–1366. [CrossRef]
35. Ozalp, C.; Pinarbasi, A.; Sahin, B. Experimental measurement of flow past cavities of different shapes. *Exp. Therm. Fluid Sci.* **2010**, *34*, 505–515. [CrossRef]
36. Jackson, T.R.; Apte, S.V.; Haggerty, R.; Budwig, R. Flow Structure and Mean Residence Times of Lateral Cavities in Open Channel Flows: Influence of Bed Roughness and Shape. *Environ. Fluid Mech.* **2015**, *15*, 1069–1100. [CrossRef]
37. Koseff, J.R.; Street, R.L. The Lid-Driven Cavity Flow: A Synthesis of Qualitative and Quantitative Observations. *J. Fluids Eng.* **1984**, *106*, 390–398. [CrossRef]
38. Ouro, P.; Juez, C.; Franca, M. Drivers for mass and momentum exchange between the main channel and river bank lateral cavities. *Adv. Water Resour.* **2020**, *137*, 103511. [CrossRef]
39. Liu, H.; Liu, M.; Ai, Y.; Huai, W. Turbulence Characteristics in Partially Vegetated Open Channels with Alternating Sparse and Dense Patches. *Phys. Fluids* **2024**, *36*, 015130. [CrossRef]
40. Chang, K.; Constantinescu, G.; Park, S.O. Comparison of Large Eddy Simulation, Detached Eddy Simulation and Unsteady RANS Numerical Predictions of Flow Past a Bottom River Cavity. In Proceedings of the 32nd IAHR World Congress, Venice, Italy, 1–6 July 2007.
41. Ouro, P.; Cea, L.; Croquer, S.; Dong, W.; Garcia-Feal, O.; Navas-Montilla, A.; Rogers, B.D.; Uchida, T.; Juez, C. Benchmark of Computational Hydraulics Models for Open-Channel Flow with Lateral Cavities. *J. Hydraul. Res.* **2024**, *62*, 441–460. [CrossRef]

42. Liu, M.; Yuan, S.; Tang, H.; Huai, W.; Yan, J. Investigation of turbulence and interfacial exchange features of the gap area within the fully developed Shallow-Submerged canopy flow. *J. Hydrol.* **2024**, *643*, 131938. [CrossRef]
43. Liu, M.; Huai, W.; Tang, H.; Wang, Y.; Yuan, S. Numerical study of mean and turbulent flow adjustments in open channels with limited near-bank vegetation patches. *Phys. Fluids* **2024**, *36*, 095117. [CrossRef]
44. Uhlmann, M. An immersed boundary method with direct forcing for the simulation of particulate flows. *J. Comput. Phys.* **2005**, *209*, 448–476. [CrossRef]
45. Gong, Y.; Stoesser, T.; Mao, J.; McSherry, R. LES of flow through and around a finite patch of thin plates. *Water Resour. Res.* **2019**, *55*, 7587–7605. [CrossRef]
46. Luo, Q.; Dolcetti, G.; Stoesser, T.; Tait, S. Water surface response to turbulent flow over a backward-facing step. *J. Fluid Mech.* **2023**, *966*, A18. [CrossRef]
47. Guo, X.; Liu, X.; Zheng, T.; Zhang, H.; Lu, Y.; Li, T. A mass transfer-based LES modelling methodology for analyzing the movement of submarine sediment flows with extensive shear behavior. *Coast. Eng.* **2024**, *191*, 104531. [CrossRef]
48. Liu, Y.; Tang, Z.; Huang, L.; Stoesser, T.; Fang, H. On the role of the Froude number on flow, turbulence, and hyporheic exchange in open-channel flow through boulder arrays. *Phys. Fluids* **2024**, *36*, 095141. [CrossRef]
49. Dai, J.; Mao, J.q.; Gong, Y.q.; Gao, H. Effect of triangular baffle designs on flow dynamics and sediment transport within standard box culverts. *J. Hydrodyn.* **2024**, *36*, 1142–1154. [CrossRef]
50. Dai, J.; Xu, Q.; Gong, Y.; Lu, Y.; Liu, X.; Mao, J. An LES Investigation of Flow Field Around the Cuboid Artificial Reef at Different Angles of Attack. *J. Mar. Sci. Eng.* **2025**, *13*, 463. [CrossRef]
51. Fadlun, E.; Verzicco, R.; Orlandi, P.; Mohd-Yusof, J. Combined immersed-boundary finite-difference methods for three-dimensional complex flow simulations. *J. Comput. Phys.* **2000**, *161*, 35–60. [CrossRef]
52. Nicoud, F.; Ducros, F. Subgrid-scale stress modelling based on the square of the velocity gradient tensor. *Flow Turbul. Combust.* **1999**, *62*, 183–200. [CrossRef]
53. Peskin, C. Fluid dynamics of heart valves: Experimental, theoretical, and computational methods. *Annu. Rev. Fluid Mech.* **1982**, *14*, 235–259. [CrossRef]
54. Christou, A.; Stoesser, T.; Xie, Z. A large-eddy-simulation-based numerical wave tank for three-dimensional wave-structure interaction. *Comput. Fluids* **2021**, *231*, 105179. [CrossRef]
55. Chorin, A.J. Numerical solution of the Navier-Stokes equations. *Math. Comput.* **1968**, *22*, 745–762. [CrossRef]
56. Cevheri, M.; McSherry, R.; Stoesser, T. A local mesh refinement approach for large-eddy simulations of turbulent flows. *Int. J. Numer. Methods Fluids* **2016**, *82*, 261–285. [CrossRef]
57. Ouro, P.; Fraga, B.; Lopez-Novoa, U.; Stoesser, T. Scalability of an Eulerian–Lagrangian large-eddy simulation solver with hybrid MPI/OpenMP parallelisation. *Comput. Fluids* **2019**, *179*, 123–136. [CrossRef]
58. Zhang, Y.; Wang, L.; Li, X.; Chen, H. Effects of vegetation density on flow, mass exchange and sediment transport in lateral cavities. *J. Hydrol.* **2024**, *631*, 129567.
59. Jarrin, N.; Benhamadouche, S.; Laurence, D.; Prosser, R. A synthetic-eddy-method for generating inflow conditions for large-eddy simulations. *Int. J. Heat Fluid Flow* **2006**, *27*, 585–593. [CrossRef]
60. Adzic, F.; Stoesser, T.; Liu, Y.; Xie, Z. Large-eddy simulation of supercritical free-surface flow in an open-channel contraction. *J. Hydraul. Res.* **2022**, *60*, 628–644. [CrossRef]
61. Fang, H.; Bai, J.; He, G.; Zhao, H. Calculations of Nonsubmerged Groin Flow in a Shallow Open Channel by Large-Eddy Simulation. *J. Eng. Mech.* **2014**, *140*, 04014016. [CrossRef]
62. Meile, T.; Boillat, J.L.; Schleiss, A.J. Flow Resistance Caused by Large-Scale Bank Roughness in a Channel. *J. Hydraul. Eng.* **2011**, *137*, 1588–1597. [CrossRef]
63. Samantaray, S.S.; Das, D. High Reynolds number incompressible turbulent flow inside a lid-driven cavity with multiple aspect ratios. *Phys. Fluids* **2018**, *30*, 075103. [CrossRef]
64. Dong, W.; Uchida, T. Role of three-dimensional vortex motions on horizontal eddies in an open-channel cavity. *Environ. Fluid Mech.* **2024**, *24*, 539–566. [CrossRef]

Journal of
*Marine Science
and Engineering*

MDPI

Article

An LES Investigation of Flow Field Around the Cuboid Artificial Reef at Different Angles of Attack

Jie Dai [1], Qianshun Xu [2], Yiqing Gong [3], Yang Lu [4], Xinbo Liu [5,*] and Jingqiao Mao [1,*]

1 College of Water Conservancy and Hydropower Engineering, Hohai University, Nanjing 210098, China; jiedai@hhu.edu.cn
2 Yellow River Engineering Consulting Co., Ltd., Zhengzhou 450003, China; xuqianshun@yrec.cn
3 Institute of Water Science and Technology, Hohai University, Nanjing 211106, China; yqgong@hhu.edu.cn
4 Department of River and Harbor Engineering, Nanjing Hydraulic Research Institute, Nanjing 210024, China; luyang@nhri.cn
5 China Yangtze Power Co., Ltd., No. 1 Jianshe Road, Yichang 443002, China
* Correspondence: liuxinbo@ctg.com.cn (X.L.); maojq@hhu.edu.cn (J.M.)

Abstract: The placement of artificial reefs (ARs) significantly influences local hydrodynamics and nutrient transport, both of which are crucial for enhancing marine ecosystems and improving marine habitats. Large eddy simulations (LESs) are performed to study the flow field around a cuboid artificial reef (CAR) with three inflow angles (α = 0°, 45°, and 90°). The numerical method is successfully validated with experimental data, and a reasonable grid resolution is chosen. The results demonstrate that the case with an inflow angle of 45° exhibits superior flow field performance, including the largest recirculation bubble length and the maximum volumes for both the upwelling and wake regions. Stronger turbulence is also observed around the CAR at this inflow angle, attributed to the intensified shear layer. The instantaneous flow features torn horseshoe vortices and rollers shed from the shear layer, which further develop into hairpin vortices.

Keywords: artificial reef; large eddy simulation; flow separation; coherent structures; upwelling region; wake region

1. Introduction

The marine ecological environment and biological resources are increasingly threatened by accelerated climate change, deteriorating water quality, and modern fishing practices [1–4]. Recently, artificial reefs (ARs), human-made submerged structures, have garnered considerable interest due to their notable successes in the preservation of marine ecosystems and enhancing oceanic environments [5–7]. The presence of ARs in marine environments significantly alters the local hydrodynamics, providing shelter for marine species and modifying the circulation of sediments and nutrients [8–10]. Therefore, a thorough understanding of the flow dynamics around ARs is essential for the restoration and enhancement of marine ecosystems.

Generally, two key regions are formed in the surrounding area of ARs when AR units are placed on the seabed: the upwelling region and the wake region [11–13]. The upwelling region is located upstream of ARs and is defined as the area where the vertical velocity, $\overline{w}$, is greater than or equal to 10% of the inlet velocity, U_0 [14,15]. This region promotes the exchange of materials between different water layers and facilitates the accumulation of organic matter [16]. The second region, referred to as the wake region, forms downstream of ARs and is characterized by a streamwise velocity, $\overline{u}$, that is negative [17,18]. This low-momentum wake region provides shelter, foraging, and spawning spaces for fish species. Extensive research has been conducted, and design parameters, such as shapes and configurations, have been found to significantly affect the flow characteristics of these two regions around ARs. Specifically, Kim et al. [19] numerically investigated the relationship between 24 representative ARs and their wake regions, finding that the AR type and reference plane strongly govern the size of the wake length. Wang et al. [18] used the ANSYS Fluent 2012R1 to compare the flow field performances between Menger-type ARs and traditional ARs. The numerical results suggested that the Menger-type ARs outperform traditional ARs in terms of flow characteristics in both the upwelling and wake regions. Two different configurations of ARs were investigated both experimentally and numerically, and the AR with a hole diameter of 0.25 m demonstrated better performance than the one with a hole diameter of 0.5 m in terms of improving the circulation of water and nutrients [13]. Maslov et al. [20] proposed a novel design for artificial reefs (ARs), comparing the performances of droplet and circular cross sections using four evaluation indices. Their findings revealed that the droplet cross section exhibited superior flow field performance and higher efficiency compared with the circular cross section. Moreover, the arrangement and deployment of ARs are effective strategies for altering the hydraulic characteristics around ARs. Liu et al. [11] demonstrated that the interference effect between the ARs persists when the transverse distance between the ARs is less than twice the length of the ARs. The low-velocity zone between the ARs in the longitudinal direction is positively correlated with the distance between them. Shu et al. [21] experimentally examined the impact of spacing between M-shaped artificial reefs (ARs) on turbulent intensity. Their results indicated that the low-turbulence region was situated upstream of the ARs, while the downstream area displayed increased turbulence.

The aforementioned research efforts are highly valuable for the design and optimization of ARs, and these investigations primarily rely on numerical simulations rather than laboratory experiments. This preference is due to the cost-effectiveness and time efficiency of numerical simulations, making them more practical compared with laboratory experiments or field monitoring. In past studies, many computational fluid dynamics (CFD) cases have focused on Reynolds-averaged Navier–Stokes (RANS) models, such as the k–ϵ model [11,13,18,22] and the k–ω model [20]. The flow through ARs is highly complex due to interactions between the flow and obstacles, which induce turbulence and the formation of three-dimensional coherent structures. These turbulent flow structures influence the swimming behavior, aggregation, and migration of fish species, as well as the transport rates of sediments and nutrients [23–25]. However, the limitation of the RANS model is that it may lose the instantaneous turbulent flow fields due to the averaging of turbulent information [26,27]. Additionally, another physical process related to flow separation requires further investigation, as the vortices and eddy currents formed in the wake region by flow separation are also factors influencing fish aggregation [28–30]. Unsteady Reynolds-averaged Navier–Stokes (URANS) models have been found to be slightly inaccurate in predicting the flow separation and reattachment [31–33]. Thus, a high-fidelity, detailed numerical simulation tool is essential to reveal the three-dimensional (3D) hydrodynamics of such complex flow. Recently, large eddy simulation (LES) has been widely used to study the flow through or around obstacles in open-channel and oceanic flows [34–36]. The LES method not only provides the necessary resolution of turbulent structures, but also accurately predicts regions of flow separation [37–39].

Very few studies have explored the turbulence structures and flow separation around artificial reefs (ARs), particularly under varying inflow angles, even though these factors play a crucial role in the flow dynamics around or through these submerged structures. Zhang et al. [40] evaluated the impact of the layout configuration of ARs using CFD methods and proposed an optimal spacing distance. Their findings revealed that the inflow angle has the most significant influence on the flow field around ARs, surpassing the effects of the longitudinal and transverse intervals between ARs.

This study employs large eddy simulation (LES) to ascertain the hydrodynamics and turbulence of flow through artificial reefs (ARs) at different angles of attack. This paper is structured as follows: Section 2 describes the numerical methods and setup, including the validation case. Section 3 presents and analyzes the results, encompassing time-averaged and instantaneous flow and turbulence parameters, as well as upwelling and wake region characteristics. Section 4 concludes with a summary and final remarks.

2. Methodology

In this study, high-resolution large eddy simulations were conducted using the open-source code Hydro3D v6.0, which has been widely applied and validated for simulating complex and challenging flow scenarios [41–44]. The code solves spatially filtered Navier–Stokes equations for incompressible viscous flow in a Eulerian framework,

$$\frac{\partial u_i}{\partial x_i} = 0 \tag{1}$$

$$\frac{\partial u_i}{\partial t} + \frac{\partial u_i u_j}{\partial x_j} = -\frac{\partial p}{\partial x_i} + v\frac{\partial^2 u_i}{\partial x_i x_j} - \frac{\partial \tau_{ij}}{\partial x_j} \tag{2}$$

where u_i and u_j are filtered velocity vectors (i and j = 1, 2, and 3 represent the x-, y-, and z-axis directions, respectively); similarly, x_i and x_j represent the spatial location vectors in the three directions; p and v are the pressure and kinematic viscosity of the fluid, respectively; and τ_{ij} represents the sub-grid scale stress.

In Hydro3D, the sub-grid scale stress tensor is calculated using the wall-adapting local eddy (WALE) viscosity model, as proposed by Nicoud and Ducros [45]. Fourth- and second-order central differences schemes are used to approximate convective and diffusive terms, respectively. Time advancement is performed using a fractional step method, with a three-step Runge–Kutta scheme to obtain second-order accuracy. In the final step, a Poisson pressure correction equation is applied to couple the pressure field with the velocity field. Hydro3D is based on finite differences, and the filtered equations were solved on a uniform Cartesian grid. The immersed boundary method (IBM) is employed to enforce the no-slip condition on submerged bodies, such as the cuboid artificial reef (CAR) in this study.

The setup of LES is analogous to the experimental study carried out by Tan et al. [46]. The experiments were conducted in a flume with dimensions of 15 × 1 × 1.3 m (length × width × height). A 1:10 scale model was constructed to represent the prototype AR. Specifically, the model was made from Plexiglas, with a length (L) of 0.1 m, width (W) of 0.2 m, and height (H) of 0.1 m. Cut openings were created on the side surfaces of the model, as shown in Figure 1. In the experimental setup, the water depth (h) was 1 m, and the inlet velocity (U_0) was set at 0.22 m/s. An Acoustic Doppler Velocimetry (ADV) system was employed to collect velocity data, with measurement profiles located 0.1 m above the center of the CAR.

The computational domain is shown in Figure 2a, with $L_x = 20.5L$ and $L_y = 8W$ in the streamwise and spanwise directions, respectively. The wall-normal height of the domain (L_z) is $5H$. The computational domain size is selected to minimize unnecessary computations, while the center is positioned $5.5L$ downstream of the inlet to ensure full flow development. A prescribed inflow condition is employed at the inlet of the domain. This is obtained by conducting simulations without the cuboid AR, using periodic boundary conditions. The instantaneous velocity at one cross section of the domain is saved for 5 flow-through periods ($T_f = L_x/U_0$) after the flow had fully developed and is then used as the inlet condition for all simulations. Convective boundary conditions are applied at the outlet of the domain, and periodic boundary conditions are imposed at the sidewalls. The no-slip boundary condition is applied at the bottom of the domain, while a free-slip boundary condition is used at the top of the domain. In order to investigate the effect of the angles (α) of inflow on the flow structure around the CAR, the placement angles (α) of the CAR are varied to 0°, 45°, and 90°, as shown in Figure 2b. Table 1 summarizes the flow conditions for three cases, including the inlet velocity (U_0), water depth (h), and inflow angle (α).

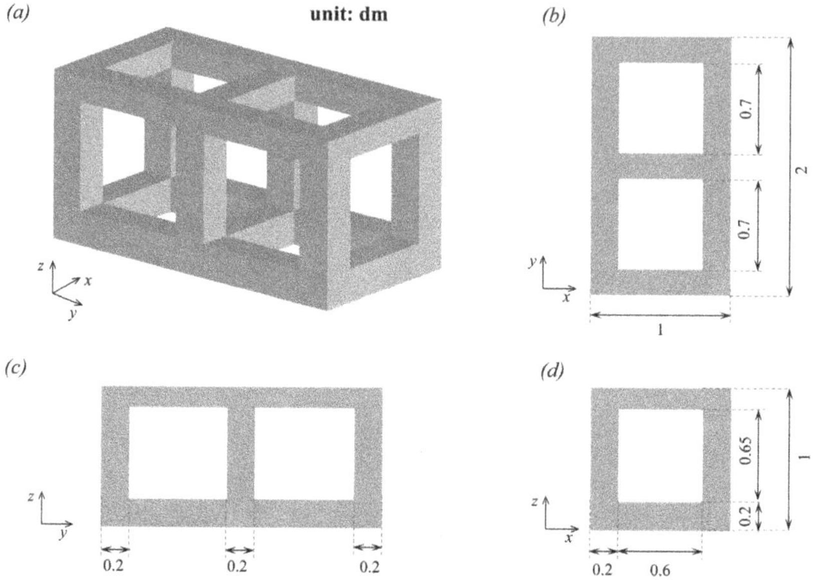

Figure 1. Schematic diagram of the model in the experiment: (**a**) from the view of generic angle, (**b**) side view along the z-axis, (**c**) side view along the x-axis, and (**d**) side view along the y-axis.

Table 1. Detailed flow conditions of the analyzed cases.

Case	U_0 (m/s)	h (m)	α (°)
1	0.22	0.5	0
2	0.22	0.5	45
3	0.22	0.5	90

The computational mesh is refined through two levels of local mesh refinement, as illustrated in Figure 2a, with a finer resolution grid set in the vicinity of the CAR (-0.35 m $< x <$ 0.95 m, -0.3 m $< y <$ 0.3 m, 0 m $< z <$ 0.25 m). Two different grid resolutions are selected for the analysis of grid sensitivity, and the sizes of the mesh are listed in Table 2. For Mesh 1, there are a total of approximately 8.94 million grid points. The finer grid resolution (Mesh 2) was discretized with 55.75 million computational points. All simulations are run for 10 T_f to obtain the fully developed flow after the prescribed inflow imposed at the inlet. First-order statistics are collected after an additional 10 T_f, and second-order statistics commence after a further 20 T_f.

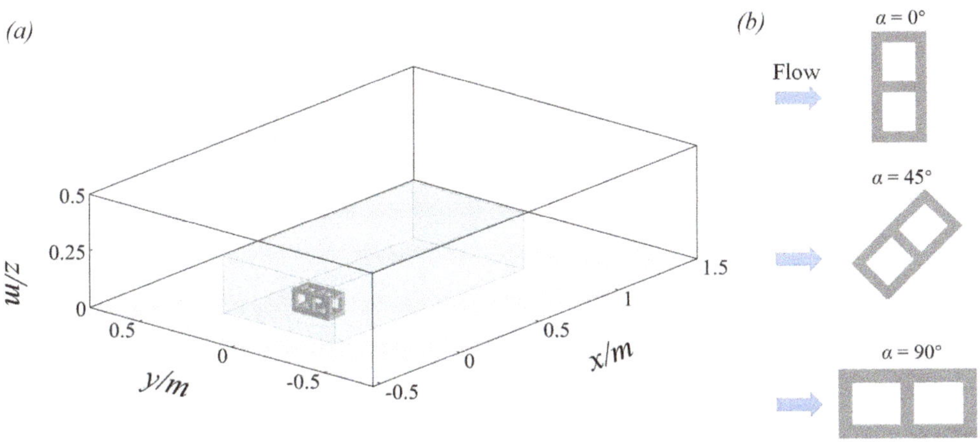

Figure 2. (**a**) Sketch of the computational domain where the blue region represents the area of local mesh refinement and (**b**) the inflow with different angles (0°, 45°, and 90°).

Table 2. Grid resolutions used for grid sensitivity analysis.

Mesh	$\Delta x = \Delta y$	Δz
1	0.004	0.002
2	0.002	0.002

Figure 3a illustrates the locations of the validation points used in the experiment, while Figure 3b shows the profile of the normalized time-averaged streamwise velocity, comparing the results from the two LES simulations with the experimental data. Overall, a satisfying match is observed between the LES results and experimental data. The normalized streamwise velocity (U_m/U_0) experiences a slight decline before gradually increasing from $x = -0.15$ m, peaking at approximately $x = 0.15$ m, and then slowly decreasing to 1.02. The results from Mesh 1 (dashed line) show slight discrepancies, particularly in the range $0.1 < x < 0.4$ m, probably due to insufficient resolution. These discrepancies are reduced in the results from a finer mesh (Mesh 2). The wall-normal distance of the first computational point (Δz^+) along the length of the computational domain is presented in Figure 3c. The average value along the streamwise direction is approximately $\overline{\Delta z^+} = 5.78$, which is less than 12 [47], indicating that the near-wall grids remain within the viscous sublayer and ensuring that the simulations are wall resolved. A sampling point downstream of the CAR is chosen, and the time series of the instantaneous velocity data are recorded. The power spectral density (PSD) of the velocity fluctuations at this point is shown in Figure 3d. The results indicate that the energy cascade along the inertial subrange follows the $-5/3$ Kolmogorov law, indicating a physically realistic energy transfer between scales. An increased dissipation rate is found at higher frequencies due to the sub-grid scale model. Therefore, the resolution of Mesh 2 is chosen to conduct the subsequent simulations.

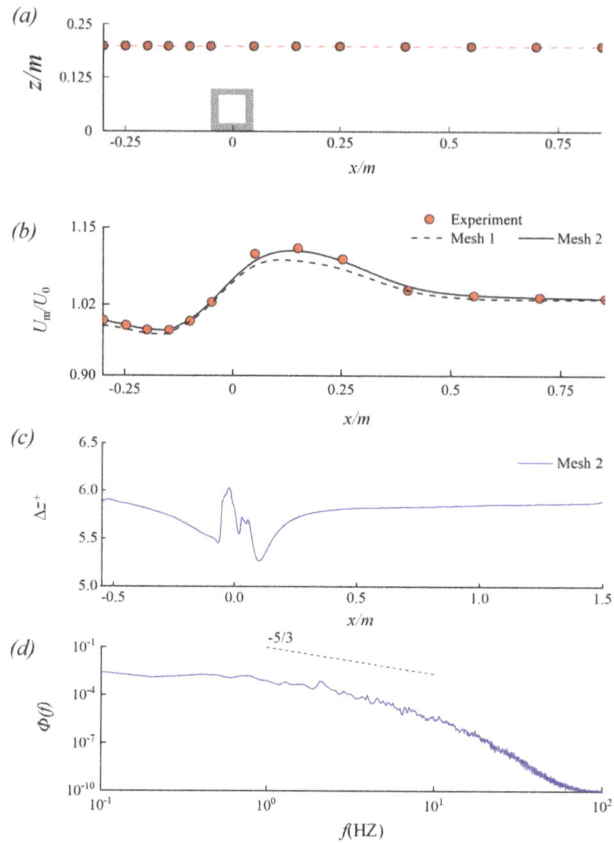

Figure 3. (**a**) Locations of the validation points, (**b**) validation between LES and experiment, (**c**) the wall-normal grid spacing of the simulation, and (**d**) power spectral density of the velocity fluctuation at the sampling point.

3. Results and Discussions

3.1. Time-Averaged Flow and Flow Separation

Figure 4 presents the contours of the time-averaged velocity distribution (left) and streamlines (right) at the plane $z/H = 0.5$ for the three cases. The presence of the CAR notably alters the incoming flow, creating distinct low-velocity and high-velocity zones due to the blockage effect of the CAR. As shown in Figure 4a, high-velocity zones are located on either side of the CAR, with the maximum velocity (U_{max}) reaching approximately 0.228 m/s. In Case 2 and Case 3, U_{max} is approximately 0.224 m/s and 0.205 m/s, respectively. Case 3 exhibits the lowest U_{max}, which can be attributed to the weakest blockage effect among the three cases. The low-velocity zones, indicated by the blue regions in Figure 4a–c, exhibit significant variations in both extent and intensity. These zones are observed behind the square cylinders of the CAR due to the sheltering effect. In Case 1 and Case 3 ($\alpha = 0°$ or $90°$), the development of the low-velocity zones is relatively limited, especially in the transverse direction. In contrast, the distribution of low-velocity zones in Case 2 ($\alpha = 45°$) shows considerable differences, in both the streamwise and transverse directions. Additionally, the minimum velocities (U_{min}) in Case 1 and Case 3 are -0.0233 m/s and -0.0236 m/s, respectively, while in Case 2, U_{min} drops to -0.0562 m/s.

In order to highlight flow separation and recirculation, the two-dimensional (2D) streamlines together with the contours of velocity are plotted in Figure 4d–f. Several recirculation zones occur downstream of these square cylinders, consistent with the studies previously reported by Liu et al. and Zheng et al. [11,48]. In Case 1, the maximum length of the recirculation zone is approximately 0.055 m, located at the center of the CAR. It is interesting to note that the sizes of the recirculation bubble in Case 2 increase dramatically. More specifically, the maximum length of the recirculation bubble in Case 2 (α = 45°) is approximately 0.139 m, which is approximately 2.5 times that of Case 1. In Case 3, the characteristics of the recirculation zones are similar to those in Case 1, with the maximum length of the recirculation zone being approximately 0.0478 m, slightly smaller than that of Case 1.

Both the time-averaged velocity distribution and streamline plots indicate that the case with an inflow angle of 45° demonstrates the best performance, as it results in the largest low-velocity and recirculation zones. This is attributed to the fact that when the inflow angle equals 45°, it allows for the largest area of incoming flow, leading to the formation of a larger deceleration area in front of and inside the CAR.

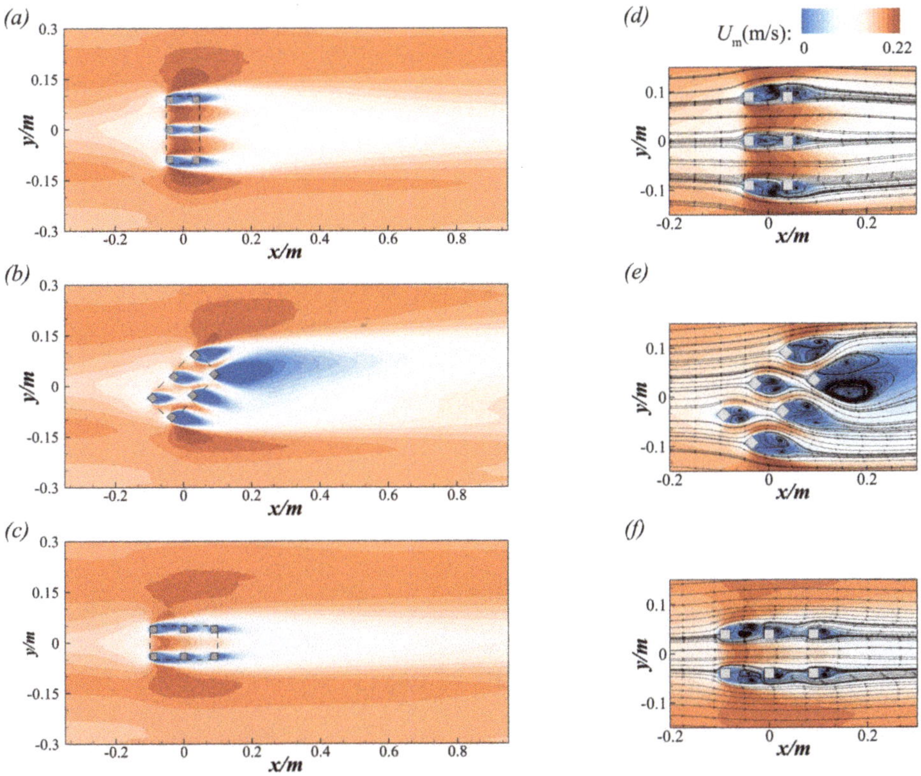

Figure 4. Contours of the time-averaged velocity distribution (**left**) and streamlines (**right**) at the plane $z/H = 0.5$: (**a**,**d**) for the case with $\alpha = 0$; (**b**,**e**) for the case with $\alpha = 45$; (**c**,**f**) for the case with $\alpha = 90$.

3.2. Turbulence Parameters

Figure 5 compares the distribution of the Reynolds shear stress (RSS), calculated as $-\rho\overline{u'v'}$, at z/H = 0.5 across the three cases, highlighting the spatial distribution of turbulence within these cross sections. High-level RSS regions are primarily observed inside and downstream of the CAR in all simulations, where the velocity gradients are most pronounced. In Case 1, the high-level RSS regions originate from the location at $x = 0$ and develop downstream, as shown in Figure 5a. The regions where $RSS = \pm 1\ \mathrm{N/m^2}$ are highlighted by dashed lines, and these areas are primarily concentrated on both sides of the CAR. Peaks in RSS occur on both sides of the CAR, with a maximum value of $\pm 2.1\ \mathrm{N/m^2}$. The distribution of RSS at the same plane of the inflow angle (α) equal to 45° is presented in Figure 5b. The region of high RSS remains visible downstream of the CAR; however, it expands both streamwise and spanwise. The maximum value of RSS in Case 2 is $\approx \pm 1.4\ \mathrm{N/m^2}$, which is slightly lower than that in Case 1. The distribution of RSS in Case 3 shows symmetry along $y = 0$, similar to that in Case 1, as depicted in Figure 5c. Notably, the region where $RSS = \pm 1\ \mathrm{N/m^2}$ is observed only inside and on the sides of the CAR, with a peak value of approximately $\pm 1.8\ \mathrm{N/m^2}$.

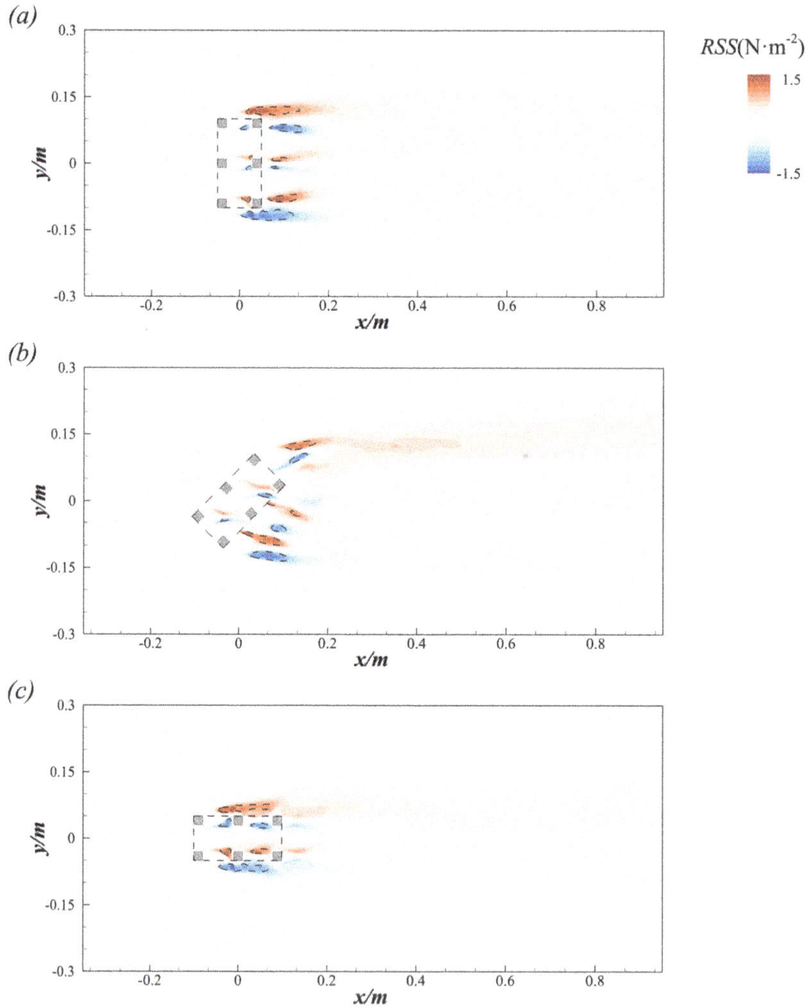

Figure 5. Contours of the Reynolds shear stress at the plane $z/H = 0.5$: (**a**) for the case with $\alpha = 0$; (**b**) for the case with $\alpha = 45$; (**c**) for the case with $\alpha = 90$.

Vertical profiles of turbulent kinetic energy ($TKE = 0.5(\overline{u'}^2 + \overline{v'}^2 + \overline{w'}^2)$) at four locations downstream at the plane $y/W = 0$ are presented in Figure 6 for the three simulated cases. In Case 1 and Case 3, a local peak value of TKE can be observed in the curves. More specifically, the profiles of TKE exhibit multiple peaks at $z = 0.022$ m, 0.077 m, and 0.107 m. This is attributed to the shear layer at the corresponding location. For Case 1, the peak value of TKE occurs approximately at $z = 0.115$ m due to the shear layer near the upper side of the CAR. A clear peak could not be identified in Case 3 with an inflow angle ($\alpha = 90°$), as shown in Figure 6a. Further downstream, local peaks become less pronounced due to the weakening of the shear layer, as depicted in Figure 6b–d. Notably, Case 2 exhibits a higher level of TKE compared with Case 1 and Case 3 at most locations. For example, 0.3 m downstream of the CAR, the TKE values at $z = 0.1$ m are 0.00104, 0.00182, and 0.00113 m^2/s^2 for Case 1, Case 2, and Case 3, respectively, as shown in Figure 6d.

The turbulent dynamics around ARs are crucial, as they significantly influence the mixing and transport of nutrients within the surrounding area [48]. Moreover, RSS and TKE are critical metrics for assessing the impact of turbulence on fish behavior [49,50]. High-levels of RSS or TKE can lead to disorientation or even physical harm to fish [51]. In turbulent environments, thresholds of $RSS < 10$ N/m^2 and $TKE < 0.5$ m^2/s^2 are typically considered favorable for fish species [52]. Consequently, the design and placement of ARs should carefully account for both factors.

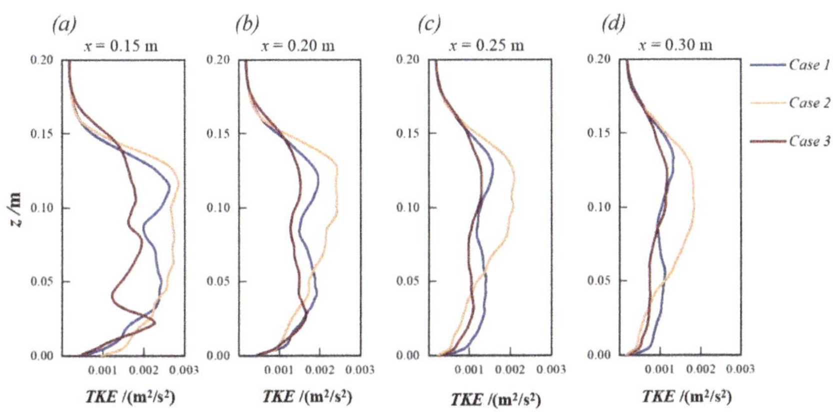

Figure 6. Vertical profiles of turbulent kinetic energy at four locations downstream of the CAR at the plane $y/w = 0$: (**a**) for $x = 0.15$ m, (**b**) for $x = 0.20$ m, (**c**) for $x = 0.25$ m, and (**d**) for $x = 0.30$ m.

3.3. Instantaneous Flow Structures

The flow through and around the CAR exhibits strong three-dimensional characteristics and is primarily influenced by different scales of turbulent structures, whose configuration is largely determined by the angle of inflow. Figure 7 presents contours of instantaneous vertical vorticity ($\omega_z \pm 15$) at the plane $z/H = 0.5$ at varying inflow angles. This plot illustrates the formation of the shear layer around the CAR. In Case 1, the shear layer separates from the leading edge of the upstream square cylinder, wraps around the downstream square cylinder, and continues to develop further downstream. This flow pattern resembles the flow around two cylinders arranged in tandem [53]. The vortex in the wake region downstream of the CAR gradually dissipates, following a pattern similar to a classic von Kármán vortex street. As the inflow angle varies from $\alpha = 0°$ to $\alpha = 45°$, the development of the shear layer changes. Part of the shear layer separates and moves directly downstream, while another portion detaches and evolves through the gap between the square cylinders inside the CAR. It is interesting to note that the wake vortex expands spanwise while moving downstream due to more significant shear deformation inside the CAR, as shown in Figure 7b. In the case of $\alpha = 90°$, the spanwise development of the vortex is significantly restricted, confined to the range of -0.05 m $< y <$ 0.05 m, where the CAR obstructs the main flow (Figure 7c).

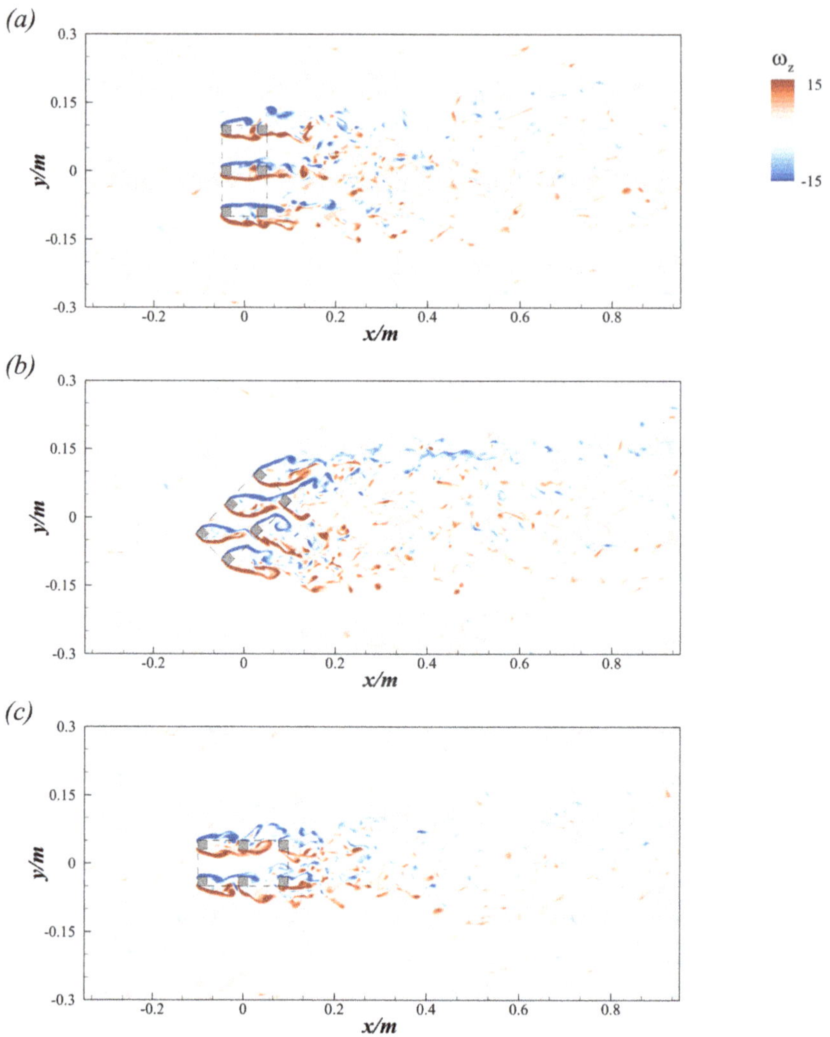

Figure 7. Contours of instantaneous vertical vorticity ($\omega_z \pm 15$) at the plane $z/H = 0.5$: (**a**) for the case with $\alpha = 0$; (**b**) for the case with $\alpha = 45$; (**c**) for the case with $\alpha = 90$.

To better visualize the 3D turbulent structures around the CAR, the Q criterion method, a second-generation vortex identification technique, is used. Figure 8 presents the the isosurfaces of the Q-criterion, contoured with instantaneous streamwise velocity (u) for three different inflow angles (α = 0°, 45°, and 90°). Different types of the turbulence structures can be classified. A spanwise roller is found shed from the upper side of the CAR, i.e., quasi-2D structures that rotate about the y-axis (Figure 8a). The roller subsequently forms a hairpin vortex and is transported downstream by the main current. Similar rollers are also found on the sides of the CAR, where they subsequently evolve into vertical vortices. The formation of these rollers is driven by flow separation and the roll-up of the shear layer. Unlike the flow around typical bluff bodies, such as a cube [54], a complete horseshoe vortex cannot be identified in front of the CAR. Instead, only the legs of torn horseshoe vortices are observed through the gap inside the CAR. Similarly, the aforementioned vortex structures are also found in the cases of $\alpha = 45°$ and $90°$, as shown in Figure 8b,c. The main differences lie in the length, scale, and direction of these vortices.

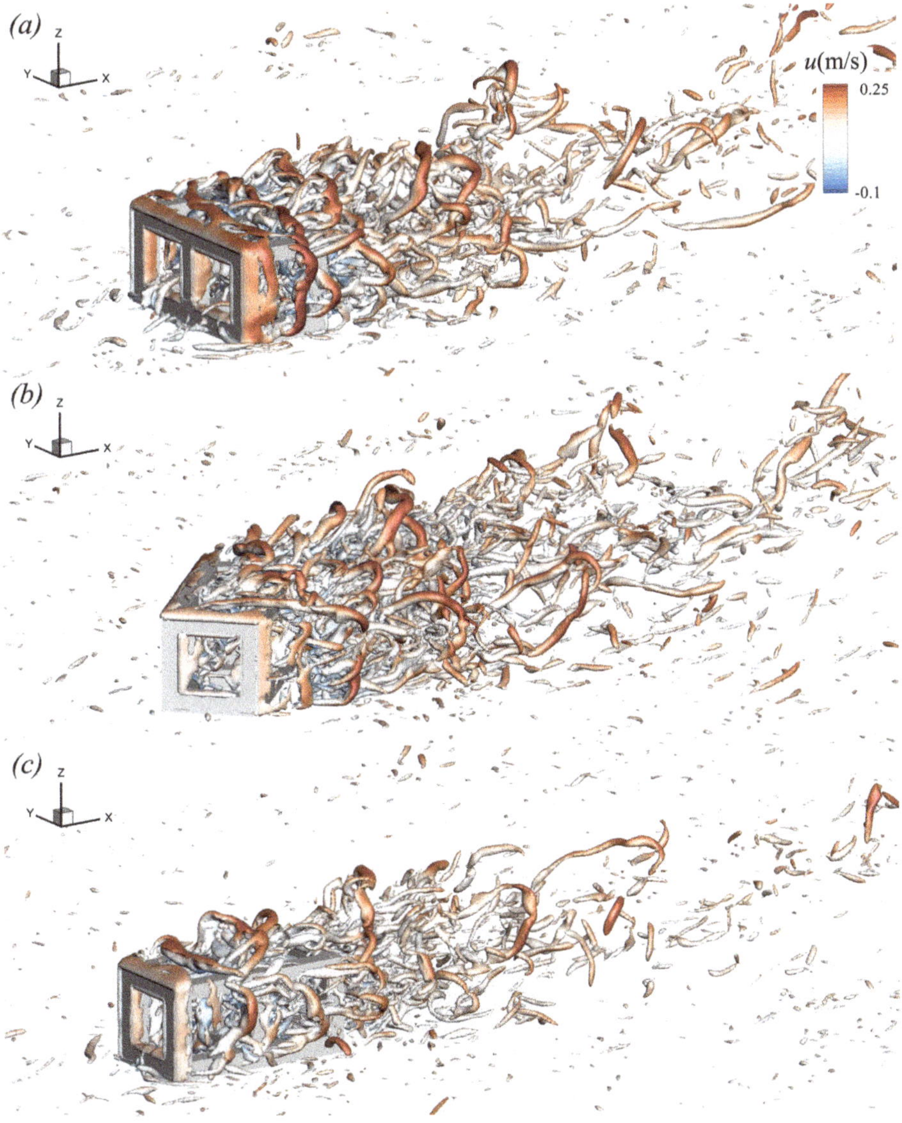

Figure 8. Isosurfaces of Q criterion for three cases, colored with instantaneous streamwise velocity: (**a**) for the case with $\alpha = 0$; (**b**) for the case with $\alpha = 45$; (**c**) for the case with $\alpha = 90$.

3.4. Upwelling and Wake Region Parameters

To quantify the effect of inflow angles (0°, 45°, and 90°) on the upwelling and wake regions, the isosurfaces of these regions are plotted in Figure 9. The upwelling region is indicated by the blue area in Figure 9a, mainly occurring on the upper front of the CAR. This is attributed to the blocking effect of the CAR, leading to the flow lifting and moving around. The wake region mainly exists downstream of the CAR, where flow separation occurs.The distribution of the upwelling and wake regions is almost symmetrical along the centerline of the CAR, as shown in Figure 9d. As the inflow angle increases, the characteristics of the upwelling and wake regions change significantly, in both size and shape, as depicted in Figure 9b,e. The upwelling region nearly covers the full extent of both the streamwise and spanwise of the CAR. At the same time, part of the wake region extends further downstream. In Case 3 (α = 90°), the size of these two regions decreases significantly, as depicted in Figure 9c,f.

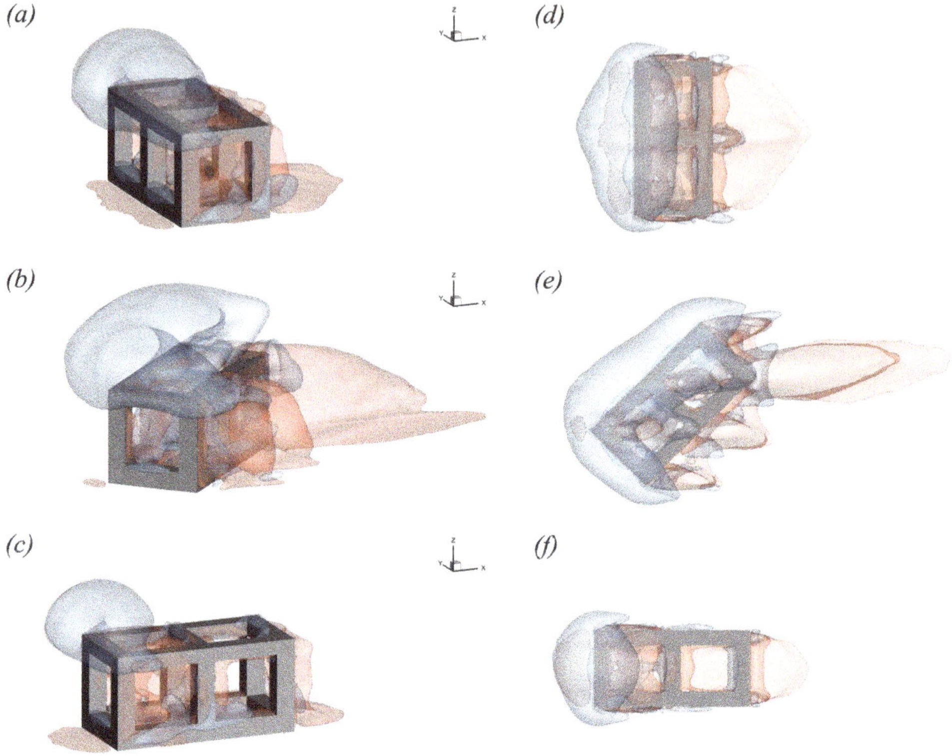

Figure 9. Side view (**left**) and top view (**right**) of distribution of the upwelling and wake region: (**a**,**d**) for the case with $\alpha = 0$; (**b**,**e**) for the case with $\alpha = 45$; (**c**,**f**) for the case with $\alpha = 90$.

The volumes of the upwelling ($V_{upwelling}$) and wake region (V_{wake}) are summarized in Table 3. As shown in Table 3, Case 2, with an inflow angle of $\alpha = 45°$, exhibits the largest volumes of both the upwelling and wake regions compared with Cases 1 and 3 ($V_{upwelling} = 2.18 \times 10^{-1}$ m^3 and $V_{wake} = 4.09 \times 10^{-1}$ m^3). $V_{upwelling}$ and V_{wake} in Case 1 decrease by approximately 35% and 15%, respectively, compared with Case 2. As expected, Case 3 ($\alpha = 90°$) exhibits the smallest volumes for both the upwelling and wake regions, with $V_{upwelling} = 7.78 \times 10^{-2}$ m^3 and $V_{wake} = 3.22 \times 10^{-1}$ m^3, showing a decrease of approximately 64% and 21%, respectively, compared with Case 2.

Table 3. Volumes of the upwelling ($V_{upwelling}$) and wake region (V_{wake}).

Case	α (°)	$V_{upwelling}$ (m^3)	V_{wake} (m^3)
1	0	1.42×10^{-1}	3.49×10^{-1}
2	45	2.18×10^{-1}	4.09×10^{-1}
3	90	7.78×10^{-2}	3.22×10^{-1}

4. Conclusions

The present study has investigated the flow field around a cuboid artificial reef (CAR) with three inflow angles (α = 0°,45°, and 90°), utilizing the open-source large eddy simulation (LES) code. First, the LES has been validated with the velocity data in the experiment, and a good agreement has been obtained. It examines variations of inflow angles on hydrodynamics and turbulence structures around the CAR. Key findings from this study include the following:

(1) Time-averaged velocity distribution and streamline plots indicated that flow separation occurs inside and downstream of the CAR. Case 2 has the maximum length of the recirculation bubble, 2.5 times and 2.9 times greater than that of Case 1 and Case 3, respectively.
(2) The high-level region of the Reynolds shear stress (RSS) is located where flow separation occurs. This region extends the most in both the streamwise and spanwise directions when the inflow angle is 45°. Additionally, Case 2 exhibits higher levels of turbulence kinetic energy (TKE) compared with Case 1 and Case 3 at most vertical locations downstream of the CAR.
(3) The instantaneous flow around the CAR is characterized by significant shear layers prevailing from the leading edge of square cylinders. Isosurfaces of the Q-criterion reveal various types of turbulent structures, including torn horseshoe vortices and hairpin vortices, with the latter formed from the rollers shed by the shear layer.
(4) Case 2, with an inflow angle of $\alpha = 45°$, exhibits the largest volumes for both the upwelling and wake regions. Compared with Case 2, $V_{upwelling}$ and V_{wake} in Case 1 ($\alpha = 0°$) decrease by approximately 35% and 15%, respectively. Case 3 ($\alpha = 90°$) shows the smallest volumes for both regions, with reductions of approximately 64% and 21% for $V_{upwelling}$ and V_{wake}, respectively, compared with Case 2.

Author Contributions: Conceptualization, J.D., Q.X. and J.M.; methodology, J.D., Q.X. and Y.G.; formal analysis, Y.L. and X.L.; writing—original draft preparation, J.D., Y.G. and Y.L.; writing—review and editing, X.L. and J.M.; visualization, J.D., Q.X. and Y.L.; supervision, X.L. and J.M.; project administration, X.L. and J.M. All authors have read and agreed to the published version of the manuscript.

Funding: This work was supported by the National Key Research and Development Program of China (2023YFC3209200), Hubei Provincial Natural Science Foundation of China (2023AFD201), and National Natural Science Foundation of China (52279013, 52279055, and U2040205).

Institutional Review Board Statement: Not applicable.

Informed Consent Statement: Not applicable.

Data Availability Statement: Data can be obtained from the first author upon request.

Acknowledgments: The authors gratefully acknowledge the support from the fundings listed above. We also appreciate the anonymous reviewers who gave comments to revise the paper.

Conflicts of Interest: Author Qianshun Xu was employed by the company Yellow River Engineering Consulting Co., Ltd., and author Xinbo Liu was employed by the company China Yangtze Power Co., Ltd. The remaining authors declare that the research was conducted in the absence of any commercial or financial relationships that could be construed as potential conflicts of interest.

References

1. Au, D. The application of histo-cytopathological biomarkers in marine pollution monitoring: A review. *Mar. Pollut. Bull.* **2004**, *48*, 817–834. [CrossRef] [PubMed]
2. Hoegh-Guldberg, O. Coral reef ecosystems and anthropogenic climate change. *Reg. Environ. Change* **2011**, *11*, 215–227. [CrossRef]

3. Wu, W.; Yang, Z.; Tian, B.; Huang, Y.; Zhou, Y.; Zhang, T. Impacts of coastal reclamation on wetlands: Loss, resilience, and sustainable management. *Estuar. Coast. Shelf Sci.* **2018**, *210*, 153–161. [CrossRef]
4. Guo, X.; Fan, N.; Zheng, D.; Fu, C.; Wu, H.; Zhang, Y.; Song, X.; Nian, T. Predicting impact forces on pipelines from deep-sea fluidized slides: A comprehensive review of key factors. *Int. J. Mining Sci. Technol.* **2024**, *34*, 211–225. [CrossRef]
5. Lee, M.O.; Otake, S.; Kim, J.K. Transition of artificial reefs (ARs) research and its prospects. *Ocean Coast. Manag.* **2018**, *154*, 55–65. [CrossRef]
6. Lima, J.S.; Zalmon, I.R.; Love, M. Overview and trends of ecological and socioeconomic research on artificial reefs. *Mar. Environ. Res.* **2019**, *145*, 81–96. [CrossRef]
7. Bull, A.S.; Love, M.S. Worldwide oil and gas platform decommissioning: A review of practices and reefing options. *Ocean Coast. Manag.* **2019**, *168*, 274–306. [CrossRef]
8. Simon, T.; Pinheiro, H.T.; Joyeux, J.C. Target fishes on artificial reefs: Evidences of impacts over nearby natural environments. *Sci. Total Environ.* **2011**, *409*, 4579–4584. [CrossRef] [PubMed]
9. Walles, B.; Troost, K.; van den Ende, D.; Nieuwhof, S.; Smaal, A.C.; Ysebaert, T. From artificial structures to self-sustaining oyster reefs. *J. Sea Res.* **2016**, *108*, 1–9. [CrossRef]
10. Paxton, A.B.; Peterson, C.H.; Taylor, J.C.; Adler, A.M.; Pickering, E.A.; Silliman, B.R. Artificial reefs facilitate tropical fish at their range edge. *Commun. Biol.* **2019**, *2*, 168. [CrossRef] [PubMed]
11. Liu, T.L.; Su, D.T. Numerical analysis of the influence of reef arrangements on artificial reef flow fields. *Ocean Eng.* **2013**, *74*, 81–89. [CrossRef]
12. Kim, D.; Woo, J.; Na, W.B. Intensively stacked placement models of artificial reef sets characterized by wake and upwelling regions. *Mar. Technol. Soc. J.* **2017**, *51*, 60–70. [CrossRef]
13. Santiago Caamaño, L.; Lamas Galdo, M.I.; Carballo, R.; López, I.; Cartelle Barros, J.J.; Carral, L. Numerical and experimental analysis of the velocity field inside an artificial reef. Application to the Ares-Betanzos estuary. *J. Mar. Sci. Eng.* **2022**, *10*, 1827. [CrossRef]
14. Liu, Y.; Zhao, Y.P.; Dong, G.H.; Guan, C.T.; Cui, Y.; Xu, T.J. A study of the flow field characteristics around star-shaped artificial reefs. *J. Fluids Struct.* **2013**, *39*, 27–40. [CrossRef]
15. Jiao, L.; Yan-Xuan, Z.; Pi-Hai, G.; Chang-Tao, G. Numerical simulation and PIV experimental study of the effect of flow fields around tube artificial reefs. *Ocean Eng.* **2017**, *134*, 96–104. [CrossRef]
16. Wu, J.; Zhang, S.; Sun, M.; Chen, Y. Experiment on the distribution of different artificial reef models for Paralichthys olivaceus. *Mar. Fish.* **2004**, *26*, 394–398.
17. Kim, D.; Woo, J.; Yoon, H.S.; Na, W.B. Efficiency, tranquillity and stability indices to evaluate performance in the artificial reef wake region. *Ocean Eng.* **2016**, *122*, 253–261. [CrossRef]
18. Wang, X.; Liu, X.; Tang, Y.; Zhao, F.; Luo, Y. Numerical analysis of the flow effect of the menger-type artificial reefs with different void space complexity indices. *Symmetry* **2021**, *13*, 1040. [CrossRef]
19. Kim, D.; Woo, J.; Yoon, H.S.; Na, W.B. Wake lengths and structural responses of Korean general artificial reefs. *Ocean Eng.* **2014**, *92*, 83–91. [CrossRef]
20. Maslov, D.; Pereira, E.; Duarte, D.; Miranda, T.; Ferreira, V.; Tieppo, M.; Cruz, F.; Johnson, J. Numerical analysis of the flow field and cross section design implications in a multifunctional artificial reef. *Ocean Eng.* **2023**, *272*, 113817. [CrossRef]
21. Shu, A.; Qin, J.; Rubinato, M.; Sun, T.; Wang, M.; Wang, S.; Wang, L.; Zhu, J.; Zhu, F. An experimental investigation of turbulence features induced by typical artificial M-shaped unit reefs. *Appl. Sci.* **2021**, *11*, 1393. [CrossRef]
22. Jiang, Z.; Liang, Z.; Zhu, L.; Liu, Y. Numerical simulation of effect of guide plate on flow field of artificial reef. *Ocean Eng.* **2016**, *116*, 236–241. [CrossRef]
23. Liao, J.C. A review of fish swimming mechanics and behaviour in altered flows. *Philos. Trans. R. Soc. B Biol. Sci.* **2007**, *362*, 1973–1993. [CrossRef]
24. Fang, H.; Liu, Y.; Stoesser, T. Influence of boulder concentration on turbulence and sediment transport in open-channel flow over submerged boulders. *J. Geophys. Res. Earth Surf.* **2017**, *122*, 2392–2410. [CrossRef]
25. Muhawenimana, V.; Wilson, C.; Ouro, P.; Cable, J. Spanwise cylinder wake hydrodynamics and fish behavior. *Water Resour. Res.* **2019**, *55*, 8569–8582. [CrossRef]
26. Yusuf, S.N.A.; Asako, Y.; Sidik, N.A.C.; Mohamed, S.B.; Japar, W.M.A.A. A short review on rans turbulence models. *CFD Lett.* **2020**, *12*, 83–96.
27. Zhao, C.; Ouro, P.; Stoesser, T.; Dey, S.; Fang, H. Response of flow and saltating particle characteristics to bed roughness and particle spatial density. *Water Resour. Res.* **2022**, *58*, e2021WR030847. [CrossRef]
28. Jiang, Z.; Liang, Z.; Liu, Y.; Tang, Y.; Huang, L. Particle image velocimetry and numerical simulations of the hydrodynamic characteristics of an artificial reef. *Chin. J. Oceanol. Limnol.* **2013**, *31*, 949–956. [CrossRef]
29. Kim, D.; Woo, J.; Na, W.B.; Yoon, H.S. Flow and structural response characteristics of a box-type artificial reef. *J. Korean Soc. Coast. Ocean Eng.* **2014**, *26*, 113–119. [CrossRef]

30. Kim, D.; Jung, S.; Na, W.B. Evaluation of turbulence models for estimating the wake region of artificial reefs using particle image velocimetry and computational fluid dynamics. *Ocean Eng.* **2021**, *223*, 108673. [CrossRef]
31. Jalalabadi, R.; Stoesser, T.; Ouro, P.; Luo, Q.; Xie, Z. Free surface flow over square bars at different Reynolds numbers. *J. Hydro-Environ. Res.* **2021**, *36*, 67–76. [CrossRef]
32. He, K.; Su, X.; Gao, G.; Krajnović, S. Evaluation of LES, IDDES and URANS for prediction of flow around a streamlined high-speed train. *J. Wind Eng. Ind. Aerodyn.* **2022**, *223*, 104952. [CrossRef]
33. Sheidani, A.; Salavatidezfouli, S.; Stabile, G.; Rozza, G. Assessment of URANS and LES methods in predicting wake shed behind a vertical axis wind turbine. *J. Wind Eng. Ind. Aerodyn.* **2023**, *232*, 105285. [CrossRef]
34. Ouro, P.; Juez, C.; Franca, M. Drivers for mass and momentum exchange between the main channel and river bank lateral cavities. *Adv. Water Resour.* **2020**, *137*, 103511. [CrossRef]
35. Luo, Q.; Dolcetti, G.; Stoesser, T.; Tait, S. Water surface response to turbulent flow over a backward-facing step. *J. Fluid Mech.* **2023**, *966*, A18. [CrossRef]
36. Guo, X.; Luo, Q.; Stoesser, T.; Hajaali, A.; Liu, X. Evolution of high-density submarine turbidity current and its interaction with a pair of parallel suspended pipes. *Phys. Fluids* **2023**, *35*, 086608. [CrossRef]
37. Liu, Y.; Stoesser, T.; Fang, H.; Papanicolaou, A.; Tsakiris, A.G. Turbulent flow over an array of boulders placed on a rough, permeable bed. *Comput. Fluids* **2017**, *158*, 120–132. [CrossRef]
38. Alzabari, F.; Wilson, C.A.; Ouro, P. Hydrodynamics of in-stream leaky barriers for natural flood management. *Water Resour. Res.* **2024**, *60*, e2024WR038117. [CrossRef]
39. Guo, X.; Liu, X.; Zheng, T.; Zhang, H.; Lu, Y.; Li, T. A mass transfer-based LES modelling methodology for analyzing the movement of submarine sediment flows with extensive shear behavior. *Coastal Eng.* **2024**, *191*, 104531. [CrossRef]
40. Zhang, J.; Zhu, L.; Liang, Z.; Sun, L.; Nie, Z.; Wang, J.; Xie, W.; Jiang, Z. Numerical study of efficiency indices to evaluate the effect of layout mode of artificial reef unit on flow field. *J. Mar. Sci. Eng.* **2021**, *9*, 770. [CrossRef]
41. Gong, Y.; Stoesser, T.; Mao, J.; McSherry, R. LES of flow through and around a finite patch of thin plates. *Water Resour. Res.* **2019**, *55*, 7587–7605. [CrossRef]
42. Luo, Q.; Stoesser, T.; Cameron, S.; Nikora, V.; Zampiron, A.; Patella, W. Meandering of instantaneous large-scale structures in open-channel flow over longitudinal ridges. *Environ. Fluid Mech.* **2023**, *23*, 829–846. [CrossRef]
43. Liu, Y.; Tang, Z.; Huang, L.; Stoesser, T.; Fang, H. On the role of the Froude number on flow, turbulence, and hyporheic exchange in open-channel flow through boulder arrays. *Phys. Fluids* **2024**, *36*, 095141. [CrossRef]
44. Dai, J.; Mao, J.q.; Gong, Y.Q.; Gao, H. Effect of triangular baffle designs on flow dynamics and sediment transport within standard box culverts. *J. Hydrodyn.* **2025**, 1–13. . [CrossRef]
45. Nicoud, F.; Ducros, F. Subgrid-scale stress modelling based on the square of the velocity gradient tensor. *Flow Turbul. Combust.* **1999**, *62*, 183–200. [CrossRef]
46. Tan, S.f.; Wang, X.z.; Zhang, M.; Liu, C.g.; Wang, A.m.; Xu, X.f.; Xu, P. Flow field effect of cuboid frame artificial reef. *Chin. J. Appl. Mech.* **2025**, *online first*. (In Chinese)
47. Rodi, W.; Constantinescu, G.; Stoesser, T. *Large-Eddy Simulation in Hydraulics*; CRC Press: Boca Raton, FL, USA, 2013.
48. Zheng, Y.; Kuang, C.; Zhang, J.; Gu, J.; Chen, K.; Liu, X. Current and turbulence characteristics of perforated box-type artificial reefs in a constant water depth. *Ocean Eng.* **2022**, *244*, 110359. [CrossRef]
49. Farzadkhoo, M.; Kingsford, R.T.; Suthers, I.M.; Felder, S. Flow hydrodynamics drive effective fish attraction behaviour into slotted fishway entrances. *J. Hydrodyn.* **2023**, *35*, 782–802. [CrossRef]
50. Wu, Y.; Liu, G.; Yang, J.; Xu, J.; Ke, S.; Li, D.; Chen, X.; Shi, X.; Lin, C. Avoidance behavior of grass carp (Ctenopharyngodon idella) shoals to low-frequency sound stimulation. *Aquat. Sci.* **2025**, *87*, 21. [CrossRef]
51. Marriner, B.A.; Baki, A.B.; Zhu, D.Z.; Cooke, S.J.; Katopodis, C. The hydraulics of a vertical slot fishway: A case study on the multi-species Vianney-Legendre fishway in Quebec, Canada. *Ecol. Eng.* **2016**, *90*, 190–202. [CrossRef]
52. Quaranta, E.; Katopodis, C.; Revelli, R.; Comoglio, C. Turbulent flow field comparison and related suitability for fish passage of a standard and a simplified low-gradient vertical slot fishway. *River Res. Appl.* **2017**, *33*, 1295–1305. [CrossRef]
53. Palau-Salvador, G.; Stoesser, T.; Rodi, W. LES of the flow around two cylinders in tandem. *J. Fluids Struct.* **2008**, *24*, 1304–1312. [CrossRef]
54. McSherry, R.J.; Chua, K.V.; Stoesser, T. Large eddy simulation of free-surface flows. *J. Hydrodyn. Ser. B* **2017**, *29*, 1–12. [CrossRef]

Journal of
Marine Science
and Engineering

MDPI

Article

Mechanical Behavior of Marine Soft Soil with Different Water Contents Under Cyclic Loading

Yajun Liu, Heng Zhang, Yindong Sun, Ke Wu * and Wenbin Xiao

School of Civil Engineering, Shandong University, Jinan 250061, China
* Correspondence: wuke@sdu.edu.cn

Abstract: This study integrates macroscopic dynamic triaxial tests with microscopic discrete element simulations to comprehensively examine the dynamic deformation characteristics of marine soft soils under cyclic loading. Unlike previous research that typically focuses solely on experimental or numerical methods, this approach combines both techniques to enable a holistic analysis of soil behavior. The dynamic triaxial tests assessed macroscopic responses, including strain evolution and energy dissipation, under varying dynamic stress ratios, confining pressures, and water contents. Concurrently, discrete element simulations uncovered the microscopic mechanisms driving these behaviors, such as particle rearrangement, porosity variations, and shear zone development. The results show that (1) The strain range of marine soft soils increases significantly with higher dynamic stress ratios, confining pressures, and water contents; (2) Cumulative dynamic strain and particle displacement intensify at water contents of 50% and 55%. However, at a water content of 60%, the samples exhibit significant damage characterized by the formation of shear bands throughout the entire specimen; (3) As water content increases, energy dissipation in marine soft soils accelerates under lower confining pressures but increases more gradually under higher confining pressures. This behavior is attributed to enhanced particle packing and reduced pore space at elevated confining pressures. This integrated methodology not only enhances analytical capabilities but also provides valuable engineering insights into the dynamic response of marine soft soils. The findings offer essential guidance for the design and stabilization of marine soft soil infrastructure in coastal urban areas.

Keywords: mechanical behavior; marine soft soil; cyclic loading; dynamic triaxial test; particle flow analysis

1. Introduction

Engineers and builders face unique challenges when developing underground spaces and constructing rail transit systems in coastal cities. Marine soft soils, commonly found in these regions, pose significant obstacles to underground engineering due to their distinctive properties, such as low shear strength, high compressibility, and high sensitivity to moisture [1,2]. The construction of subways on these soft soil foundations is particularly vulnerable to issues such as foundation instability and structural damage, raising serious concerns about the safety and reliability of such projects. Therefore, a comprehensive understanding of the dynamic deformation characteristics of marine soft soils under cyclic loading [3], along with the mechanisms driving changes in soil properties, is crucial for effectively managing the dynamic behavior of these soils [4].

Currently, dynamic triaxial test analysis is widely used to study the dynamic characteristics of marine soft soil. In dynamic triaxial testing, cyclic or vibrating loads are applied to soil samples to simulate the effects of dynamic loads, such as those generated by earthquakes and traffic vibrations [5,6]. Additionally, numerical simulation methods are frequently employed to analyze the stress–strain behavior of marine soft soils. Dynamic triaxial testing is particularly useful for evaluating soil behavior under cyclic loading, allowing investigation into responses such as strength and stiffness reduction, principal stress rotation, excess pore water pressure generation, and deformation accumulation in marine soft clays [7]. A series of undrained triaxial tests under low-frequency cyclic loading reveal that marine soft clay accumulates plastic deformation and pore pressure under cyclic loads, with higher deformation and pore pressure observed under conditions of elevated confining pressure and stress ratio [8]. Dynamic triaxial tests have also been employed to examine the properties of freeze–thaw soils under cyclic loading, with scanning electron microscopy (SEM) analyses illustrating the weakening effects of freeze–thaw cycles and dynamic loading on soft soils [9]. Using numerical simulations with the "cyclic flow (CM) model"—an elasto-plastic constitutive model—and a coupled soil–water finite element–finite difference (FE–FD) code called DBLEAVES, cyclic triaxial tests of marine soft clays were simulated to assess soil behavior under cyclic conditions [10]. In another study, the variable confining pressure (VCP) mechanical behavior of marine soft clay was examined under long-term traffic loading (50,000 cycles) through two series of monotonic shear tests. Post-cycling stress paths, pore water pressures, and shear strengths were compared with those from standard monotonic shear tests without cyclic loading history [11]. To develop a more accurate stiffness degradation model for marine soft clay, dynamic triaxial tests were conducted at varying stress levels, and cumulative displacement and stiffness degradation were modeled in ABAQUS using the USDFLD subroutine [12]. For more realistic assessment of cyclic triaxial behavior in saturated soils, a series of bidirectional cyclic triaxial tests were designed to account for drainage conditions and variable confining pressures during cyclic and intermittent loading phases [13]. Large-scale interfacial shear tests and numerical simulations were also conducted, showing that the pile-soil interfacial shear strength exhibits a hardening–stabilizing behavior with peak strength and slightly elevated stabilized strength under cyclic loading [14]. Triaxial tests on marine clay have demonstrated how the cyclic stress ratio (CSR) and initial confining pressure (p′(0)) influence the dynamic properties of marine soft soil [15]. Additionally, microstructural evolution during loading and unloading in natural marine clays was studied using undrained triaxial tests, with mercury intrusion porosimetry (MIP) and field-emission scanning electron microscopy (FESEM) used to examine pore distribution and the arrangement of soil particles and aggregates [16]. Based on 2-D PFC, non-circular particle with direction was constructed and specimens with different distribution of long axis orientation were established. Loads were applied in vertical direction. Changes in fabric such as the long axis orientation, particle contact direction, and particle contact force in the process of loading were analyzed [17]. Based on PFC3D 5.0 software and theory, combined with the indoor triaxial test results to calibrate the basic micro parameters of numerical simulation, and the conventional triaxial numerical simulation of loess numerical samples under 200 kPa confining pressure was conducted [18]. Using the PFC numerical simulation method, the particle contact model was modified by considering the particle interaction and low gravity effect, and the research on the microscopic parameters of the particle contact model of the lunar soil and its macroscopic mechanical properties was carried out [19].

In conclusion, dynamic triaxial testing is widely regarded as one of the most effective methods for studying the mechanical properties of marine soft soils. Many researchers have employed this method to investigate the macroscopic mechanical behavior of marine soft soils under cyclic loading. However, it has limitations in capturing microscopic phenomena, such as particle collisions and rolling interactions. While numerical simulations are commonly used to model macroscopic behavior, microstructural changes in marine soft soils are often analyzed using techniques like scanning electron microscopy, revealing a gap in integrating macroscopic and microscopic analyses. Particle flow analysis, a numerical technique, offers significant advantages for studying the micromechanical behavior of geotechnical materials by simulating particle interactions and their effects on complex system behavior. Despite its potential, this method has not been extensively applied to marine soft soils. To address this gap, the present study integrates dynamic triaxial testing with particle flow analysis to explore the dynamic behavior of marine soft soils under varying water content conditions. This combined approach provides a comprehensive understanding of marine soft soil behavior under cyclic loading from both macroscopic and microscopic perspectives, highlighting the utility of particle flow analysis in capturing the evolution of particle deformation under dynamic conditions. The findings offer valuable theoretical and technical insights for marine geotechnical engineering research and design.

2. Analysis Methods

2.1. Geotechnical Dynamic Triaxial Test Analysis Method

The dynamic triaxial test is a method used to evaluate the dynamic properties of soils by simulating the stress conditions that soils experience in their natural environment. This test applies controlled vertical and horizontal stresses to soil samples, allowing cyclic or vibratory loads to replicate the effects of dynamic forces such as earthquakes and traffic vibrations. The dynamic triaxial test is particularly effective for assessing soil behavior under dynamic loading, including characteristics of the stress–strain hysteresis curve. Key factors such as water content, confining pressure, and dynamic stress ratio directly influence the dynamic behavior of marine soft soil. Consequently, this study utilizes the dynamic triaxial test as a macro-scale analysis method to investigate the mechanical behavior of marine soft soil under cyclic loading, focusing on the morphology of the stress–strain hysteresis curve across varying water content, confining pressure, and dynamic stress conditions.

The testing equipment used in this study is a fully digital, pneumatic, closed-loop repetitive loading triaxial apparatus, as shown in Figure 1. This computer-controlled system allows for precise control over stress or strain rate loading tests, with test frequency and loading magnitude carefully managed to ensure accuracy. The triaxial chamber accommodates specimens with diameters of 1.4 inches and 2.8 inches, with maximum pressures of 100 kPa in the inner chamber and 200 kPa in the outer chamber.

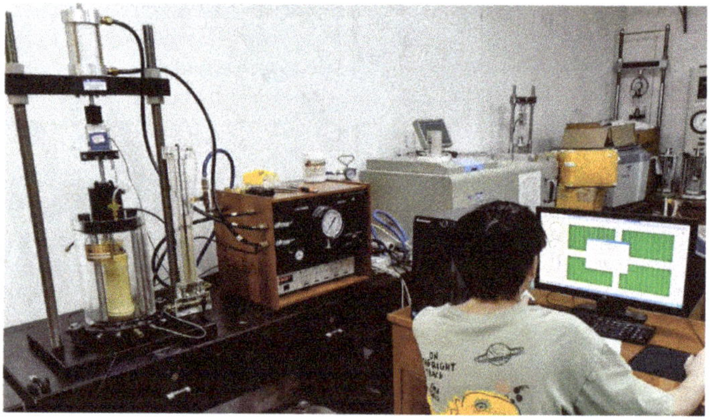

Figure 1. Triaxial apparatus.

Shenzhen, located in the Pearl River Delta region of Guangdong Province, China, is characterized by extensive distributions of marine soft soils, which pose significant challenges to local engineering and construction projects. This type of soil is particularly prevalent in the construction of rail transit systems within the Shenzhen area. For this study, marine soft soil specimens were collected from the shield section of the tunnel between Left Battery Station and Prince Bay Station along Shenzhen Urban Rail Transit Line 12. The fundamental hysico-mechanical properties of these specimens are summarized in Table 1.

Table 1. Basic Physical and Mechanical Indicators of Soil.

Liquid Limit (%)	Plastic Limit (%)	Plasticity Index (%)	Specific Gravity (kg/m^3)	Particle Composition (%)		
				Sand Grains (0.02–2 mm)	Powder Grain (0.002–0.02 mm)	Clay Grain (<0.002 mm)
63.62	37.37	26.25	1.69	18.8%	58.5%	22.7%

The tests were conducted in accordance with China's Geotechnical Laboratory Test Regulations (GB/T 50123-1999) [20] and the United States' Standard Test Method for Consolidated Drained Triaxial Compression Test for Soils (ASTM D7181-11) [21]. The specimen preparation process involves the following steps: (1) soil sample collection and air drying: Soil samples are collected and placed in a well-ventilated area to air dry. The purpose of air drying is to remove excess moisture to prevent its influence on the soil's properties. (2) Crushing and sieving: Once air-dried, the soil is placed on a rubber sheet, crushed using a wooden mallet, and sieved through a 2 mm aperture sieve. This ensures uniformity of soil particles, which is critical for maintaining consistency across tests. (3) Water content adjustment: The water content of the air-dried soil is measured, and the required amount of water is calculated based on the target design water contents (e.g., 40%, 45%, 50%, 55%, 60%). Water is evenly sprayed onto the soil, ensuring uniform moisture distribution. The soil is then placed in a sealed plastic bag to equilibrate for 24 h. (4) Specimen preparation: The soil is compacted to 95% of its maximum dry density. Five specimens are prepared for each water content level, yielding a total of 25 specimens. The mass error for each specimen is carefully controlled within ±2 g. (5) Preservation: The prepared specimens are immediately wrapped in plastic wrap to prevent moisture loss, ensuring stable water content throughout the testing process.

To investigate the dynamic properties of marine soft soils, parameters such as water content, dynamic stress ratio, and confining pressure were systematically varied and analyzed. The water content was adjusted to specific levels, and cylindrical specimens measuring 140 mm in height and 70 mm in diameter were prepared accordingly. Prior to testing, each specimen was subjected to K_0-consolidation with $K_0 = 0.7$ for 24 h to simulate the in situ stress conditions, and confining pressure was maintained at a constant throughout loading. The test utilized moisture contents of 40%, 45%, 50%, 55%, and 60%; confining pressures of 50, 100, and 200 kPa; and cyclic stress ratios of 0.15, 0.25, and 0.35.

A stress-controlled loading method was employed, wherein axial dynamic loads were applied in a sinusoidal pattern and precisely monitored through computer control to ensure accuracy and consistency. Before initiating the test, a static bias stress σ_s was applied to each specimen for bias consolidation. After consolidation, loading proceeded according to the test program, with sinusoidal dynamic stress applied to simulate tunnel traffic loading. Test termination was governed by the number of load cycles: for specimens that remained intact during loading, testing ceased after 1000 cycles; for specimens that failed during loading, testing stopped upon failure. The test was conducted at a load frequency of 0.1 Hz. The dynamic stress ratio, defined as the ratio of dynamic stress to confining pressure, was halved due to the double-amplitude loading method. The test program is outlined in Table 2 [22].

The dynamic properties of marine soft soil are investigated by analyzing key factors such as water content, confining pressure, and dynamic stress ratio. The relevant parameters are defined as follows: static bias stress σ_s, representing the initial static stress applied to the specimen; static bias stress ratio η, defined as the ratio of static stress components; dynamic stress σ_d, referring to the cyclic or dynamic stress applied; water content w, indicating the moisture level in the soil; confining pressure σ_c, representing the surrounding compressive stress; dynamic stress ratio η_d, defined as the ratio of dynamic to static stress; and the vibration number N, representing the number of load cycles applied. For this test, the failure criterion is defined as an axial strain of 5%, meaning a specimen is considered to have failed when its axial strain reaches or exceeds 5%.

Table 2. Experimental plan and control parameters.

Specimen Number	Water Content (%)	Surrounding Pressure (kPa)	Deviator Stress (kPa)	Dynamic Stress (kPa)	Dynamic Stress Ratio
CT-1	40	50	22	15	0.15
CT-2		100	43	30	0.15
CT-3		100	43	50	0.25
CT-4		100	43	70	0.35
CT-5		200	86	60	0.15
CT-6	45	50	22	15	0.15
CT-7		100	43	30	0.15
CT-8		100	43	50	0.25
CT-9		100	43	70	0.35
CT-10		200	86	60	0.15
CT-11	50	50	22	15	0.15
CT-12		100	43	30	0.15
CT-13		100	43	50	0.25
CT-14		100	43	70	0.35
CT-15		200	86	60	0.15
CT-16	55	50	22	15	0.15
CT-17		100	43	30	0.15
CT-18		100	43	50	0.25
CT-19		100	43	70	0.35
CT-20		200	86	60	0.15
CT-21	60	50	22	15	0.15
CT-22		100	43	30	0.15
CT-23		100	43	50	0.25
CT-24		100	43	70	0.35
CT-25		200	86	60	0.15

2.2. Energy Loss Analysis Method

The dynamic response of the soil mass is characterized not only by changes in displacement and stress but also by energy dissipation. The extent of energy loss provides insights into the soil's adaptability and hysteresis under dynamic loading. Consequently, the study of energy dissipation is crucial for evaluating the long-term stability of marine soft soil under cyclic loading, as it helps determine whether the soil experiences plastic deformation or damage. In the stress–strain behavior of marine soft soil subjected to cyclic loading, the area of the hysteresis loop corresponds to the energy dissipated during each loading cycle. Energy dissipation is typically quantified using two primary methods:

(1) Calculation of hysteresis loop area

Energy dissipation is quantified as the area enclosed by the hysteresis loop in a single loading cycle. This area represents the energy lost due to internal friction and material dissipation within the soil sample during each loading and unloading cycle.:

$$W = \oint \sigma d\varepsilon \tag{1}$$

where σ is the stress during cyclic loading; ε is the corresponding strain; $\oint$ is the closed integral of the hysteresis loop.

(2) Discrete calculation methods

When the stress–strain curve is represented as discrete data points, energy dissipation can be calculated through approximate integration methods:

$$W \approx \sum_{i=1}^{n} \frac{(\sigma_i + \sigma_{i+1})}{2} \cdot (\varepsilon_{i+1} - \varepsilon_i) \tag{2}$$

The experimentally obtained (σ, ε) data were input into the equation to calculate the energy dissipation for each loading cycle.

Based on the stress–strain hysteresis curve of marine soft soil, the discrete calculation method was applied, and a Python 3.11 was developed to calculate energy dissipation. This approach allows for the generation of characteristic energy loss curves under varying water content conditions, which can be used to assess the stability of marine soft soil.

2.3. Discrete Particle Flow Analysis Method

Particle flow code (PFC) is a widely used simulation software based on the discrete element method (DEM), designed to accurately predict the physical behavior of particle systems under various conditions, such as flow, compression, separation, and mixing. PFC models the interactions between particles and their motion laws, enabling detailed simulations of complex particle systems [23,24]. In this study, PFC was employed to investigate the mechanical behavior of marine soft soil under dynamic loading from a microscopic perspective.

The study commenced with a comprehensive analysis of both macroscopic and microscopic mechanical properties of marine soft soils, focusing on key parameters such as particle size distribution, mineral composition, and porosity. The findings from this analysis are extensively discussed in the literature [5]. The paper also presents the particle distribution characteristics of the specimens used in the dynamic triaxial tests, as detailed in Table 1. By integrating data from the literature [5] with the information in Table 1, and referencing the microstructural characteristics of marine soft soils outlined in the study by [20], a particle flow calculation model for marine soft soil was developed using DEM.

The specimen in the dynamic triaxial test measures 70 mm in diameter and 140 mm in height, and a two-dimensional numerical model of the same size was established for simulation. Wall units were used to define the triaxial tester boundaries, with rigid boundaries applied to create the model structure. Four walls were generated in the simulation: the top and bottom walls represent the axial loading plates, while the left and right walls simulate the rubber sleeve surrounding the specimen. To replicate the rubber sleeve effect, the top and bottom walls were assigned higher stiffness values, while the left and right walls were assigned lower stiffness values.

The study is based on the following assumed conditions: (1) marine soft soil exhibits homogeneity in particle distribution and mineralogy; (2) the behavior of marine soft soil is isotropic; and (3) the particle distribution is characterized using the particle coordination number, which serves as a key parameter for evaluating the structural stability of particle assemblies and the strength of inter-particle interactions. In the context of marine soft soil research, the coordination number not only reflects the physical contacts between particles but also has a direct correlation with the macroscopic mechanical properties of the material, such as compressibility, shear strength, and permeability. By integrating these methodologies and assumptions, the micromechanical characterization of marine soft soils is effectively realized.

In constructing the model, the radius expansion method was applied, initially generating smaller particles within the model range and then enlarging them to meet initial porosity requirements. A direct replication of particle gradation from the physical test would involve millions of particles, significantly impacting computational efficiency. Thus, the radius expansion method was optimized to produce 22,440 particles with an initial porosity of 0.25, a particle diameter range of 0.3–1 mm, and a density of 1.69 g/cm^3. This is shown in Figure 2.

In the PFC particle flow model, particle unit parameters cannot be directly assigned to replicate the macroscopic properties of soil. Instead, a calibration process is required to establish a correspondence between the model's microscopic parameters and the actual physical and mechanical properties of the soil. This involves iterative adjustment of the model's microscopic parameters to ensure that the numerical simulation results align with the soil's macroscopic physical and mechanical behavior, a process known as macro–micro parameter calibration.

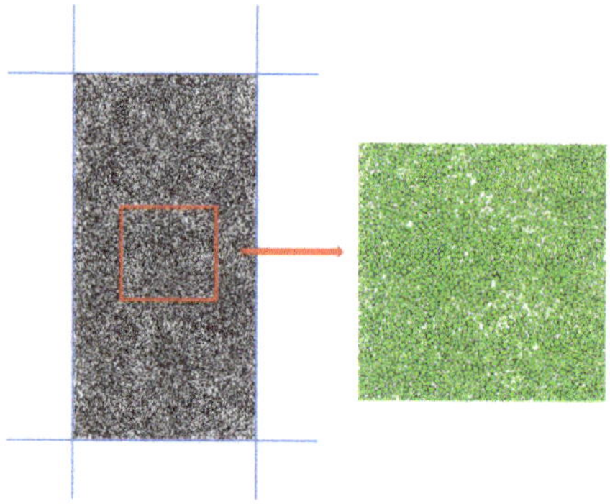

Figure 2. Numerical simulation model sample.

Key microscopic parameters in the linear contact bond model include friction coefficient, linear contact effective modulus Ec, the ratio of normal to tangential contact stiffness between particles, normal bond strength σc, and tangential bond strength τc. By fine-tuning these parameters, the numerical results can more closely match those obtained from laboratory tests. The calibration of fine-scale parameters is crucial for accurately simulating the mechanical properties of materials in numerical models. However, this process can be complex and requires adjustments to different models based on actual working conditions. Both domestic and international scholars have conducted extensive research on fine-scale parameter calibration, identifying key principles that can serve as guidelines. In this study, we follow the calibration principles outlined in the literature [23,24] and perform the calibration of fine-scale parameters for marine soft soil using the functional relationship between fine-scale and macro-mechanical parameters proposed in [25]. Based on the basic properties of marine soft soils presented in reference [5], the calibrated micro-parameter values for marine soft soils are provided in Table 3, corresponding to the macro-mechanical properties obtained from reference [22].

Table 3. Calibration results of microscopic parameters.

Water Content	Coefficient of Friction	Linear Contact Effective Modulus (MPa)	Stiffness Ratio	Normal Bond Strength (kPa)	Tangential Bond Strength (kPa)
40%	0.17	50	2	30	30
45%	0.15	40	2	25	25
50%	0.14	20	2	20	20
55%	0.13	10	2	10	10
60%	0.11	8	2	1	1

In PFC, axial loading of the specimen is simulated by applying velocity to the upper and lower walls [26]. Since the walls in PFC are massless entities to which Newton's second law does not apply, it is not possible to apply force directly to them. In this study, an axial stress control method is used, where sinusoidal stress is converted into velocity, which is then applied to the walls. This conversion is achieved through a servo mechanism, as detailed below:

$$v = G(\sigma^{req} + \sigma_z) \tag{3}$$

where σ^{req} is the axial stress; σ_z is between the particle and the upper and lower walls.

Axial loading stress:

$$\sigma^{req} = \Delta q sin(2\pi f t) + \sigma \tag{4}$$

where Δq is the stress amplitude; f is the loading frequency; t is the loading time; σ is circumferential pressure.

In the simulation, the sinusoidal stresses were set to match those used in the laboratory tests, with a loading frequency of 10 Hz applied to improve computational efficiency. After generating the specimen particles, they were pressurized using a servo-mechanism programmed in the FISH language, and each specimen underwent bias consolidation with $K_0 = 0.7$ before cyclic loading. The perimeter pressure was maintained constant throughout the loading process. Parameters of the cyclic loading, including frequency, amplitude (in terms of force or displacement), loading form (unidirectional or bidirectional), and the number of cycles, were specified via programming scripts (e.g., FISH language) or through the software's built-in functions. Once bias consolidation was complete, a sinusoidal load matching the magnitude of the laboratory test was applied to the specimen, with the loading process concluding after 200 cycles. The specimen's response—including particle displacement, deformation, stress, and strain—was recorded in real time using PFC's monitoring tools.

2.4. Analytical Method Validation

To verify the consistency between the macroscopic test results and the microscopic particle flow analysis, axial deformation curves for varying water contents were obtained using both methods under the conditions of $\sigma_c = 100$ kPa and $\eta_d = 0.15$. This is shown in Figure 3.The results from both approaches indicate that the strain in marine soft soil increases rapidly at the initial stage of loading, followed by a gradual reduction in the growth rate until stabilization, and eventually exhibits a slow linear increase. The cumulative strain changes observed in the discrete element simulations closely align with those from the experimental tests, with an error margin within 7%, demonstrating the reliability of the discrete element numerical simulation method. These findings confirm that the discrete element numerical model developed in this study is suitable for simulating dynamic triaxial tests of marine soft soil and for advancing research into its micromechanical properties.

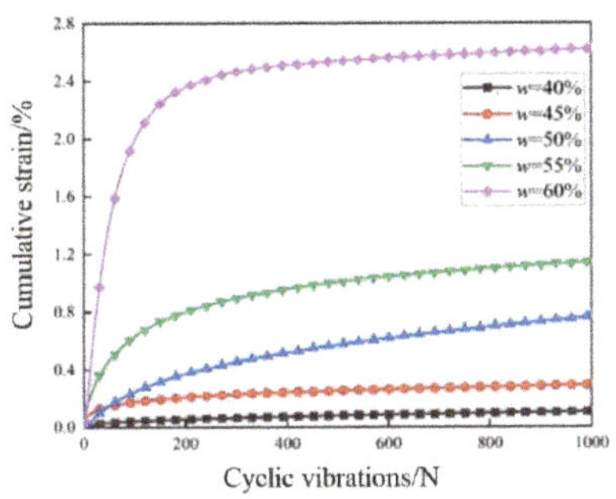

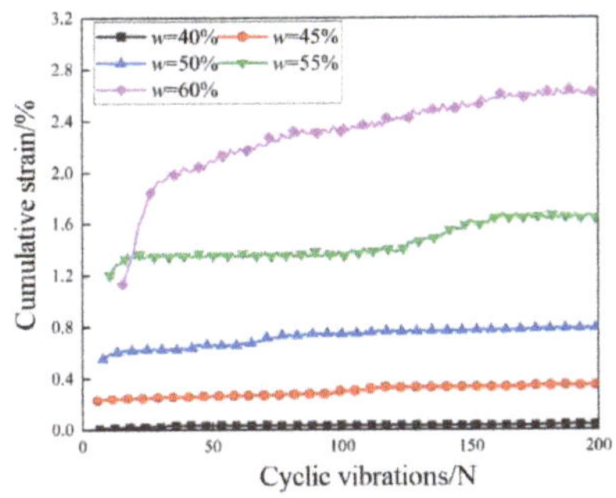

Figure 3. The curve between strain and number of cycles obtained from triaxial specimen and discrete element analysis. (**a**) Test results. (**b**) Discrete element analysis results.

3. Results

3.1. Morphological Characteristics of Stress–Strain Hysteresis Curves in Marine Soft Soils

The stress–strain hysteresis curve is a cyclic loading and unloading development curve that illustrates the relationship between dynamic stress and dynamic strain at each point during cyclic loading. In dynamic triaxial tests of marine soft soils, the initial cycles reveal the soil's initial stiffness—its fundamental capacity to resist deformation under dynamic loading—and provide early indicators of nonlinear behavior, such as strain softening or hardening. These initial cycles also elucidate the soil's energy dissipation mechanisms, offering valuable insights into its behavior and stability under dynamic conditions. This analysis can further identify early signs of soil damage, providing a basis for implementing timely preventive measures. Accordingly, this study analyzes the first 10 cycles of the hysteresis curve to gain a deeper understanding of soil behavior in the early stages of dynamic loading, offering critical theoretical support for developing soil dynamics models and predicting soil performance in engineering applications.

Water contents of w = 40%, w = 45%, and w = 50% were selected for testing, with confining pressures of σ_c = 50 kPa, σ_c = 100 kPa, and σ_c = 200 kPa, and dynamic stress ratios of η_d = 0.15, η_d = 0.25, and η_d = 0.35. Figure 4 illustrates the stress–strain hysteresis curves for the specimens during the first 10 cycles under these test conditions. The results indicate that the hysteresis curves for all specimens exhibit a progression from loose to tight as the number of cycles increases. Notably, the strain range of the specimens expanded significantly with increased water content, and the hysteresis loops lengthened and tilted toward the axial strain direction, suggesting that higher water content intensified soil softening. Additionally, the area of the hysteresis loop increased with each cycle, further highlighting the soil's enhanced energy dissipation capacity under high water content conditions, which, in turn, amplified the degree of softening.

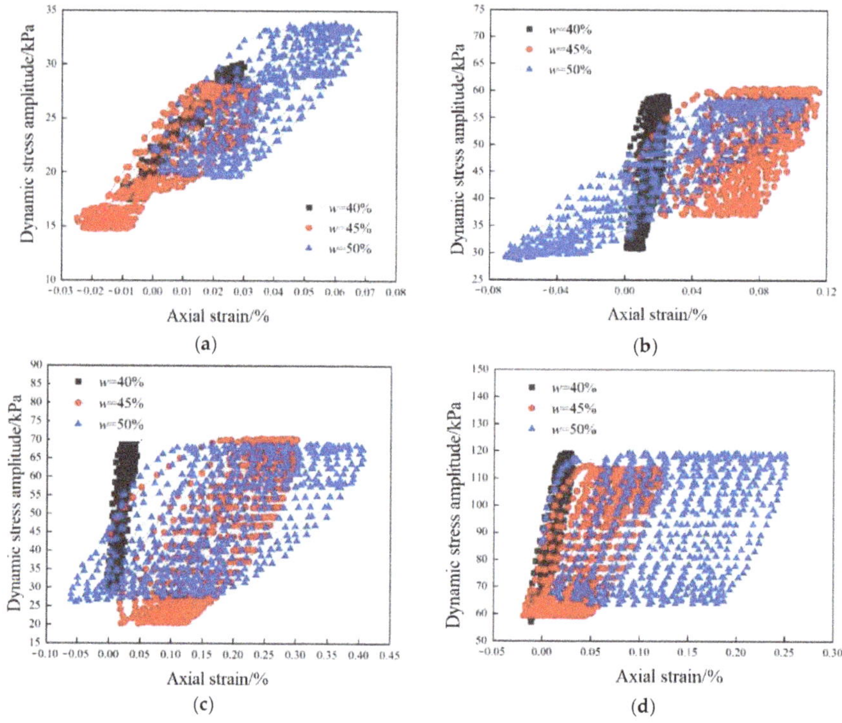

Figure 4. Stress–strain curves of the first 10 cycles under different moisture content conditions. (**a**) σ_c = 50 kPa, η_d = 0.15. (**b**) σ_c = 100 kPa, η_d = 0.15. (**c**) σ_c = 100 kPa, η_d = 0.25. (**d**) σ_c = 200 kPa, η_d = 0.15.

The water content w was varied from 40% to 60%, with a confining pressure of σ_c = 100 kPa and dynamic stress ratios η_d of 0.15, 0.25, and 0.35. Figure 5 shows the stress–strain hysteresis curves for the specimens over the first 10 cycles under these test conditions. At a given water content, the hysteresis curves displayed a progression from loose to tight as the dynamic stress ratio increased, leading to a significant expansion of the strain range and an increase in the area of the hysteresis loop per cycle. This indicates that higher dynamic stress ratios amplify the degree of softening in the soil. Additionally, as water content increased, the hysteresis curves shifted to the right, with maximum strain values of 0.12%, 0.25%, 0.41%, and 4.5% observed in specimens subjected to 10 cycles of vibration at a dynamic stress ratio of 0.35 across different water contents. The cumulative strain over the 10 cycles increased with rising water content, making the softening effect on the soil more pronounced.

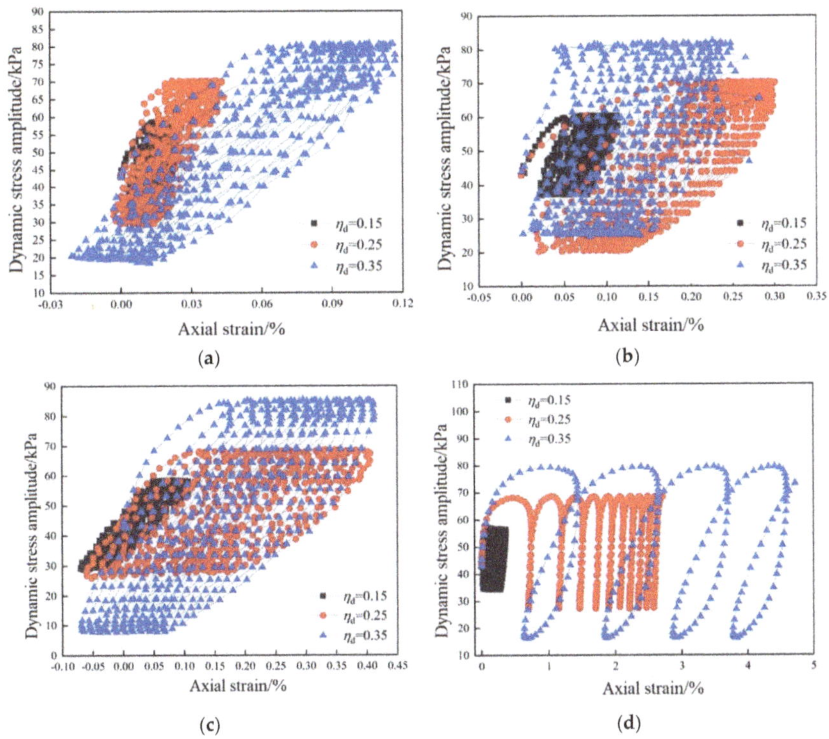

Figure 5. Stress–strain curves of the first 10 cycles under different dynamic stress ratios. (**a**) σ_c = 50 kPa, w = 40%. (**b**) σ_c = 100 kPa, w = 45%. (**c**) σ_c = 100 kPa, w = 50%. (**d**) σ_c = 200 kPa, w = 60%.

The water content w was selected as 40%, 45%, 50%, 55%, and 60%, with confining pressures σ_c of 50 kPa, 100 kPa, and 200 kPa, and a dynamic stress ratio η_d = 0.15. Figure 6 shows the stress–strain hysteresis curves of the specimens over the first 10 cycles under these conditions. At a constant dynamic stress ratio, a higher confining pressure corresponds to a greater applied dynamic stress. The development of stress–strain hysteresis loops under different confining pressures follows a sparse-to-dense-to-compact pattern over 10 cycles. As accumulated plastic strain increases, the hysteresis loop gradually becomes denser. Higher confining pressures yield larger hysteresis loop areas, broader strain ranges, and enhanced energy dissipation capacity, indicating that absorbed and dissipated energy increase with confining pressure. Furthermore, as water content rises, the hysteresis curves shift to the right, with maximum strain values of 0.03%, 0.12%, 0.24%, and 1.0% observed for specimens at 200 kPa across different water contents after 10 cycles. The cumulative strain over 10 cycles also increases with higher water content, making the soil softening effect more pronounced.

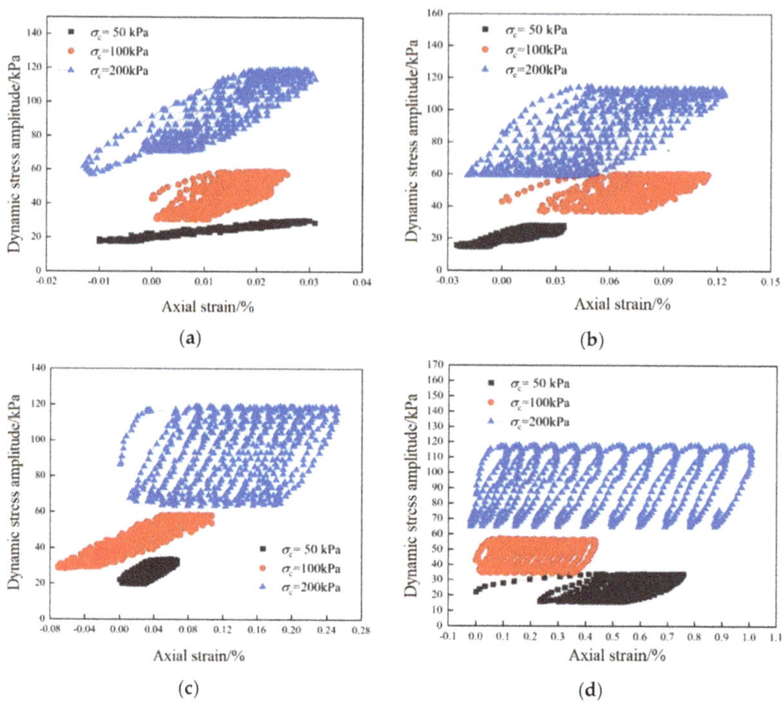

Figure 6. Stress–strain curves of the first 10 cycles under different confining pressures. (**a**) w = 40%, η_d = 0.15. (**b**) w = 45%, η_d = 0.15. (**c**) w = 50%, η_d = 0.25. (**d**) w = 45%, η_d = 0.15.

3.2. Characterization of Energy Loss in Marine Soft Soil

Figure 7 presents a schematic diagram illustrating the relationship between energy dissipation in marine soft soil under a dynamic stress ratio of 0.15, as influenced by changes in water content and perimeter pressure. From the figure, the following observations can be made:

(1) Effect of water content on energy loss: At low water content, friction and particle contact in the marine soft soil are stronger, leading to higher energy dissipation and greater soil stiffness, with energy loss primarily occurring through friction between soil particles. As water content increases to a medium level, the water gradually fills the pore spaces within the soil, enhancing the lubricating effect and causing the energy dissipation mechanism to shift toward viscous dissipation. At high water content, the soil becomes softer, and the proportion of water increases, potentially reaching a higher level of energy loss. In this condition, the soil may exhibit behavior akin to liquefaction, and the energy dissipation mechanism becomes more complex. However, the energy loss may decrease as the deformation mode becomes more water-dominated, reducing friction and, thus, frictional dissipation.

(2) Effect of perimeter pressure on energy loss: When the perimeter pressure is 50 kPa, the larger pore spaces in the soil result in more noticeable water effects, leading to more significant soil deformation and higher energy dissipation. As water content increases, energy loss rises more gradually, but water content still plays an important role. At 100 kPa perimeter pressure, the soil is more compacted, enhancing particle contact and increasing stiffness, which reduces energy loss. The stronger contact force and higher stiffness limit soil deformation, leading to more efficient energy transfer but less energy dissipation compared to lower perimeter pressures. For soil with medium water content, the effect of perimeter pressure on energy dissipation becomes more

pronounced. At 200 kPa perimeter pressure, the soil becomes denser, and particle contact reaches its maximum, resulting in increased stiffness and significantly reduced energy dissipation. At this point, energy is primarily transferred through elastic deformation, with minimal energy loss, particularly at higher water content.

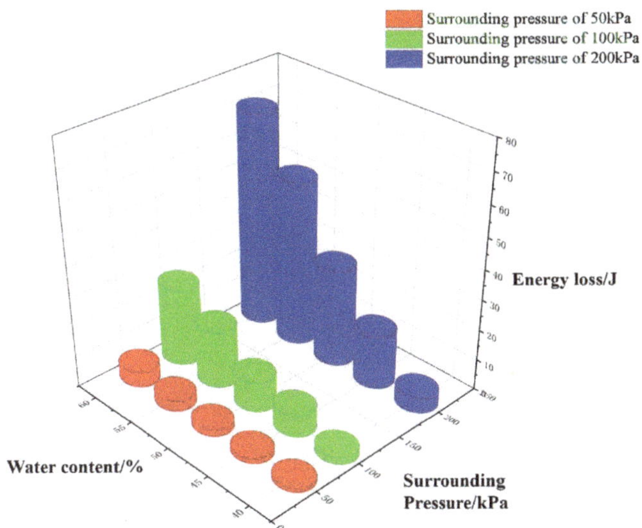

Figure 7. Schematic diagram of energy dissipation in marine soft soil under a dynamic stress ratio of 0.15.

In summary, at lower perimeter pressures (50 kPa), energy dissipation increases more noticeably with water content, as the soil is more prone to deformation and the energy dissipation mechanism shifts. At higher perimeter pressures (200 kPa), the effect of water content on energy dissipation is less significant, as the soil becomes more compact and the stronger particle contact reduces sensitivity to changes in water content. The curve at 100 kPa perimeter pressure lies between these extremes, showing a smoother change in energy dissipation.

Figure 8 presents a schematic diagram showing the relationship between energy dissipation, water content, and dynamic stress ratio for marine soft soil under a perimeter pressure of 100 kPa. From the figure, it can be observed that when the dynamic stress ratio is 0.15, the increase in energy dissipation with water content is relatively gradual, primarily driven by friction and viscous dissipation. At a dynamic stress ratio of 0.25, the energy dissipation curve rises smoothly, likely representing an intermediate case. However, when the dynamic stress ratio is 0.23, the energy dissipation increases rapidly with rising water content, which is closely related to the larger dynamic stress and corresponding soil deformation. In summary, higher dynamic stress ratios lead to more significant energy loss. At higher stress ratios, soil deformation is more pronounced, and increased water content results in a rapid rise in energy dissipation. At lower dynamic stress ratios, energy dissipation increases more slowly, primarily influenced by the soil's stiffness and the lubricating effect of water. The overall trend is that energy dissipation intensifies with increasing water content, especially at higher dynamic stress ratios.

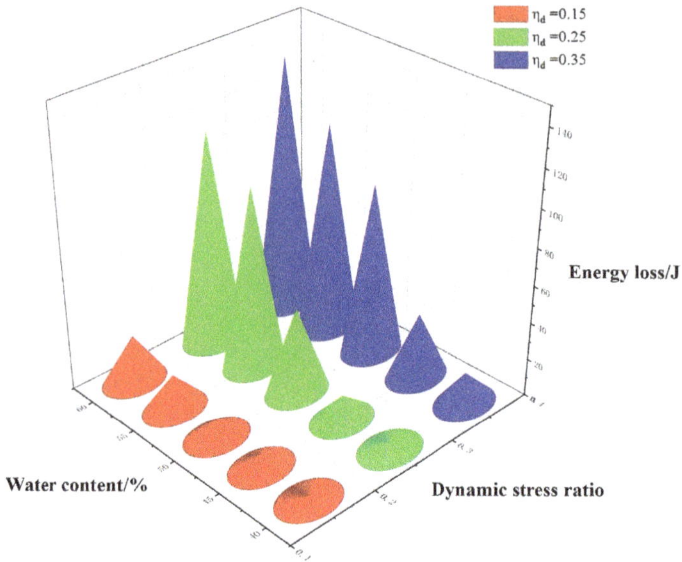

Figure 8. Schematic diagram of energy dissipation in marine soft soil at 100 kPa perimeter pressure condition.

3.3. Displacement Fields

PFC 5.0 software can monitor particle motion, plot displacement vectors, and observe the initiation and development of damage and shear zone formation within the specimen. Figures 9–11 show the particle displacement vector distribution in marine soft soil specimens. The displacement field trends and their responses under different conditions reveal the complex behavior of granular materials, particularly marine soft soils, under cyclic loading.

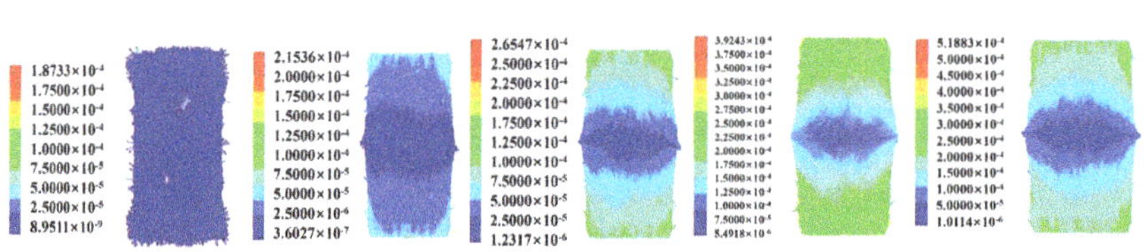

(a) CT-1 (b) CT-2 (c) CT-3 (d) CT-4 (e) CT-5

Figure 9. Particle displacement distribution of samples with a moisture content of 40.

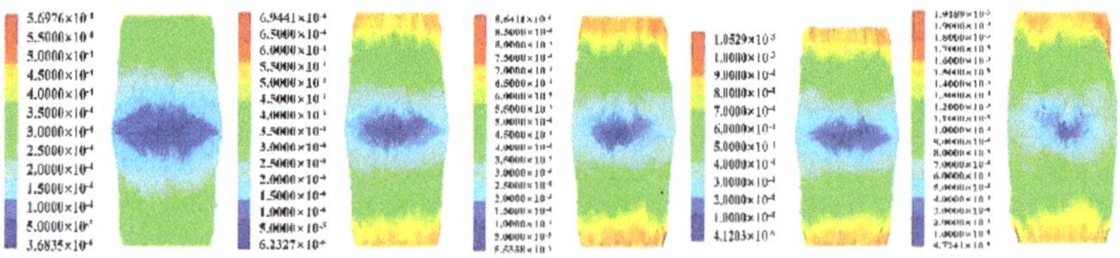

(a) CT-11 (b) CT-12 (c) CT-13 (d) CT-14 (e) CT-15

Figure 10. Particle displacement distribution of samples with a moisture content of 50.

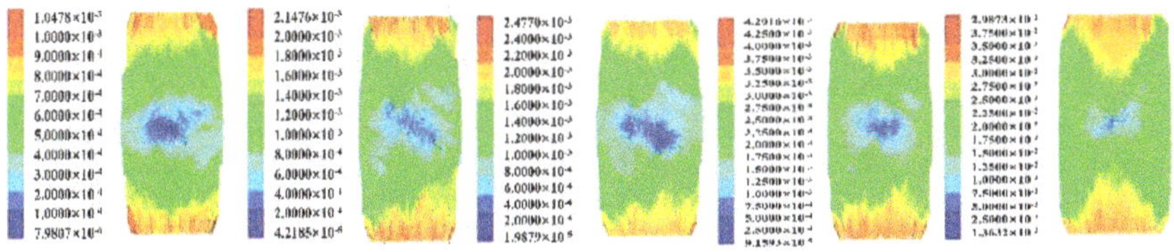
(**a**) CT-21 (**b**) CT-22 (**c**) CT-23 (**d**) CT-24 (**e**) CT-25

Figure 11. Particle displacement distribution of samples with a moisture content of 60.

At lower water contents, specimens exhibit limited accumulated dynamic strain, and particle motion is more random. As water content increases, especially at 50% and 55%, cumulative dynamic strain grows, particle displacement becomes more apparent, and initial shear zones start to form within the specimen, which remains intact. At a water content of 60%, however, the specimen shows significant damage, with the shear zone running through its entirety. At low water content, the lack of lubrication between particles restricts movement, resulting in minimal accumulated dynamic strain and preventing the formation of effective shear bands. As water content rises, the lubrication between particles increases, facilitating displacement and rearrangement, which supports shear band formation. Once particles begin relative motion and accumulate dynamic strain, localized stress concentration occurs between particles, leading to shear band formation. As water content continues to increase, particles slip more easily, creating more prominent shear bands that eventually extend throughout the specimen. Additionally, as the dynamic stress ratio rises, the shear band becomes increasingly visible, eventually spanning the entire specimen, with its inclination angle shifting from 0° to 45° and its width expanding. This increase in dynamic stress amplitude imposes greater cyclic loading on the specimen, promoting shear zone development and expansion. The gradual tilt of the shear band to 45° suggests that, under dynamic loading, the shear zone aligns with the direction of maximum shear stress, while the widening of the shear zone indicates the expansion of the damage region. Increased particle displacement under rising confining pressure shows that higher circumferential pressure significantly enhances relative particle movement. However, this increase in displacement is not accompanied by prominent shear band development, as high confining pressure suppresses shear band formation. The heightened peripheral pressure raises inter-particle contact forces, which facilitate relative displacement but also stabilize the sample as a whole, resulting in more uniform particle interaction. This uniform interaction reduces localized shear stress concentrations, hindering shear band formation. Under high circumferential pressure, particles undergo more deformation, absorbing shear deformation through particle rearrangement, which weakens the localized characteristics of shear bands. This particle deformation and rearrangement mechanism promotes an even distribution of particle displacement, further inhibiting concentrated shear zone development under high confining pressure.

3.4. Porosity Evolution

Using the macroscopic dynamic triaxial test method, this study investigated the porosity evolution characteristics of marine soft soil under varying conditions of water content, dynamic stress ratio, and confining pressure. Figure 12 shows the porosity variation curves with respect to the number of cycles across different water content levels. As loading progresses, porosity within the specimen generally reaches a dynamic equilibrium state after an initial adjustment period, remaining relatively stable as the number of cycles increases. This indicates that particle interactions and arrangements tend toward a stable structural configuration following rapid adaptation in the early stages, leading to minimal porosity changes during subsequent cyclic loading.

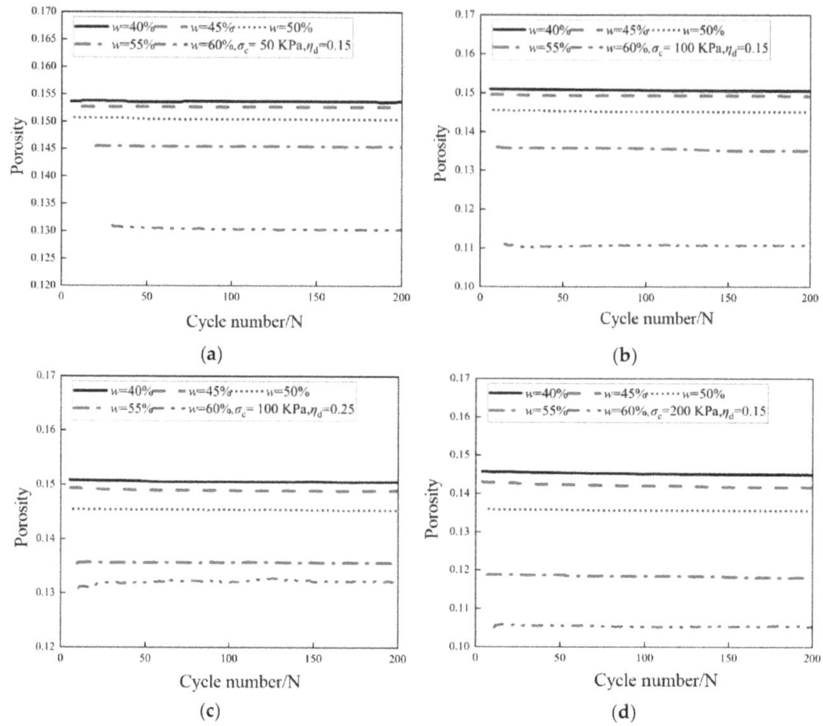

Figure 12. Deformation curve of porosity with cycle times under different water content conditions. (a) σ_c = 50 kPa, η_d = 0.15. (b) σ_c = 100 kPa, η_d = 0.15. (c) σ_c = 100 kPa, η_d = 0.25. (d) σ_c = 200 kPa, η_d = 0.15.

Further analysis of the effect of water content on porosity revealed that as moisture content increased from 40% to 60%, the porosity of the samples gradually decreased, with the magnitude of this decrease becoming more pronounced at higher moisture levels. This trend underscores the critical role of water in granular materials, particularly through its lubricating effects and the enhancement of capillary water bridging. These mechanisms facilitate tighter particle packing, thereby reducing porosity. The basis for this analysis lies in the behavior of the thin water film that forms on the surface of marine soft soil particles as water content begins to increase. This film enhances lubrication between particles, reducing direct friction and facilitating closer particle packing. However, as water content continues to rise, excess water reduces effective particle-to-particle contact. Under cyclic loading, particle rearrangement tends to result in tighter packing at lower water content levels. In contrast, at higher water content, excessive lubrication between particles and the weakening of capillary water bridging effects hinder effective rearrangement and packing.

Figure 13 presents the deformation curve of porosity with respect to the number of cycles. The curve indicates that porosity within the specimen shows a relatively stable, fluctuating downward trend during continuous cyclic loading, suggesting that the particle structure reaches a dynamic equilibrium after an initial phase of rapid adaptation. Notably, the effect of increasing dynamic stress ratio on porosity does not follow a uniform trend under the same confining pressure conditions. Specifically, at a water content of 40%, porosity gradually decreased with increasing dynamic stress ratio. However, at a water content of 45%, while porosity decreased across all dynamic stress ratios, the magnitude of this decrease diminished as the dynamic stress ratio increased.

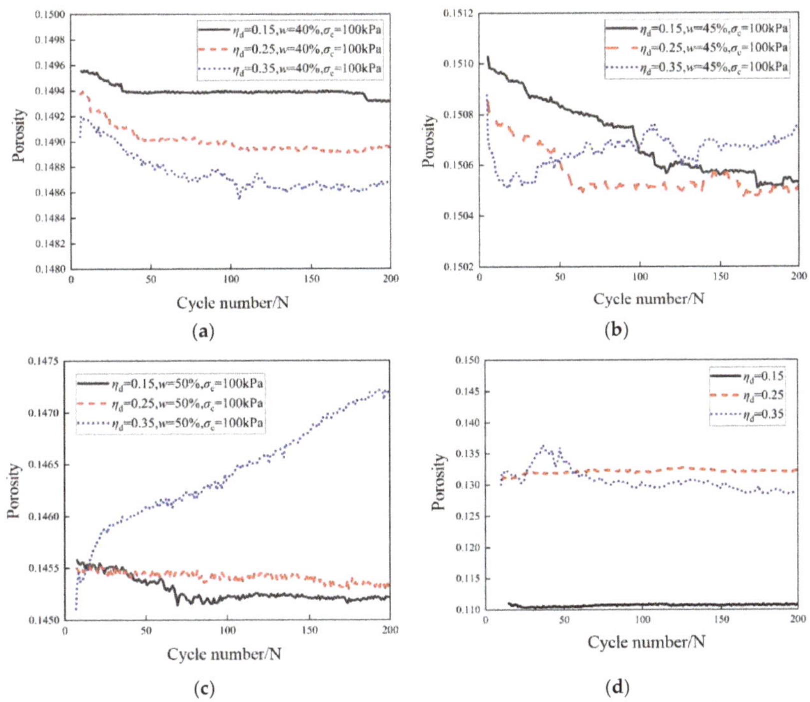

Figure 13. Deformation curve of porosity with number of cycles under different dynamic stress ratios. (**a**) w = 40%, σ_c = 100 kPa. (**b**) w = 45%, σ_c = 100 kPa. (**c**) w = 50%, σ_c = 100 kPa. (**d**) w = 60%, σ_c = 100 kPa.

In particular, at a dynamic stress ratio of 0.35, the porosity was observed to be higher than that of specimens with a dynamic stress ratio of 0.15 in the later stages of cyclic loading. This difference became more pronounced as water content increased, ultimately resulting in consistently higher porosity in specimens with a dynamic stress ratio of 0.35 compared to those with a ratio of 0.15 throughout the cyclic loading process. This phenomenon may be attributed to the more frequent and intense interactions between particles at higher dynamic stress ratios, leading to a saturation of particle rearrangement and structural adjustments, which limits further reductions in porosity.

At higher moisture contents, the presence of water significantly enhances lubrication between particles, reducing inter-particle friction and preventing close particle packing, thereby influencing porosity trends. As the dynamic stress ratio increases, the lubrication layer formed by moisture between particles intensifies due to more frequent interactions and collisions, particularly under high moisture conditions. This reduces direct contact and close packing, thus maintaining or even increasing porosity to some extent.

Figure 14 presents the variation curves of porosity with the number of cycles under different dynamic stress ratios. During continuous cyclic loading, although the porosity within the specimen exhibited a slight fluctuating downward trend, it remained relatively stable overall. This suggests that the internal structure of the granular material gradually adapted to the effects of cyclic loading, reaching a dynamic equilibrium state after an initial phase of rapid adjustment. The gradual decrease in porosity with increasing confining pressure can be explained by two mechanisms: enhanced contact pressure between particles and particle rearrangement leading to densification. Higher confining pressure directly increases the inter-particle pressure, promoting closer packing and thereby reducing porosity. Additionally, particle rearrangement under higher confining pressure further optimizes the stress transfer path between particles, achieving a more compact structure. Under consistent dynamic stress ratios and confining pressure, porosity was observed to decrease progressively with increasing water content, with higher water contents resulting in greater reductions in porosity. This trend is primarily attributed to the enhanced lubrication effect between particles as water content increases. A thicker water film forms on particle surfaces with higher moisture, reducing direct contact and friction, thus facilitating particle rearrangement and tighter packing during loading, which ultimately decreases porosity.

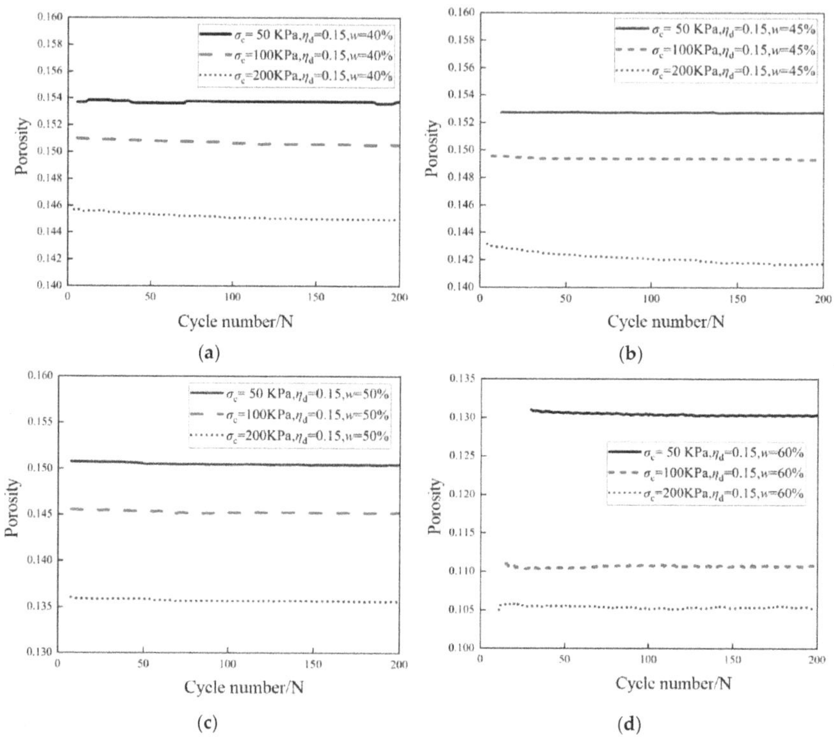

Figure 14. Deformation curve of porosity with number of cycles. (**a**) w = 40%, η_d = 0.15. (**b**) w = 45%, η_d = 0.15. (**c**) w = 50%, η_d = 0.15. (**d**) w = 60%, η_d = 0.15.

4. Conclusions

In coastal cities, engineers and builders face significant challenges due to the presence of marine soft soils when developing underground spaces and constructing rail transit systems. Understanding the mechanical behavior of marine soft soils under dynamic loading is essential for reinforcing these specialized formations or designing structures with enhanced durability in such conditions. This study investigates the morphological characteristics of stress–strain hysteresis curves in marine soft soil under cyclic loading, employing a combination of macroscopic dynamic triaxial testing and microscopic particle flow analysis. The research highlights particle displacement patterns and porosity evolution in marine soft soil under varying conditions of water content, dynamic stress ratio, and confining pressure. Key findings include:

(1) Strain behavior and energy dissipation: A detailed analysis of the first 10 cycles of the hysteresis curve reveals that the strain range of the specimen increases significantly with higher dynamic stress ratios, confining pressures, and water content. The hysteresis loop extends and tilts toward the axial strain direction, with a corresponding increase in the overall loop area, indicating enhanced energy dissipation. This trend

suggests that higher dynamic stress ratios, confining pressures, and water content exacerbate strain softening in marine soft soil during the initial cycles.

(2) Particle displacement and shear band formation: At lower water contents, increased dynamic stress ratios lead to greater particle displacement but do not result in significant shear band formation. However, as water content increases to 50%, particle displacement becomes more pronounced, and initial shear bands begin to form, indicating that moderate water content provides sufficient lubrication while facilitating orderly particle movement. Further increases in water content to 60% result in wider and more pronounced shear zones, signifying extensive specimen damage. Although higher confining pressure enhances particle displacement, it does not lead to marked shear band formation, likely due to the stabilizing effect of high confining pressure, which promotes more uniform stress distribution and suppresses shear band development.

(3) Energy dissipation: As water content increases, energy dissipation in marine soft soil accelerates at lower confining pressures but increases more gradually at higher confining pressures. At higher perimeter pressures, energy dissipation stabilizes, and the influence of water content on energy loss diminishes. Comparison of curves at different pressures shows that at lower pressures, energy dissipation increases more significantly, whereas at higher pressures, the rate of energy loss levels off. Additionally, higher dynamic stress ratios lead to greater energy dissipation in marine soft soils.

(4) Porosity and moisture content: At low moisture content, porosity is minimally affected by the dynamic stress ratio, indicating that insufficient moisture prevents substantial changes in particle interactions. As moisture content reaches moderate levels, however, porosity becomes more responsive to changes in dynamic stress ratio, likely due to moisture-induced lubrication that reduces direct particle contact and promotes pore formation. At high moisture content, porosity change becomes more pronounced, especially with increasing dynamic stress ratio. This is due to increased lubrication between particles, which reduces friction and expands pore spaces. Conversely, higher confining pressure typically reduces porosity by promoting tighter particle packing and reducing pore space.

In practical applications, marine soft soil is subjected to complex environmental conditions and construction-induced disturbances, leading to greater variations in load amplitude, increased surrounding pressures, and other influencing factors. These conditions significantly complicate its mechanical behavior under dynamic loading, making the stress–strain state more prone to instability and failure. While this study simplifies external loading conditions to examine the mechanical characteristics of marine soft soil under cyclic loading, the findings effectively capture the soil's mechanical response and particle transport mechanisms under dynamic conditions. These results provide valuable theoretical parameters and practical insights for the design and construction of engineering projects involving marine soft soil.

Author Contributions: Conceptualization, investigation, resources, K.W.; methodology, writing—original draft preparation, Y.L.; software, writing—review and editing, W.X.; validation, formal analysis, data curation, H.Z.; writing—review and editing, Y.S. All authors have read and agreed to the published version of the manuscript.

Funding: This work was supported by National Natural Science Foundation of China (52179106).

Institutional Review Board Statement: Not applicable.

Informed Consent Statement: Informed consent was obtained from all subjects involved in the study.

Data Availability Statement: The data presented in this study are available on request from the corresponding author.

Conflicts of Interest: The authors declare no conflict of interest.

References

1. Kim, A.R.; Chang, I.; Cho, G.C.; Shim, S.H. Strength and Dynamic Properties of Cement-Mixed Korean Marine Clays. *KSCE J. Civ. Eng.* **2018**, *22*, 1150–1161. [CrossRef]
2. Huancollo, H.J.M.; Saboya, F.; Tibana, S.; McCartney, J.S.; Borges, R.G. Thermal Triaxial Tests to Evaluate Improvement of Soft Marine Clay through Thermal Consolidation. *Geotech. Test. J.* **2023**, *46*, 579–597. [CrossRef]
3. Emmanuel, E.; Anggraini, V.; Raghunandan, M.E.; Asadi, A.; Bouazza, A. Improving the engineering properties of a soft marine clay with forsteritic olivine. *Eur. J. Environ. Civ. Eng.* **2019**, *26*, 519–546. [CrossRef]
4. Nguyen, T.N.; Nguyen, T.D.; Bui, T.S. Geotechnical Properties of Soft Marine Soil at Chan May Port, Vietnam. *Inz. Miner.-J. Pol. Miner. Eng. Soc.* **2021**, *2*, 207–215. [CrossRef]
5. Wu, K.; Xu, W.; Liu, Y.; Liu, D.; Dou, Z.; Liu, Z.; Chen, R.; Hao, D. Mechanical properties of marine soft soil and its influence on urban underground track transport engineering. *J. Water Resour. Archit. Eng.* **2024**, *22*, 122–131.
6. Viens, L.; Bonilla, L.F.; Spica, Z.J.; Nishida, K.; Yamada, T.; Shinohara, M. Nonlinear earthquake response of marine sediments with distributed acoustic sensing. *Geophys. Res. Lett.* **2022**, *49*, e2022GL100122. [CrossRef]
7. Lu, Y.; Fu, W.; Xue, D. Deformation Characteristics of Soft Marine Soil Tested under Cyclic Loading with Low Frequency. *Adv. Civ. Eng.* **2020**, *2020*, 8875315. [CrossRef]
8. Qiao, F.; Bo, J.S.; Qi, W.H.; Wang, L.; Chang, C.Y.; Zhang, Z.P.; Wang, J. Study on the dynamic characteristics of soft soil. *RSC Adv.* **2020**, *10*, 4630–4639. [CrossRef] [PubMed]
9. Ding, Z.; Kong, B.W.; Wei, X.J.; Zhang, M.Y.; Xu, B.L.; Zhao, F.J. Laboratory Testing to Research the Micro-Structure and Dynamic Characteristics of Frozen-Thawed Marine Soft Soil. *J. Mar. Sci. Eng.* **2019**, *7*, 85. [CrossRef]
10. Cai, H.; Zhang, Q.; Ye, G.L. Numerical simulation on undrained cylic triaxial test of soft marine clay considering end restrictions of soil specimen. *Ocean Eng.* **2021**, *216*, 108100. [CrossRef]
11. Tong, J.H.; Wang, Q.; Zhang, W.K.; Guo, L.; Ni, D.Y.; Zhai, J.L.; Yuan, D.Y.; Wang, H.Z. Post-cyclic undrained behavior of marine soft clay subjected to partially drained cyclic loading. *Mar. Georesources Geotechnol.* **2023**, *42*, 898–911. [CrossRef]
12. Wang, Y.; Qi, Z.; Wei, T.Z.; Bao, J.J.; Zhang, X.; Zhou, Y.S. Numerical Study on the Responses of Suction Pile Foundations under Horizontal Cyclic Loading Considering the Soil Stiffness Degradation. *J. Mar. Sci. Eng.* **2023**, *11*, 2336. [CrossRef]
13. Yang, J.Q.; Cui, Z.D.; Xi, B.L.; Song, W.L. Experimental study on cyclic triaxial behaviors of saturated soft soil considering time intermittent and variable confining pressure. *Soil Dyn. Earthq. Eng.* **2024**, *179*, 108508. [CrossRef]
14. Zhou, J.; Liu, C.J.; Zhu, K.F.; Shen, P.P. Macroscopic behaviour and mesoscopic mechanism of shear strength characteristics of pile-soil interface under cyclic loading. *Ocean Eng.* **2024**, *297*, 116998. [CrossRef]
15. Zhao, L.Y.; Shan, Z.G.; Wang, M.Y.; Ni, W.D.; Jin, H.X.; Shi, L. Undrained deformation behavior of marine clay under two-way cyclic loading. *Mar. Georesources Geotechnol.* **2023**, *42*, 465–474. [CrossRef]
16. Wang, X.Y.; Sun, H.; Niu, F.J. Microstructure study of natural marine clay in loading and unloading processes. *Eng. Geol.* **2023**, *327*, 107363. [CrossRef]
17. Li, F.; Zhang, K.Y.; Luo, X.J. Micro-study of soil's inherent anisotropy with PFC simulation. *Appl. Mech. Mater.* **2013**, *275–277*, 383–386.
18. Cao, B. Sensitivity analysis of loess triaxial mesoscopic parameters based on PFC3D. In Proceedings of the 6th International Conference on Energy Science and Chemical Engineering, Dali, China, 17–19 July 2020; Volume 565, p. 012070.
19. Lin, C.-X.; Ling, D.-S.; Zhong, S.-Y. Application of particle flow code numerical simulation in research of geotechnical behavior of lunar soil. *J. Zhejiang Univ. (Eng. Sci.)* **2015**, *49*, 1679–1691.
20. *GB/T50123-2019*; Standard for Geotechnical Testing Method. National Standard of the People's Republic of China: Beijing, China, 2019.
21. *ASTM D7181-11*; Method for Consolidated Drained Triaxial Compression Test for Soils. American Society for Testing and Materials: Conshohocken, PA, USA, 2011.
22. Xiao, W.; Wu, K.; Xu, W.; Liu, Y.; Lu, H.; Chen, R. Experiment and analysis on dynamic characteristics of marine soft clay. *Mar. Georesources Geotechnol.* **2024**, *5*, 1–22. [CrossRef]
23. Ni, X.; Yin, X.; Cai, Z. Study on the meso-mechanism of consolidation of soft soil based on PFC-COMSOL. *J. Hebei Univ. Eng. (Nat. Sci. Ed.)* **2017**, *34*, 30–36.
24. Ji, S.T.; Karlovsek, J. Optimized differential evolution algorithm for solving DEM material calibration problem. *Eng. Comput.* **2023**, *39*, 2001–2016. [CrossRef]
25. Geng, X. Analysis of Microscopic Damage Mechanism of Marine Silt Cement-solidified Soil Based on PFC3D. Master's Thesis, Wuhan University of Engineering, Wuhan, China, 2022.
26. Tang, Y.; Chen, H.J.; Xiong, L.X.; Xu, Z.Y. Conventional triaxial loading and unloading test and PFC numerical simulation of rock with single fracture. *Arch. Civ. Eng.* **2024**, *70*, 233–254. [CrossRef]

Journal of
*Marine Science
and Engineering*

MDPI

Article

Centrifugal Test Study on the Vertical Uplift Capacity of Single-Cylinder Foundation in High-Sensitivity Marine Soil

Mingzhe Wei [1], Yanghui Ye [2], Wei Zhao [1,*], Zehao Wang [1], Fuhao Ge [1] and Tingkai Nian [1,*]

1 State Key Laboratory of Coastal and Offshore Engineering, Dalian University of Technology, Dalian 116024, China; wmz724@mail.dlut.edu.cn (M.W.); wang_zh@mail.dlut.edu.cn (Z.W.); 2544407283@mail.dlut.edu.cn (F.G.)
2 Economic and Technical Research Institute, State Grid Jiangxi Electric Power Co., Ltd., Nanchang 330096, China; jxlnyyh@163.com
* Correspondence: zhaoweiyt@163.com (W.Z.); tknian@dlut.edu.cn (T.N.)

Abstract: Offshore wind power is a new type of clean energy with broad development prospects. Accurate analysis of the uplift capacity of offshore wind turbine foundations is a crucial prerequisite for ensuring the safe operation of wind turbines under complex hydrodynamic conditions. However, current research on the uplift capacity of suction caissons often neglects the high-sensitivity characteristics of marine soils. Therefore, this paper first employs the freeze–thaw cycling procedure to prepare high-sensitivity saturated clay. Subsequently, a single−tube foundation for wind turbines is constructed within a centrifuge through a penetration approach. Ten sets of centrifuge model tests with vertical cyclic pullout are conducted. Through comparative analysis, this study explores the pullout capacity and its variation patterns of suction caisson foundations in clay with different sensitivities under cyclic loading. This research indicates the following: (1) The preparation of high-sensitivity soil through the freeze−thaw procedure is reliable; (2) the uplift capacity of suction caissons in high−sensitivity soil rapidly decreases with increasing numbers of cyclic loads and then tends to stabilize. The cumulative displacement rate of suction caissons in high-sensitivity soil is fast, and the total number of pressure–pullout cycles required to reach non-cumulative displacement is significantly smaller than that in low-sensitivity soil; (3) the vertical cyclic loading times and stiffness evolution patterns of single-tube foundations, considering the influence of sensitivity, have been analyzed. It was found that the secant stiffness exhibits a logarithmic function relationship with both the number of cycles and sensitivity. The findings of this study provide assistance and support for the design of suction caissons in high-sensitivity soils.

Keywords: offshore wind turbines; freeze–thaw; high-sensitivity soil; centrifuge model test; cyclic loading

1. Introduction

Wind, solar, and hydropower are currently the most advanced renewable energy sources, with wind energy projected to experience the fastest growth in production among renewables [1]. Since 2020, offshore wind energy has gradually become a key strategic industry within the global wind sector [2]. As the energy landscape shifts from land-based to oceanic sources, seabed stability assessment has emerged as a critical research area [3]. For example, Nian et al. have explored earthquake-induced seabed instability and examined how temperature affects the rheological properties of mudflows [4,5]. Common offshore wind turbine foundations include monopile, gravity-based, jacket, and floating foundations [6,7]. However, as wind power moves toward deeper waters, monopile foundations face challenges such as escalating costs and construction complexities [8]. In contrast, cylindrical foundations, which feature lower production costs, simpler design, and recyclability, are more adaptable to complex marine hydrodynamic conditions and varying seabed strata [9], leading to extensive research and application in recent years [10–12].

The application of cylindrical foundations dates back to 2002, when Denmark first installed a cylindrical foundation with a diameter of 12 m and a height of 6 m [13]. A key design consideration for these foundations is their response under cyclic loading. Kou et al. conducted indoor model tests to investigate the displacement, rotation angle, and rotation center variation characteristics of cylindrical foundations with suction caissons under different horizontal cyclic loads [14]. Lu utilized centrifuge model tests to investigate the deformation and bearing characteristics of cylindrical foundations in sandy soil under unidirectional cyclic loading [15]. Jiao conducted centrifuge model tests to study the development of pore pressure in clayey soil foundations under cyclic loading, revealing that the depth of soil liquefaction increases with the amplitude of cyclic loading. Moreover, based on pore pressure data, the author further examined the effects of cyclic loading on the bearing capacity of cylindrical foundations [16]. All in all, the bearing capacity and cyclic displacement of single−cylinder foundations have been examined mainly by scaled model tests [17–19], finite element simulations [20–23], and limiting analyses [24,25]. Some have examined the effects of foundation geometry, soil conditions, and load direction on the bearing capacity of cylinder foundations [26–28]. There are also some scholars who have studied the effects of load magnitude, load mean, load direction, and other factors on the cyclic displacement of the cylinder base [29]. The sensitivity of marine soils may be high due to rapid sedimentation on the seafloor, where the superporous water pressure within the sediments does not dissipate in time [30–32]. Although numerous studies have examined cylindrical foundations, few have addressed the high sensitivity of marine soils. There has been limited research into the effects of soil softening in high-sensitivity soils under cyclic loading, and design methods often fail to account for the influence of sensitivity on soil strength [33,34]. The settlement and sliding failure of the Phase II regulation structures at the Yangtze River Estuary in Shanghai was attributed to inadequate consideration of the high sensitivity of marine soil [35]. Therefore, it is crucial to account for the high sensitivity of marine clay in research and design processes.

Therefore, this paper proposes the use of a freeze–thaw procedure to prepare high-sensitivity saturated clay and obtain the undisturbed strength c_{ui} and remolded strength c_{ud} through vane shear tests to calculate the sensitivity $S_{\mathrm{t}} = c_{\mathrm{ui}}/c_{\mathrm{ud}}$. Subsequently, in a centrifuge, the actuator is controlled for the purpose of installing a single-cylinder foundation for wind turbines through penetration. Ten centrifuge model tests of vertical cyclic pullout were conducted. Through comparative analysis, this study explores the pullout capacity and its variation patterns of suction caisson foundations in clay with different sensitivities under cyclic loading. Additionally, a logarithmic function relationship is proposed between the secant stiffness of hysteresis loops—associated with cyclic loading and cumulative displacement—and clay sensitivity, as well as the number of loading cycles. The findings offer valuable insights to support future experimental research and engineering design applications.

2. Experimental Methods

2.1. Preparation of High-Sensitivity Soil for Centrifuge Testing

The raw material, consisting of ball clay or kaolin, is manually mixed with an appropriate amount of water until it reaches a specific consistency. It is then transferred into a vacuum mixing tank for thorough blending, ensuring that no large agglomerates remain in the soil slurry. Simultaneously, a vacuum pump evacuates air, maximizing the saturation of the slurry [36]. The well-mixed slurry is then carefully poured, layer by layer, into a standard laboratory mold lined with filter paper, geotextile, and a pressure plate. The top surface of the slurry is also covered with filter paper, geotextile, and a pressure plate to facilitate normal consolidation. Consolidation pressure is applied using standard weights placed on top of the mold. To maintain constant saturation throughout the process, a water layer is kept on the surface, and a dial indicator records soil settlement data to monitor consolidation.

Once the soil achieves normal consolidation strength, following the consolidation process, the method by Nian et al. on the effects of temperature changes on soil strength [37] is referenced. Freeze–thaw cycling is applied to regenerate the soil structure and enhance its sensitivity. This approach is used to prepare the high−sensitivity marine saturated soil needed for the experiment. A sensitivity range of 4 to 8 is considered as high sensitivity [38].

By subjecting the clay to different temperature variations and multiple freeze–thaw cycles, electric cruciform shear tests, and making corrections to the data [39], the undisturbed strength and remolded strength of the specimens were obtained; then, the sensitivity was calculated based on the results. A comparative analysis of the clay's sensitivity before and after freeze–thaw cycling was conducted, as shown in Table 1. The results indicate a significant increase in sensitivity post-freeze–thaw treatment. The soil samples, both pre- and post-freeze–thaw, were sectioned, sublimation-dried, metal-coated, and scanned using an electron microscope to observe the microstructural changes. The analysis revealed that, following the freeze–thaw process, the number of small pores in the soil decreased, while the number of large pores increased, as illustrated in Figure 1.

Table 1. Table of Sensitivity Changes in Clay Before and After Freeze–Thaw Cycles.

Freezing Temperature	Number of Cycle	S_t (Before Freezing and Thawing)	S_t (After Freezing and Thawing)	Growth Rate of Sensitivity Before and After Freeze–Thaw
−10 °C	1	1.86	5.44	+192.5%
−20 °C	1	1.48	4.97	+235.8%
−10 °C	5	1.94	7.98	+311.3%
−20 °C	5	2.07	8.16	+294.2%

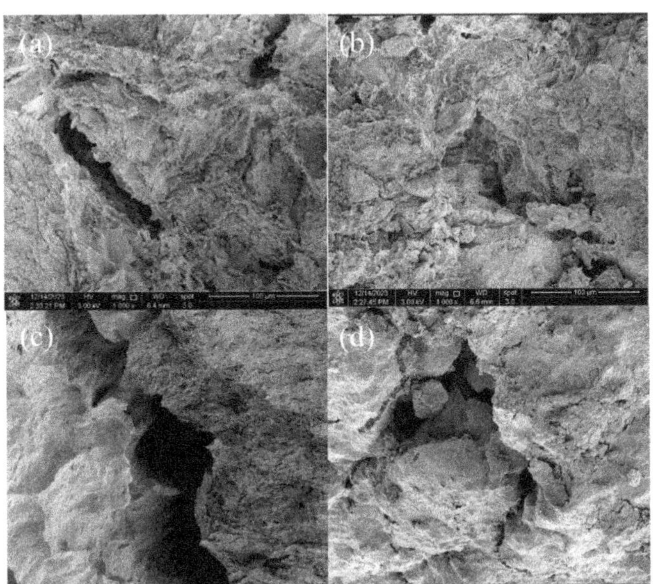

Figure 1. Microscopic images of kaolin and ball clay before and after freeze–thaw cycles: (**a**) microscopic image of ball clay before freeze–thaw; (**b**) microscopic image of kaolin before freeze–thaw; (**c**) microscopic image of ball clay after freeze–thaw; (**d**) microscopic image of kaolin after freeze–thaw.

The large pores observed after the freeze–thaw process are formed due to the suction force exerted by the freezing front on the water in the unfrozen region. The temperature differential creates matric suction in the unfrozen zone, causing pore water to migrate from warmer areas to colder ones. As the water accumulates at the freezing front, it freezes and expands into ice crystals, ultimately leading to the coalescence of smaller pores into larger ones. This transformation results in increased soil sensitivity.

Using this freeze–thaw cycling method, using kaolin clay, and with a fixed consolidation strength of 10 kPa, high-sensitivity marine saturated soil for centrifuge testing was prepared by setting different consolidation times and freeze–thaw cycles. Following the preparation, vane shear tests were conducted to measure the strength and sensitivity of the soil. The control group consisted of low-sensitivity soil, while the experimental group featured high-sensitivity soil produced via the freeze–thaw method. The relevant soil parameters are detailed in Table 2.

Table 2. Soil Strength Parameter Table.

Category	c_{ui} (kPa)	c_{ud} (kPa)	S_t
High-sensitivity soil A_1	8.08	1.11	7.31
High-sensitivity soil B_1	7.68	1.14	6.73
High-sensitivity soil C_1	8.45	1.24	6.81
High-sensitivity soil D_1	8.83	1.26	7.01
Low-sensitivity soil A_2	9.01	4.29	2.10
Low-sensitivity soil B_2	7.69	4.07	1.89
Low-sensitivity soil C_2	9.18	3.96	2.32
Low-sensitivity soil D_2	8.46	3.72	2.27

2.2. Design of the Centrifuge Test Scheme

In practical engineering, it is common to place multiple suction caisson foundations within a specific area. To ensure the safe implementation of such foundations in real-world projects, particular attention is given to investigating their impact on soil properties under relevant test conditions. According to relevant research [40], the following assumptions are typically made:

(1) Wind, wave, and other environmental loads are uniformly distributed across all suction caisson foundations within the same marine area.

(2) The overall bending moment acting on the foundation is resisted by the combined uplift and compression capacities of each suction caisson.

(3) Regarding the foundation of a multi-bucket structure, the bending moment loads on each bucket are borne by the underlying soil. The magnitude of these loads depends on factors such as the footprint of the jacket structure, the structural weight, and environmental loads. When investigating the vertical extraction of suction buckets, the bending moment loads of the individual buckets are neglected.

The prototype monopole selected for the test has a diameter of 4 m, a length of 4 m, and a thickness of 0.04 m for both the top and sidewalls. In the centrifuge model test, an acceleration value of 100 g was used, which allowed for the calculation of the model monopole dimensions: a diameter of 40 mm, a length of 40 mm, and a thickness of 4 mm. The top of the monopole is equipped with a 5 mm diameter drainage valve and is connected to the centrifuge loading rod using a threaded rod with a length of 15 mm and a diameter of 8 mm. The centrifuge device is equipped with an actuator capable of radial movement up to 600 mm and circumferential movement up to 180°, as shown in Figure 2a. The purpose of compression, extraction, and cyclic loading is achieved by externally setting relevant force control parameters and controlling the radial movement of the actuator while the centrifuge is rotating at high speed. The position of the centrifuge test model box, suction cylinder, and sensors is shown in Figure 2b. The arrangement of the vane shear test and suction cylinder placement points is shown in Figure 2c. The penetration, monotonic, and cyclic loading processes are illustrated in the schematic diagram in Figure 2d.

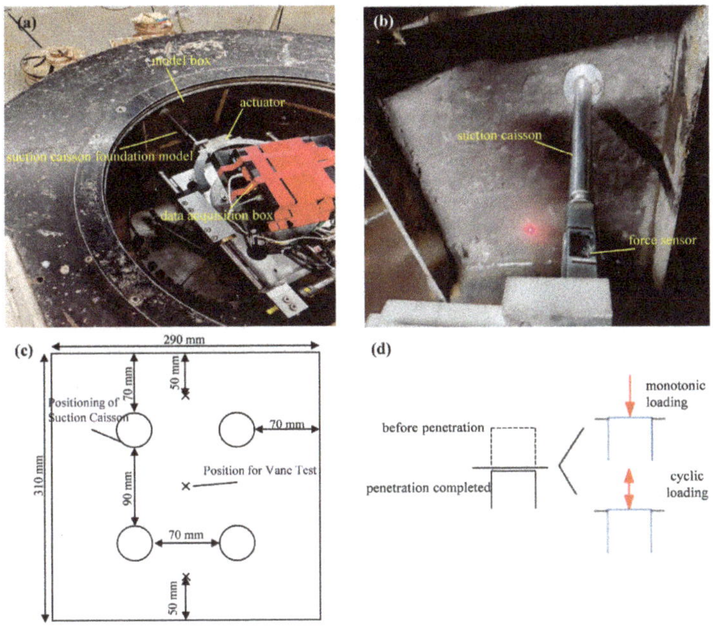

Figure 2. Centrifuge test: (**a**) centrifugal model test device; (**b**) photo of test model box and suction cylinder; (**c**) point layout drawing; (**d**) schematic diagram of penetration, monotonic, and cyclic loading.

Due to the randomness and unpredictability of horizontal cyclic loads, such as those caused by wind, waves, and currents, these loads are typically characterized by their amplitude and mean value. Firstly, the direction and symmetry characteristics of the load are determined by analyzing the relative magnitudes of the amplitude and mean value. Next, the load level is evaluated by comparing the ratio of the amplitude and mean value to the static bearing capacity. To account for the randomness of cyclic loads, they are often simplified into a series of load sequences with varying amplitudes in design. These sequences are then equivalently transformed, based on the principle of strain accumulation, into loads corresponding to the maximum amplitude within the sequence, along with an equivalent number of cycles [41]. Cyclic loads are classified into four basic types based on their direction and symmetry, with bidirectional symmetric cyclic loads being considered the most critical loading condition [42,43]. Therefore, bidirectional symmetric cyclic loading was selected for the centrifuge simulation test.

In the centrifuge model tests, to minimize boundary effects and avoid interactions between different points, the distance between the suction caisson and the edges of the model box was kept at least 1.8D, and the distance between suction caissons was ensured to be at least 2D (where D is the diameter of the caisson). Each soil box had four test locations, with the arrangement of suction caisson locations and vane shear test locations shown in Figure 2c. Out of the four test locations in each soil box, one was designated to determine the ultimate bearing capacity, while the other three were used for cyclic penetration tests.

2.3. Centrifuge Test Procedures

The centrifuge model test was conducted in the 450 g·t drum-type centrifuge at Dalian University of Technology, as shown in Figure 2a. The test acceleration used in this paper was 100 g, and the data acquisition frequency was 40–50 Hz [44,45]. The main test procedures were divided into the following 6 steps.

2.3.1. Preliminary Inspection and Equipment Installation

The suction caisson was installed, and all sensors and the data acquisition system were checked to ensure proper functioning. The actuator's direction was adjusted to confirm that the suction caisson was perpendicular to the soil surface.

2.3.2. Suction Caisson Foundation Installation and Positioning

The centrifuge was gradually accelerated to 100 g in increments of 20 g, maintaining each step for at least one minute. Upon reaching 100 g, the actuator was controlled to penetrate the suction caisson radially at a speed of 10 mm/s (given the suction cylinder's length of 40 mm, to ensure a more precise and controlled penetration to the top of the suction cylinder during the installation process, a speed of 10 mm/s is employed for a duration of 4 s) until the top of the foundation was level with the soil surface. During penetration, the valve at the top of the suction caisson was kept open to release air and water from the caisson. After penetration, the centrifuge was stopped.

2.3.3. Determination of Ultimate Bearing Capacity

After the initial penetration, the suction caisson's valve was closed, and the surrounding soil was inspected for gaps. The centrifuge was then re-accelerated to 100 g in the same incremental manner. Once stabilized, the actuator continued penetrating the soil at 1 mm/s for an additional 10 mm (0.25 times the length of the suction caisson). The resistance at this depth was considered the vertical bearing capacity V_0.

2.3.4. Determination of Cyclic Frequency

The cyclic frequency was chosen based on the load application rate. Undrained conditions were more easily achieved when the cycle duration was less than 2 s. The cyclic test frequency generally ranged from 0.2 to 0.7 Hz, representing different loading rates. Given the limitations of the centrifuge's loading precision, a frequency range between 0.4 and 0.8 was applied, with cycle durations between 1.25 and 2.5 s. Setting the test frequency between 0.4 and 0.5 Hz met both the undrained condition and equipment precision requirements.

2.3.5. Vertical Cyclic Loading Test

After determining the ultimate bearing capacity, the centrifuge was gradually slowed and stopped. The suction caisson was then positioned at the cyclic test location, and the centrifuge was accelerated uniformly back to 100 g. Once stabilized, the ultimate bearing capacity V_0, cyclic frequency (0.3–0.5 Hz), maximum and minimum cyclic loads, and the force-controlled loading frequency were input into the control built-in program for cyclic press–pull testing. The test was terminated once the preset cyclic termination conditions were met, and the process was repeated at other locations.

2.3.6. Termination Conditions for Cyclic Loading

There are two ways to determine the termination conditions for cyclic loading:

(1) The test could be terminated when the amplitude and average value of the vertical cyclic displacement of the suction caisson stabilized within a certain range, indicating no further displacement accumulation.

(2) If the sum of the amplitude and average displacement reached one-fourth of the caisson's length, the suction caisson was considered to have been pulled out, and the test was terminated.

The maximum cyclic load was calculated as the sum of the average load and the amplitude, while the minimum load was the difference between the average and the amplitude. In each test, the load amplitude ratio V_C/V_0 remained constant as the number of cycles increased (where V_0 is the ultimate bearing capacity, V_C is the load amplitude, and V_a is the average load value). The ratio V_C/V_0 ranged from 0.37 to 0.64, while V_a/V_0 ranged from 0.01 to 0.05, considering the inherent errors in force-controlled cycling.

3. Experimental Results and Analysis

The entire centrifuge test primarily focuses on the response in terms of load and displacement. The force sensor is placed between the suction bucket and the actuator, serving to connect them. The laser displacement sensor is positioned below the force sensor and is also connected to the actuator. Throughout the test, force sensors are used to record the cyclic force conditions, and displacement sensors are used to record displacement data. Finally, a relationship diagram between cyclic load and displacement is compiled. By defining the secant stiffness of the hysteresis loop, which represents the relationship between cyclic load and displacement changes, and plotting its scatter plot, a functional relationship between the secant stiffness, the number of cycles, and the soil sensitivity is fitted.

3.1. Experimental Relationship Between Normalized Vertical Displacement and Number of Cycles

By defining one cylinder length as the unit length, the normalized vertical displacement is calculated as the ratio of the cylinder's vertical displacement to the unit length L. During the test, the position changes in the suction caisson were recorded each time the set pressure and tension values were reached. Based on these recorded values, an experimental relationship between normalized vertical displacement and the number of cycles was plotted. In the diagram, tension is designated as positive and compression as negative, as illustrated in all parts of Figure 3.

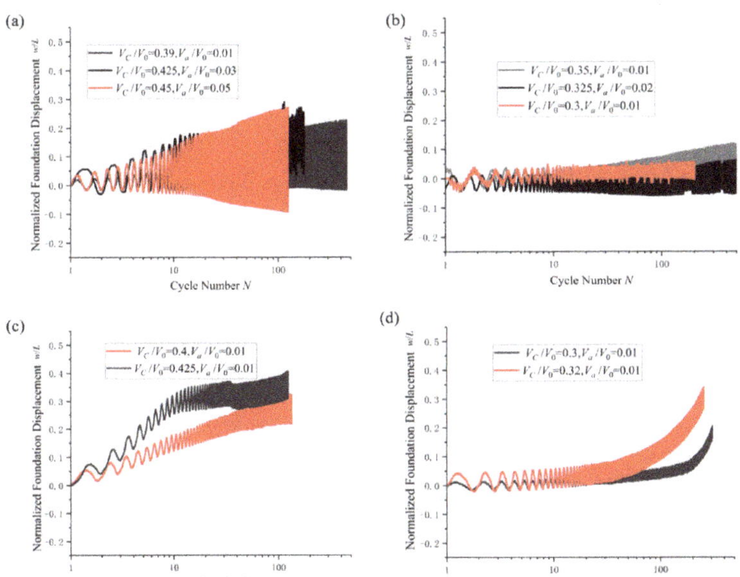

Figure 3. The experimental relationship between the normalized vertical displacement w/L and the cycle number N: (**a**) normalized vertical displacements versus number of cycles for low-sensitivity soil at $V_C/V_0 = 0.39$, $V_a/V_0 = 0.01$; $V_C/V_0 = 0.425$, $V_a/V_0 = 0.03$; $V_C/V_0 = 0.45$ and $V_a/V_0 = 0.05$ condition; (**b**) normalized vertical displacements versus number of cycles for low-sensitivity soil at $V_C/V_0 = 0.35$, $V_a/V_0 = 0.01$; $V_C/V_0 = 0.325$, $V_a/V_0 = 0.02$; $V_C/V_0 = 0.3$ and $V_a/V_0 = 0.01$ condition; (**c**) normalized vertical displacements versus number of cycles for highly sensitive soils at $V_C/V_0 = 0.4$, $V_a/V_0 = 0.01$; $V_C/V_0 = 0.425$, $V_a/V_0 = 0.01$ conditions; (**d**) normalized vertical displacements versus number of cycles for highly sensitive soils at $V_C/V_0 = 0.3$, $V_a/V_0 = 0.01$; $V_C/V_0 = 0.32$, $V_a/V_0 = 0.01$ conditions.

As shown in Figure 3, the relationship depicted is between the normalized vertical displacement w/L and the number of cycles N. The cycling process stops when the displacement reaches one-quarter of the cylinder length, at which point the cylinder is considered to have been pulled out. The upward bearing capacity of the soil during cycling is generally lower than its downward bearing capacity. As a result, under the same vertical force, the upward and downward bearing capacities of the cylinder change due to soil softening and soil structural modifications during continuous cycling. This leads to the upward displacement being greater than the downward displacement, ultimately causing the extraction of the cylinder.

In high-sensitivity soils, where the pullout bearing capacity is lower than the compression bearing capacity, the outward displacement of the cylinder continues to accumulate with each cycle. This accumulation causes the cylinder to progressively move outward, and the compression position changes constantly, making it impossible to maintain a stable compression range. In contrast, normally consolidated low-sensitivity soils show a more stable compression position during cycling and may even experience further penetration in some cases. From the perspective of forces acting on the cylinder during pullout, the primary forces involved are end resistance and frictional resistance of the soil. Since the frictional resistance remains the same during both compression and pullout, the end resistance is closely tied to the soil's strength. During cyclic loading, the soil undergoes disturbance and softening, leading to a reduction in strength. The sensitivity of the soil reflects the degree of structural integrity and the rate of strength decay after being disturbed. Therefore, under cyclic loading, high-sensitivity soils experience a rapid decrease in strength, and the soil at the cylinder location is unable to provide adequate bearing capacity. Consequently, the accumulation of upward and downward displacements is more pronounced in high-sensitivity soils compared to the control group. Furthermore, the rapid strength degradation in high-sensitivity soils after disturbance results in a pullout rate that far exceeds that of low-sensitivity soil. This is why the number of cycles for high-sensitivity soils is generally lower than that for low-sensitivity soils, and the softening rate is much faster. In summary, sensitivity is a critical factor in the response to cyclic loading, significantly influencing both displacement accumulation and bearing capacity.

3.2. Relationship Between Normalized Cyclic Loading and Normalized Displacement

By recording the position changes and vertical load magnitudes of the suction cylinder at each instance when the set pressure and tension values were reached during the experiments, a relationship between normalized vertical load and normalized vertical displacement was established (as shown in Figure 4).

Figure 4 shows the relationship between normalized cyclic loading and normalized vertical displacement for low-sensitivity soil, also known as the hysteresis loop. When the load amplitude is high, the maximum displacement of the cylinder reaches $0.25L$ within 100 cycles, as shown in Figure 4c,f. When the load amplitude is lower, the maximum displacement does not reach $0.25L$, even after 400–500 cycles, as seen in Figure 4b,d. To clearly demonstrate the change in the hysteresis loop with the number of cycles, for groups with a larger number of experimental cycles, typical hysteresis loops at different loading stages are provided in the figures. From Figure 4a,b, it can be observed that the upward displacement of the single cylinder accumulates significantly with increasing cycle number, while the downward displacement decreases slightly with increasing cycle number. In contrast to Figure 4c, the downward displacement of the single cylinder also accumulates, but the accumulation is less than that of the upward displacement. This is because the average load V_a/V_0 in Figure 4c, which is 0.05, is slightly greater than the average load V_a/V_0 in Figure 4a,b, which is 0.01. The slightly higher pressure compared to tension exacerbates the accumulation of downward displacement.

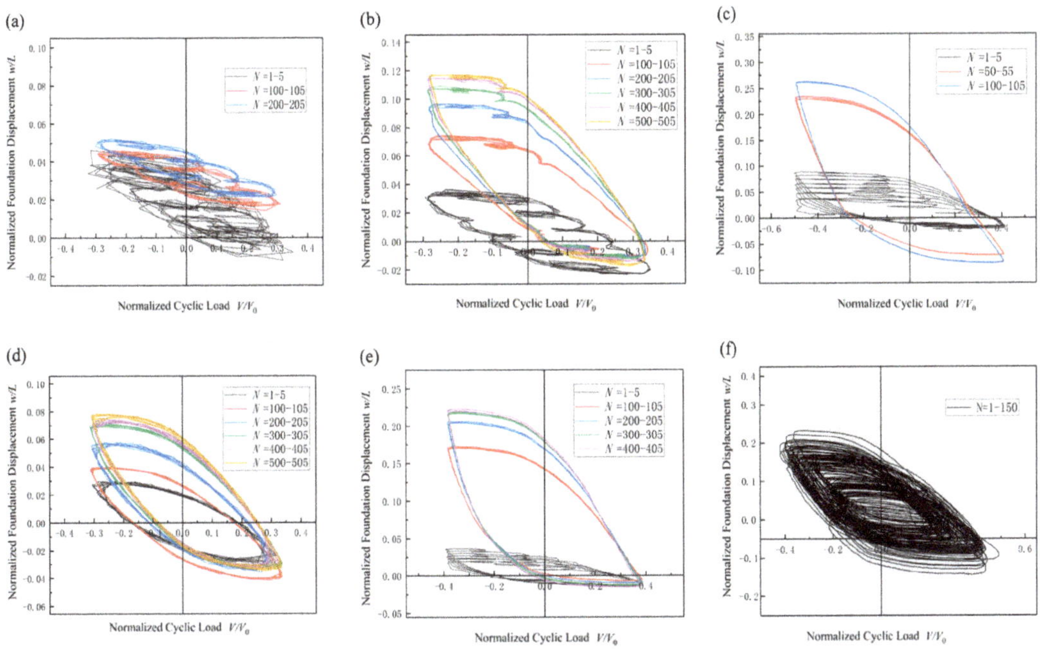

Figure 4. Normalized cyclic load and normalized displacement relationship diagram of low-sensitivity soil: (**a**) normalized cyclic load versus normalized vertical displacement for $V_C/V_0 = 0.3$ and $V_a/V_0 = 0.01$; (**b**) normalized cyclic load versus normalized vertical displacement for $V_C/V_0 = 0.35$ and $V_a/V_0 = 0.02$; (**c**) normalized cyclic load versus normalized vertical displacement for $V_C/V_0 = 0.45$ and $V_a/V_0 = 0.05$; (**d**) normalized cyclic load versus normalized vertical displacement for $V_C/V_0 = 0.325$ and $V_a/V_0 = 0.01$; (**e**) normalized cyclic load versus normalized vertical displacement for $V_C/V_0 = 0.39$ and $V_a/V_0 = 0.01$; (**f**) normalized cyclic load versus normalized vertical displacement for $V_C/V_0 = 0.425$ and $V_a/V_0 = 0.03$.

Figure 5 illustrates the relationship between normalized cyclic loading and normalized vertical displacement for high-sensitivity soil. The slope of the line connecting the highest and lowest points of the Nth hysteresis loop is defined as the secant stiffness K_N of the Nth hysteresis loop for the single cylinder, as shown in Figure 6. This secant stiffness can be obtained by calculating the ratio of the change in vertical load to the change in vertical displacement within that cycle:

$$K_N = \frac{V_{\max} - V_{min}}{W_{\max} - W_{min}} \quad (1)$$

where $V_{\max}$ and $V_{\min}$ are maximum and minimum values of vertical loads, respectively; $W_{\max}$ and $W_{\min}$ are maximum and minimum values of vertical displacement, respectively. By further segmenting the hysteresis loops representing the cyclic loading–displacement relationship for both low-sensitivity soil and high−sensitivity soil, their secant stiffnesses are calculated and plotted as scatter plots, as shown in Figure 7.

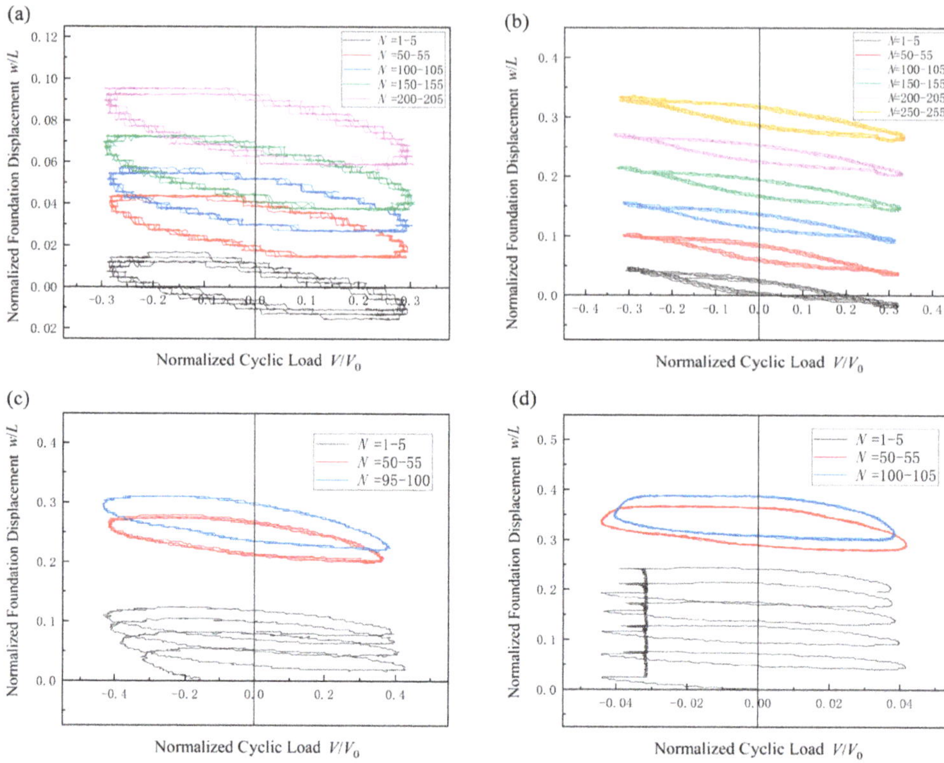

Figure 5. Normalized cyclic load and normalized displacement relationship diagram of high-sensitivity soil: (**a**) normalized cyclic load versus normalized vertical displacement for $V_C/V_0 = 0.3$; $V_a/V_0 = 0.01$; (**b**) normalized cyclic load versus normalized vertical displacement for $V_C/V_0 = 0.32$; $V_a/V_0 = 0.01$; (**c**) normalized cyclic load versus normalized vertical displacement for $V_C/V_0 = 0.4$; $V_a/V_0 = 0.01$; (**d**) normalized cyclic load versus normalized vertical displacement for $V_C/V_0 = 0.425$; $V_a/V_0 = 0.01$.

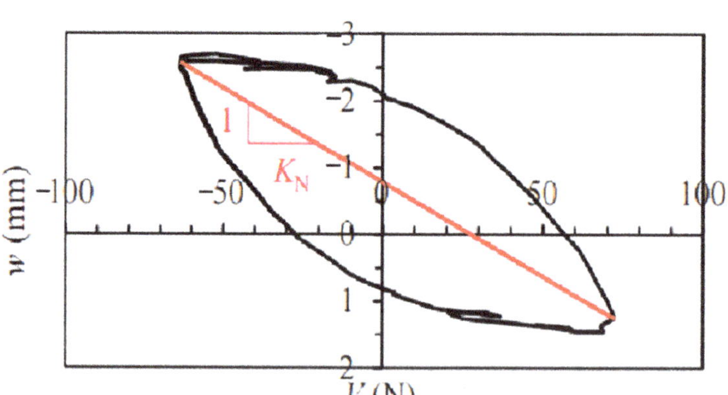

Figure 6. Definition of secant stiffness.

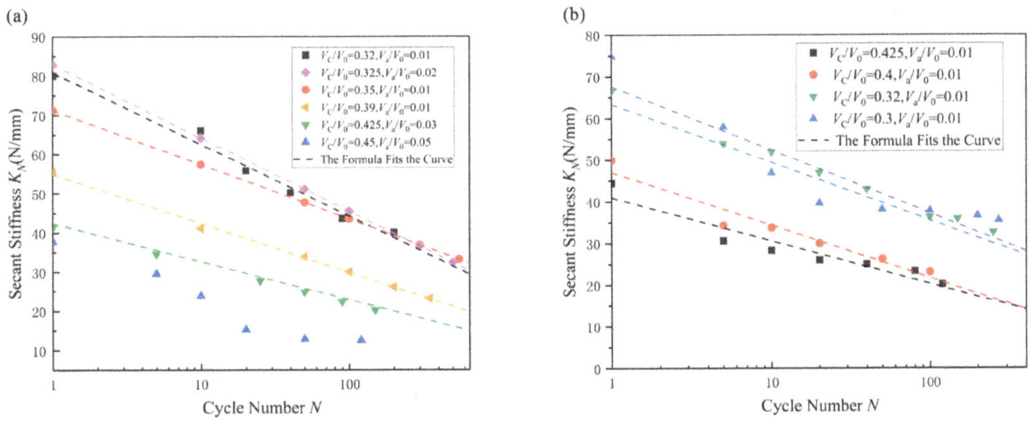

Figure 7. Scatter plot of secant stiffness: (**a**) scatter plot of cut-line stiffness of low-sensitivity soil; (**b**) scatter plot of stiffness of high-sensitivity earth cut line.

Based on Figure 7, it can be observed that the hysteresis loop stiffness K_N decreases as V_C/V_0 increases, and K_N also decreases as the number of cycles N increases [46]. The relationship between the cyclic secant stiffness K_N and the number of cycles N can be fitted as follows:

$$K_N = K_1 + A_k lnN \tag{2}$$

where K_1 and A_k are the coefficient variables. Next, by fitting the load magnitude and initial secant stiffness for both low-sensitivity soil and high-sensitivity soil, the following parameters can be obtained:

$$K_1 = (78.06\, lnS_t - 381.91)\zeta_b + 213.37{S_t}^{-0.206} \tag{3}$$

where ζ_b is load magnitude, S_t is level of sensitivity, the R-squared (R^2) value for the data fit with low-sensitivity soil is 94.9%, and the R-squared (R^2) value for the data fit with high-sensitivity soil is 98.3%. Furthermore, the parameter A_k can be estimated and fitted with the load magnitude to obtain the following relationship:

$$A_k = \left(54.33{S_t}^{-0.688}\right)\zeta_b + 6.045\, lnS_t - 22.212 \tag{4}$$

The R-squared (R^2) value for the data fit with low-sensitivity soil is 92.3%, and the R-squared (R^2) value for the data fit with high-sensitivity soil is 83.7%. Substituting Equations (3) and (4) into Equation (2), the relationship between the secant stiffness K_N, the number of cycles N, and the sensitivity S_t can be derived; ultimately, we arrive at Equation (5).

$$\begin{aligned} K_N = & \left[(78.06\, lnS_t - 381.91)\zeta_b + 213.37{S_t}^{-0.206}\right] \\ & + \left[\left(54.33{S_t}^{-0.688}\right)\zeta_b + 6.045\, lnS_t - 22.212\right] \times lnN \end{aligned} \tag{5}$$

Finally, the data were substituted into the formula to obtain the dashed straight line in Figure 7. Comparing it with the original data scatter plot, it can be seen that the error is not significant.

4. Conclusions

This study utilized the freeze–thaw method to prepare high-sensitivity soil and conducted centrifuge model tests on vertically symmetrical cyclic loading of offshore wind turbine suction caissons. The high-sensitivity soil served as the experimental group, while low-sensitivity soil was used as the control group. The primary goal was to compare and investigate the cyclic load–displacement response under vertically symmetrical loading, focusing on the effect of soil sensitivity. The following conclusions were drawn from the experiments:

- Under symmetric cyclic loading, the uplift bearing capacity of the suction caisson was found to be lower than its compressive bearing capacity. As the number of cycles increased, the caisson continued to be pulled upward. The average load value was influenced by both the cyclic amplitude and soil sensitivity. Higher cyclic amplitude and greater soil sensitivity resulted in larger average load values. Soils with high sensitivity exhibited stronger structural properties, contributing to higher average load values.
- During the experiments, high-sensitivity soils showed a rapid reduction in strength under the disturbance of cyclic loading, causing a swift decrease in both uplift and compressive bearing capacities. As a result, the suction caisson experienced greater displacement to reach the set load value. In contrast, the low-sensitivity soil group exhibited minimal strength degradation, resulting in a higher number of cycles and a longer displacement accumulation period. The compressive position remained stable within a certain range, and a near-steady state was observed under low cyclic amplitudes.
- Secant stiffness, defined as the secant slope of the hysteresis loop relating normalized cyclic load to normalized displacement, was investigated. The vertical cumulative displacement and stiffness evolution of the suction caisson were examined, taking into account the influence of soil sensitivity. A logarithmic function relationship between secant stiffness, number of cycles, and sensitivity was established.
- The centrifuge test was accelerated to a maximum of 100 g, simulating long-duration cyclic loading according to the time-scaling law. However, in real-world conditions, the intervals between uplift and compression cycles are shorter, and various factors, such as temperature fluctuations and reconsolidation, may affect the response. These factors were not accounted for in the experiment due to inherent limitations and errors in model testing. Due to limitations such as the environmental conditions in the laboratory and errors in experimental equipment, using the same parameters may lead to some variations in results, resulting in non-reproducibility of the experiments. We kindly seek the readers' understanding for this.

Author Contributions: Conceptualization, W.Z. and T.N.; methodology, M.W. and F.G.; experimental, M.W., F.G., and Y.Y.; formal analysis, Z.W.; investigation, W.Z. and T.N.; resources, W.Z.; data curation, M.W. and Y.Y.; writing—original draft preparation, M.W. and Z.W.; writing—review and editing, Z.W., T.N., and W.Z.; visualization, Z.W.; supervision, T.N.; project administration, T.N.; funding acquisition, T.N. All authors have read and agreed to the published version of the manuscript.

Funding: This research was funded by National Natural Science Foundation of China, grant numbers 52079020 and 42377185, and the open research fund program of Zhoushan Field Scientific Observation and Research Station for Marine Geo-hazards.

Institutional Review Board Statement: Not applicable.

Informed Consent Statement: Not applicable.

Data Availability Statement: Data can be obtained from the first author upon request.

Acknowledgments: The authors gratefully acknowledge the support from the funding listed above. We also appreciate the anonymous reviewers who gave comments to revise the paper.

Conflicts of Interest: Author Yanghui Ye was employed by the company State Grid Jiangxi Electric Power Co., Ltd. The remaining authors declare that the research was conducted in the absence of any commercial or financial relationships that could be construed as a potential conflict of interest.

References

1. Cozzi, M.E.L.; Goodson, T. *Global Energy Review*; IEA: Paris, France, 2021; Available online: https://www.iea.org/reports/global-energy-review-2021 (accessed on 1 April 2021).
2. Wang, S.; Liu, J.; Liu, Z.; Li, N. Analysis of Current Situation and Future Development of Offshore Wind Power Industry. *South Energy Constr.* **2023**, *10*, 103–112.
3. Guo, X.; Fan, N.; Zheng, D.; Fu, C.; Wu, H.; Zhang, Y.; Song, X.; Nian, T. Predicting impact forces on pipelines from deep-sea fluidized slides: A comprehensive review of key factors. *Int. J. Min. Sci. Technol.* **2024**, *34*, 211–225. [CrossRef]
4. Nian, T.; Guo, X.; Zheng, D.; Xiu, Z.; Jiang, Z. Susceptibility assessment of regional submarine landslides triggered by seismic actions. *Appl. Ocean Res.* **2019**, *93*, 101964. [CrossRef]
5. Nian, T.; Guo, X.; Fan, N.; Jiao, H.; Li, D. Impact forces of submarine landslides on suspended pipelines considering the low-temperature environment. *Appl. Ocean Res.* **2018**, *81*, 116–125. [CrossRef]
6. Byrne, B.W.; Houlsby, G.T. Foundations for offshore wind turbines. *Philos. Trans. R. Soc. A* **2003**, *361*, 2909–2930. [CrossRef]
7. Liu, R.; Zhou, L.; Lian, J.; Ding, H. Behavior of monopile foundations for offshore wind farms in sand. *J. Waterw. Port. C ASCE* **2016**, *142*, 04015010. [CrossRef]
8. Bae, K.T.; Choo, Y.W.; Youn, H.J.; Kim, J.Y.; Kwon, O.S. Geotechnical perspective on offshore wind plan, strategy, projects and research in Korea. In Proceeding of the TC 209 Workshop, Seoul, Republic of Korea, 20 September 2017; Volume 19, pp. 13–20.
9. Wang, K.; Zhu, Y.; Yu, M.; Peng, M. Study on the Compressive Bearing Capacity Mode of Suction Caisson Foundations for Offshore Wind Turbines. *J. Eng. Geol.* **2022**, *30*, 1753–1761. (In Chinese)
10. Bienen, B.; Gaudin, C.; Cassidy, M.J.; Rausch, L.; Purwana, O.A.; Krisdani, H. Numerical modelling of a hybrid skirted foundation under combined loading. *Comput. Geotech.* **2012**, *45*, 127–139. [CrossRef]
11. Kim, D.J.; Choo, Y.W.; Kim, J.H.; Kim, S.; Kim, D.S. Investigation of monotonic and cyclic behavior of tripod suction bucket foundations for offshore wind towers using centrifuge modeling. *J. Geotech. Geoenviron.* **2014**, *140*, 04014008. [CrossRef]
12. Wang, X.; Yang, X.; Zeng, X. Centrifuge modeling of lateral bearing behavior of offshore wind turbine with suction bucket foundation in sand. *Ocean Eng.* **2017**, *139*, 140–151. [CrossRef]
13. Zhang, Y.; Wang, C.; Li, D.; Qi, Y. Numerical Simulation of Composite Bearing Characteristics of Skirted Suction Foundations in Sandy Soil. *Sci. Technol. Eng.* **2021**, *21*, 1515–1521. (In Chinese)
14. Kou, H.; Zhou, N.; Yang, D.; Feng, J.; Tian, H. Experimental Study on Horizontal Bearing Characteristics of Suction Bucket Foundations for Offshore Wind Turbines in Sandy Soil Foundations. *J. Eng. Geol.* **2021**, *29*, 1752–1758. (In Chinese)
15. Lu, X.; Wang, Y.; Zhang, J.; Sun, G.; Shi, Z. Centrifuge Tests on Deformation of Bucket Foundations under Horizontal Dynamic Loads. *Chin. J. Geotech. Eng.* **2005**, *27*, 789–791. (In Chinese)
16. Jiao, B.; Shi, Z.; Lu, X.; Zhang, J. Centrifuge Experimental Study on the Horizontal Dynamic Response of Bucket Foundations in Silty Soil. *Eng. Mech.* **2010**, *27*, 131–141. (In Chinese)
17. Choo, Y.W.; Kim, D.; Park, J.H.; Kwak, K.; Kim, J.H.; Kim, D.S. Lateral response of large-diameter monopiles for offshore wind turbines from centrifuge model tests. *Geotech. Test. J.* **2014**, *37*, 107–120. [CrossRef]
18. Clukey, E.C.; Aubeny, C.P.; Murff, J.D. Comparison of analytical and centrifuge model tests for suction caissons subjected to combined loads. *Offshore Mech. Arct. Eng.* **2004**, *126*, 364–367. [CrossRef]
19. Ding, H.; Liu, Y.; Zhang, P.; Le, C. Model tests on the bearing capacity of wide-shallow composite bucket foundations for offshore wind turbines in clay. *Ocean Eng.* **2015**, *103*, 114–122. [CrossRef]
20. Hu, Y.; Randolph, M.F. Bearing capacity of caisson foundations on normally consolidated clay. *Soils Found.* **2002**, *42*, 71–77. [CrossRef]
21. Bransby, M.F.; Yun, G.J. The undrained capacity of skirted strip foundations under combined loading. *Géotechnique* **2009**, *59*, 115–125. [CrossRef]
22. Lee, S.H.; Vicent, S.; Kim, S.R. An experimental investigation of the cyclic response of bucket foundations in soft clay under one-way cyclic horizontal loads. *Appl. Ocean Res.* **2018**, *71*, 59–68.
23. Skau, K.S.; Chen, Y.; Jostad, H.P. A numerical study of capacity and stiffness of circular skirted foundations in clay subjected to combined static and cyclic general loading. *Géotechnique* **2018**, *68*, 205–220. [CrossRef]
24. Aubeny, C.; Han, S.W.; Murff, J.D. Suction caisson capacity in anisotropic, purely cohesive soil. *Int. J. Geomech.* **2003**, *3*, 225–235. [CrossRef]
25. Aubeny, C.; Murff, J.D. Simplified limit solutions for the capacity of suction anchors under undrained conditions. *Ocean Eng.* **2005**, *32*, 864–877. [CrossRef]
26. Kim, S.R. Evaluation of combined horizontal-moment bearing capacities of tripod bucket foundations in undrained clay. *Ocean Eng.* **2014**, *85*, 100–109.
27. Bagheri, P.; Kim, J.M. Evaluation of cyclic and monotonic loading behavior of bucket foundations used for offshore wind turbines. *Appl. Ocean Res.* **2019**, *91*, 101865. [CrossRef]
28. Barari, A.; Glitrup, K.; Christiansen, L.R.; Ibsen, L.B.; Choo, Y.W. Tripod suction caisson foundations for offshore wind energy and their monotonic and cyclic responses in silty sand: Numerical predictions for centrifuge model tests. *Soil Dyn. Earthq. Eng.* **2021**, *149*, 106813. [CrossRef]
29. Jeong, Y.H.; Kim, J.H.; Ko, K.W.; Park, H.J. Simplified estimation of rotational stiffness of tripod foundation for offshore wind turbine under cyclic loadings. *Appl. Ocean Res.* **2021**, *112*, 102697. [CrossRef]

30. Wang, Z.; Zheng, D.; Guo, X.; Gu, Z.; Yan, C.; Nian, T. A review of submarine landslides associated with rapid sedimentation. *Mar. Georesour. Geotechnol.* **2024**, 1–25. [CrossRef]
31. Jiao, H.; Guo, X.; Fan, N.; Wu, H.; Nian, T. An undrained dynamic strain-pore pressure model for deep-water soft clays from the South China Sea. *Front. Mar. Sci.* **2024**, *11*, 1377474. [CrossRef]
32. Wang, Z.; Zheng, D.; Gu, Z.; Guo, X.; Nian, T. A Methodology to Evaluate the Real-Time Stability of Submarine Slopes under Rapid Sedimentation. *J. Mar. Sci. Eng.* **2024**, *12*, 823. [CrossRef]
33. Khemakhem, M.; Chenaf, N.; Garnier, J.; Favraud, C.; Gaudicheau, P. Development of degradation laws for describing the cyclic lateral response of piles in clay. In Proceedings of the SUT Offshore Site Investigation and Geotechnics, London, UK, 12–14 September 2012; p. SUT–OSIG-12-27.
34. Liao, W.; Zhang, J.; Wu, J.; Yan, K. Response of flexible monopile in marine clay under cyclic lateral load. *Ocean Eng.* **2018**, *147*, 89–106. [CrossRef]
35. Fan, Q.; Li, N. Causes and Countermeasures of Local Damage to the Northern Guide Bank of the Yangtze Estuary Phase II Project. China Harbour and Waterfront Engineering: Beijing, China, 2004. (In Chinese)
36. Anagnostopoulos, C.; Chrysanidis, T.; Anagnostopoulou, M. Experimental data of cement grouting in coarse soils with different superplasticisers. *Data Brief* **2020**, *30*, 105612. [CrossRef] [PubMed]
37. Nian, T.; Gu, Z.; Guo, X.; Zhang, H.; Jia, Y. Effect of temperature rise on the mechanical behaviour of deep-sea clay surrounding oil and gas pipelines. *Ocean. Eng.* **2024**, *302*, 117533. [CrossRef]
38. *JGJ 83–2011;* Code for Geotechnical Investigation in Soft Soil Areas. Architecture and Building Press: Beijing, China, 2011. (In Chinese)
39. Morris, P.H.; Williams, D.J. A new model of vane shear strength testing in soils. *Géotechnique* **1994**, *44*, 771–774. [CrossRef]
40. Carbon Trust. *Suction Installed Caisson Foundations for Offshore Wind;* Design Guidelines: London, UK, 2019.
41. Andersen, K.H. Cyclic soil parameters for offshore foundation design. *Front. Offshore Geotech. III* **2015**, *5*, 5–82.
42. Kou, H.; Fang, W.; Zhou, N.; Huang, J.; Zhang, X. Dynamic response of single-bucket foundation in clay under vertical variable amplitude cyclic loadings. *Ocean Eng.* **2023**, *273*, 113973. [CrossRef]
43. Randolph, M.; Gourvenec, S. *Offshore Geotechnical Engineering;* CRC Press: Boca Raton, FL, USA, 2017.
44. Guo, X.; Nian, T.; Zhao, W.; Gu, Z.; Liu, C.; Liu, X.; Jia, Y. Centrifuge experiment on the penetration test for evaluating undrained strength of deep-sea surface soils. *Int. J. Min. Sci. Technol.* **2022**, *32*, 363–373. [CrossRef]
45. Wang, L.; Wang, Z.; Li, S.; Wang, Z.; Zhu, H. Study on the settlement behavior of underwater protective caissons based on centrifuge model tests. *Waterw. Harb.* **2024**, *45*, 448–454. (In Chinese)
46. Leblanc, C.; Byrne, B.W.; Houlsby, G.T. Response of stiff piles to random two-way lateral loading. *Géotechnique* **2010**, *60*, 715–721. [CrossRef]

Journal of
Marine Science and Engineering

MDPI

Article

Explainable Ensemble Learning Approaches for Predicting the Compression Index of Clays

Qi Ge [1], Yijie Xia [2], Junwei Shu [3], Jin Li [2] and Hongyue Sun [3,*]

[1] College of Civil Engineering, Nanjing Forestry University, Nanjing 210037, China
[2] Institute of Artificial Intelligence, Nanjing University of Information Science and Technology, Nanjing 210044, China
[3] Ocean College, Zhejiang University, Zhoushan 316021, China
* Correspondence: shy@zju.edu.cn

Abstract: Accurate prediction of the compression index (c_c) is essential for geotechnical infrastructure design, especially in clay-rich coastal regions. Traditional methods for determining c_c are often time-consuming and inconsistent due to regional variability. This study presents an explainable ensemble learning framework for predicting the c_c of clays. Using a comprehensive dataset of 1080 global samples, four key geotechnical input variables—liquid limit (LL), plasticity index (PI), initial void ratio (e_0), and natural water content w—were leveraged for accurate c_c prediction. Missing data were addressed with K-Nearest Neighbors (KNN) imputation, effectively filling data gaps while preserving the dataset's distribution characteristics. Ensemble learning techniques, including Random Forest (RF), Gradient Boosting Decision Trees (GBDT), Extreme Gradient Boosting (XGBoost), and a Stacking model, were applied. Among these, the Stacking model demonstrated the highest predictive performance with a Root Mean Squared Error (RMSE) of 0.061, a Mean Absolute Error (MAE) of 0.043, and a Coefficient of Determination (R^2) value of 0.848 on the test set. Model interpretability was ensured through SHapley Additive exPlanations (SHAP), with e_0 identified as the most influential predictor. The proposed framework significantly improves both prediction accuracy and interpretability, offering a valuable tool to enhance geotechnical design efficiency in coastal and clay-rich environments.

Keywords: coastal foundation; clays; compression index; consolidation; ensemble learning; model interpretability

1. Introduction

In recent years, the construction of geotechnical infrastructures along coastal areas with clay ground has gained significant popularity due to limited constructional space [1–4]. Clay ground poses unique challenges for the maintenance and design of structures, with consolidation being a crucial concern [5–8]. The compression index (c_c), which represents the slope of the linear segment in the e-logσ' graph, serves as a key parameter for quantifying the compressibility of clays [9]. Generally, clays with a higher c_c are more compressible and deformable under load, while those with a lower c_c resist deformation better. However, conventional c_c determination via oedometer testing is time-consuming and expensive, hindering efficient geotechnical assessment.

For this reason, many efforts have focused on associating fundamental geotechnical characteristics with the compression index [9–11]. While statistical models based on geotechnical variables like natural water content (w), liquid limit (LL), plasticity index (PI), and initial void ratio (e_0) have been proposed [12–15], they often lack generalizability across diverse datasets. Recent studies have shifted towards machine learning models [16–18], particularly ensemble learning approaches, to overcome these limitations and improve predictive accuracy. For example, Zhang et al. [19] implemented Random Forest (RF), a classic ensemble learning technique, utilizing a dataset of 311 samples with three factors: LL, PI, and e_0. Asteris et al. [20] developed extreme learning machine models to estimate c_c using the void ratio at liquid limit, LL, and PI. Long et al. [21] employed tree-based ML algorithms to establish a correlation between c_c and w, LL, PI, e_0, and specific gravity (G_s), drawing from 391 samples collected in Northern Iran. Lee et al. [22] gathered geotechnical data from soft ground along Korea's southern coastline and devised a c_c prediction model using linear regression analysis alongside three ensemble learning models. Díaz and Spagnoli [23] introduced a novel super-learner machine learning algorithm for c_c prediction, leveraging a comprehensive dataset comprising over 1000 samples.

Nevertheless, despite the promising potential of ML models to enhance conventional methods in predicting the compression index, several challenges remain: (1) many existing models are limited by either insufficient data or a narrow geographical focus, (2) the practical application of data-driven models is often hindered by inconsistencies in real-world test data, which frequently contains missing items or values due to diverse collection methods/sites, and (3) the "black box" nature of ML models [24–26] makes it difficult to discern the precise relationships between geotechnical inputs and the resulting compression index.

To address these concerns, this research aims to develop an effective and interpretable machine learning framework for predicting the compression index of diverse clay samples. A comprehensive worldwide dataset is first built to mitigate regional biases. Building upon the dataset, we focus on three major contributions: (1) We employ the imputation method of K-Nearest Neighbors (KNN), therefore effectively handling the missing data. (2) We apply various ensemble learning methods, including RF, Gradient Boosting Decision Trees (GBDT), Extreme Gradient Boosting (XGBoost), and a novel use of Stacking, to model the correlations between geotechnical inputs and the compression index, conducting a comprehensive comparison to identify the most accurate and robust model. (3) We utilize the SHapley Additive exPlanations (SHAP) algorithm to interpret the predictive outcomes, providing transparent insights into feature importance and addressing the "black box" issue of ML models. These contributions result in a robust, globally applicable tool for predicting soil compression behavior, potentially enhancing the accuracy and efficiency of coastal geotechnical design and analysis across diverse geological contexts.

2. Data Overview and Analysis

A comprehensive global dataset of clay soil parameters was compiled from peer-reviewed studies, representing 1080 samples from diverse geographical regions such as Algeria, Spain, Nigeria, Iran, India, Germany, Ireland, and Bangladesh, among others [18,23,27–34] This dataset was selected to reduce regional bias and improve the generalizability of the model across a wide range of soil conditions. It encompasses four independent variables (i.e., the input features)-LL, PI (with 169 missing values), e_0, and w-which are used to establish a relationship with the output feature, the compression index. The selection of these four variables as input features is based on prior research into data-driven models for predicting c_c [19–23]. These variables are critical for understanding the consolidation behavior of clays, making them ideal for building a robust and interpretable predictive model. The dataset exhibits substantial variability in both input and output variables due to differences in soil origin. Given its extensive and diverse range of data, the database is well suited for robust and reliable analysis. Key statistics of the dataset are presented in Table 1 and Figure 1.

Table 1 demonstrates that the database includes clays with a broad range of plasticity, from low to very high. This is reflected in the LL values, which span 181.9%, and PI values that range from 2.0% to 69.0%. In contrast, the database also includes clays with varying degrees of compressibility, with e_0 ranging from 0.3 to more than 7. These statistical metrics reveal significant variability and dispersion among the variables, underscoring the comprehensive global scope of the compiled database.

Table 1. Statistical characteristics of the dataset.

	Maximum	Minimum	Mean	Median	SD	Range
LL (%)	199.0	17.1	45.4	44.0	14.5	181.9
PI (%)	69.0	2.0	21.3	21.5	9.0	67.0
e_0	7.1	0.3	0.8	0.7	0.4	6.8
w (%)	244.1	8.0	29.2	25.9	15.2	236.1
c_c	2.2	0.01	0.2	0.2	0.2	2.2

SD: Standard Deviation.

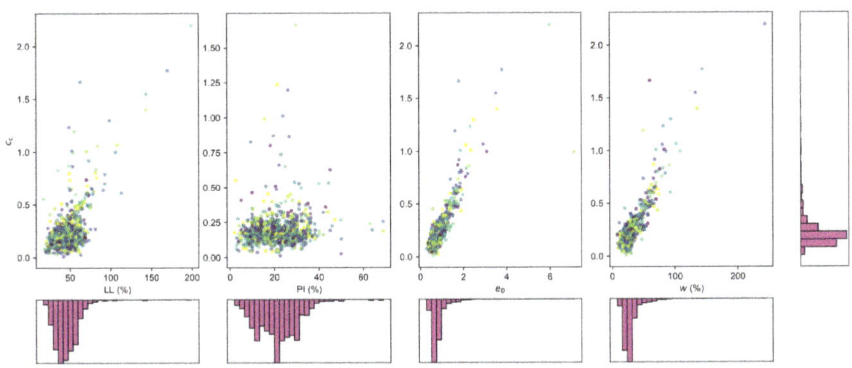

Figure 1. Scatter plots and distribution histograms of input and output features.

Figure 1 depicts the correlation between c_c and the four input features. The plots demonstrate a positive association, indicating that an increase in the input features generally corresponds to a rise in c_c. This correlation is particularly evident with e_0 and w, whereas it is less pronounced with LL and PI. These trends can be further analyzed using the Pearson correlation coefficient (r) and a correlation heatmap, as shown in Figure 2. The heatmap provides a quantitative assessment of the linear relationships between each pair of features. According to Figure 2, there is a general positive correlation between cc and all four input features, as indicated by the positive correlation values. Specifically, c_c exhibits a strong positive correlation with e_0 (r = 0.86) and w (r = 0.88), while the correlations with PI (r = 0.1) are relatively weaker. Additionally, a strong positive correlation is observed between the two groups of input features e_0 and w (r = 0.91), and LL and PI (r = 0.91).

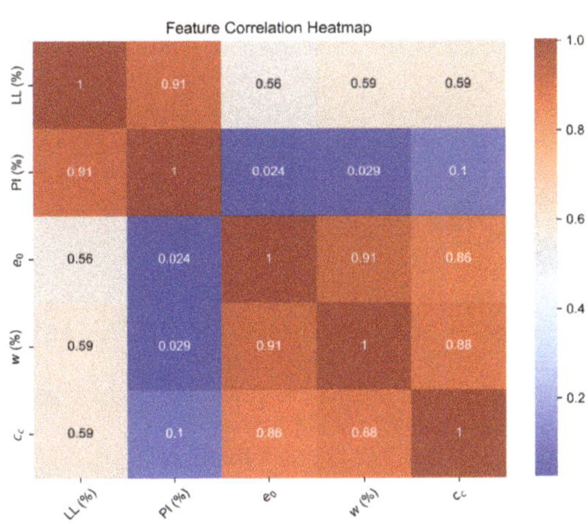

Figure 2. Correlation heatmap between different features.

3. Methodology

3.1. KNN Imputation for Missing Values

The KNN algorithm can be employed as a method for imputing missing values in the original datasets [35,36]. This technique involves selecting the k nearest neighbors of an instance with the missing data, using only the attributes with observed values. Typically, Euclidean distance is utilized to measure the dissimilarity between instances, considering only the dimensions where values are present. The missing value is then imputed by calculating the weighted average of the k nearest neighbors' values, where the weight assigned to each neighbor is inversely proportional to its distance from the instance with the missing value.

The Euclidean distance d between instances x_i and x_j is:

$$d(x_i, x_j) = \sqrt{\sum_{l=1}^{p} \left(x_{il} - x_{jl}\right)^2} \tag{1}$$

where p is the number of dimensions and x_{il} and x_{jl} are the values of the l-th dimension for instances x_i and x_j, respectively. To impute the missing value x_i^*, the weighted average is calculated as:

$$x_i^* = \frac{\sum_{k=1}^{K} \frac{x_k}{d_{ik}}}{\sum_{k=1}^{K} \frac{1}{d_{ik}}} \tag{2}$$

where x_k is the value from the k-th nearest neighbor, and d_{ik} is the distance between the instance with the missing value and its k-th nearest neighbor.

The choice of k is crucial in KNN interpolation. A small k may make the result sensitive to nearby points and noise, leading to overfitting, while a large k can smooth the result but risk underfitting by ignoring local features. Typically, odd values like 3, 5, or 7 [37] are used to avoid ties. For this study, cross-validation determined that the optimal k was 5 based on the experimental data scale.

3.2. Ensemble Learning Algorithms

We utilized ensemble learning methods, which combine several simple models to form a more powerful and accurate prediction tool. Specifically, we used:

3.2.1. Random Forest

RF is an ensemble method that combines multiple decision trees built from bootstrapped samples. It uses Bagging (Bootstrap Aggregation) to improve accuracy and reduce variance, thus combating overfitting [38]. Each tree is trained on a random subset of the data with replacement, and feature randomness is applied when splitting nodes. For prediction, each tree votes for a class, and the class with the majority of votes is chosen as the final output. Key tuning parameters for RF [39] include the number of trees in the forest (*n_estimators*), the maximum depth of the tree (*max_depth*), and the minimum number of samples required to split an internal node (*min_samples_split*).

3.2.2. Gradient Boosting Decision Tree

GBDT combines decision trees with ensemble learning techniques [40]. GBDT employs boosting and classification to enhance accuracy and reduce overfitting through regularization. Decision trees are easy to interpret and handle missing features effectively, as each node depends on a single attribute. Boosting improves GBDT performance by sequentially creating a series of models that convert multiple weak classifiers into a strong, accurate predictor. Key tuning parameters for GBDT [41] include the number of trees in the estimator (*max_iter*), the depth of the trees (*max_d*), and the learning rate.

3.2.3. Extreme Gradient Boosting

XGBoost is primarily designed to enhance speed and performance through the use of gradient-boosted decision trees [42]. It generates a predictive model by leveraging a boosting ensemble of weak classification trees, employing gradient descent to optimize the loss function. XGBoost supports three key gradient boosting techniques: Gradient Boosting, Regularized Boosting, and Stochastic Boosting. It facilitates the incorporation and tuning of regularization parameters and supports parallel processing during tree construction. Critical tuning parameters for XGBoost [43] encompass the maximum number of trees (*ntree*), the minimum loss reduction required to make a further partition on a leaf node (*gamma*), and the learning rate.

3.2.4. Stacking

Stacking, also known as Stacked Generalization, is a sophisticated nonlinear modeling technique that enhances prediction accuracy by integrating multiple base estimators [44]. This approach features a two-tier structure: the first layer (level-0) consists of the base learners, while the second layer (level-1) houses the meta-learner. The outputs from level-0 are fed into level-1, a crucial mechanism in Stacking for reducing model bias. In this research, level-0 includes three base learners derived from the GBDT, XGBoost, and RF algorithms. To refine quantile regression results, ridge regression is utilized as the meta-learner at level-1.

The above ensemble learning models were implemented in Python language (version 3.7.4) by using the sklearn.ensemble library. Hyperparameters were optimized through 5-fold cross-validation combined with a grid search.

3.3. *SHAP for Model Interpretability*

To understand how different factors affect our predictions, we used the SHAP anlysis. This method breaks down a prediction to show the impact of each feature, helping to explain which factors are most influential. Developed by Lundberg and Lee [45], SHAP is an advanced explainable artificial intelligence (XAI) tool rooted in game theory, designed to interpret the outputs of any machine learning model. The SHAP methodology assesses each input variable's contribution to the prediction, determining whether these contributions are positive or negative [46]. SHAP summary plots are utilized to rank the most critical features, while SHAP dependence plots analyze the impact of selected input features, thereby evaluating their influence. The SHAP approach formulates an explanatory model $g(\cdot)$, which is expressed as follows:

$$g(z') = \varphi_0 + \sum_{i=1}^{K} \varphi_i z'_i \tag{3}$$

where $z' \in \{0,1\}^K$, and K denotes the number of independent factors; $\varphi_i \in \mathrm{R}$ represents the SHAP value for the i-th factors; and ϕ_0 is the constant term when all independent factors are absent.

$$\varphi_i = \sum_{S \subseteq F \setminus \{i\}} \frac{|S|!(|F| - |S| - 1)!}{|F|!} \left[f_{S \cup \{i\}}\left(x_{S \cup \{i\}}\right) - f_S(x_S) \right] \tag{4}$$

where F denotes the set of all independent factors, and S represents all subsets of F that exclude the i-th factor. Additionally, x_S denotes the values of the independent factors in the subset S.

Due to the complexity of calculating SHAP values, several approximation methods have been developed, including Tree SHAP, Deep SHAP, and Kernel SHAP, the latter of which is utilized in this study [47].

3.4. Implementation Pipeline

As depicted in Figure 3, the implementation process of the proposed approach can be summarized into the following key steps:

- Collect comprehensive global data, including LL, PI, e_0, and w as input features and c_c as the output feature, to minimize local bias.
- Apply KNN to impute missing values in the dataset, then partition the data into training and testing sets (80% training, 20% testing).
- Train and fine-tune different ensemble learning models—RF, GBDT, XGBoost, and Stacking—to realize data-driven c_c prediction.
- Assess different models using Root Mean Squared Error (RMSE), Mean Absolute Error (MAE), and Coefficient of Determination (R^2) to identify the best prediction model.
- Analyze and clarify the key factors influencing the performance of the best model via the interpretable machine learning algorithm SHAP.

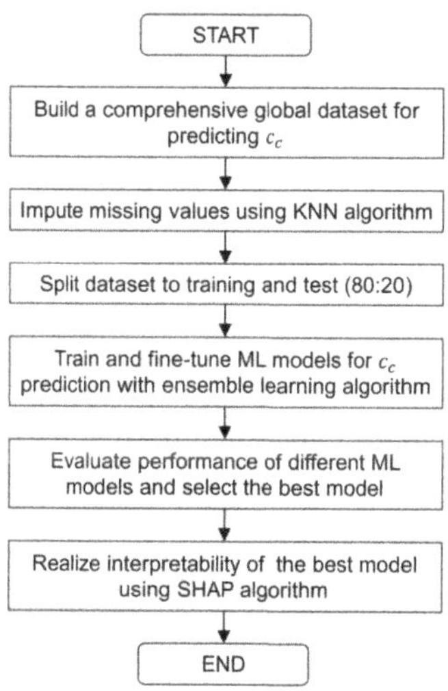

Figure 3. Flowchart of the proposed method for c_c prediction.

4. Results

4.1. Prediction Results of ML Models

The original dataset comprises 1080 records, with 169 instances having missing values for the PI variable. Initially, the KNN method was employed to generate these missing PI values. Subsequently, the dataset was randomized to prevent the potential bias of having consecutive data from the same source within either the training or testing sets, thereby ensuring more reliable model performance and evaluation. We then divided the randomized dataset into training and testing subsets using an 80/20 split. This split ratio was selected because it is widely recognized in machine learning studies as an effective balance, providing sufficient data for training to capture underlying patterns while retaining enough data for a robust evaluation of model performance on unseen data. The four ensemble learning methods underwent hyperparameter tuning to optimize their performance, employing 5-fold cross-validation in conjunction with a grid search.

Figure 4 presents scatter plots of the actual values of c_c (x-axis) versus the predicted values (y-axis) obtained from each model on the test set. As illustrated in Figure 4, the majority of data points for each ensemble learning model are clustered near the no-error line, reflecting a strong agreement between the predicted and observed values. To further assess the performance of the four ensemble learning models, a residual analysis was conducted. Residuals, defined as the differences between the predicted and actual values, are presented in Figure 5 through a histogram distribution. Analysis of this figure reveals that the residuals for all models are predominantly centered around zero and exhibit Gaussian distributions, indicating a normal distribution with minimal outliers. These observations suggest that the ensemble learning algorithms demonstrate strong generalization capabilities and do not exhibit significant prediction errors.

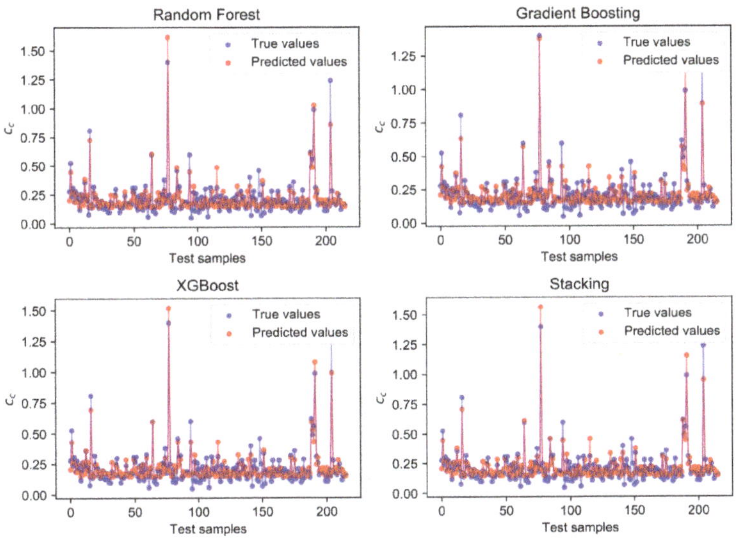

Figure 4. Prediction scatter plot (true values vs. predicted values) of all ML models.

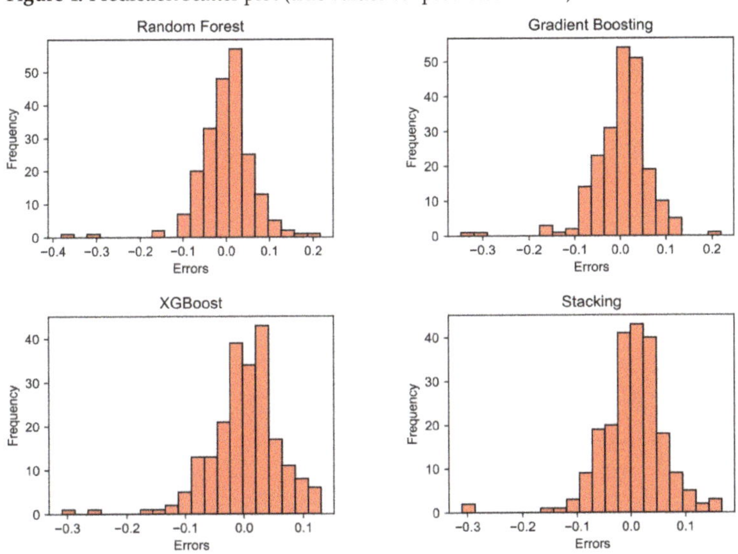

Figure 5. Residuals' histogram distribution in test sets of all ML models.

In Table 2, the RMSE, MAE, and R^2 values [48] for both the training and test datasets are presented for all ensemble learning models. These metrics reflect the ensemble learning models' outstanding predictive performance and generalizability, as evidenced by the similarity in values between the training and test datasets (e.g., for the RF model: RMSE of 0.066 and 0.062, MAE of 0.044 and 0.043, and R^2 of 0.868 and 0.843, respectively). Overfitting occurs when a model performs significantly better on the training data than on unseen data, suggesting that it has memorized the training data rather than learning general patterns.

The minimal differences between the training and test performance metrics of the models indicate their strong ability to generalize to new data and suggest that they are not prone to overfitting. Among the four ensemble models, the proposed Stacking model outperforms RF, GBDT, and XGBoost from existing research, with the test set RMSE, MAE, and R^2 values of 0.061, 0.043, and 0.848, respectively. Consequently, the Stacking model was selected as the optimal candidate for subsequent coupling with the SHAP algorithm for model interpretability analysis.

Table 2. Summary of performance metrics of all ML models.

Model	RMSE		MAE		R^2	
	Training	Test	Training	Test	Training	Test
RF	0.066	0.062	0.044	0.043	0.868	0.843
GBDT	0.061	0.062	0.042	0.044	0.890	0.844
XGBoost	0.062	0.063	0.042	0.043	0.884	0.839
Stacking	0.060	0.061	0.041	0.043	0.893	0.848

4.2. Analysis of Model Interpretability

To clarify the key factors influencing the performance of the Stacking model in predicting the compression index, we performed a factor importance analysis using SHAP. Figure 6 illustrates the importance ranking of the four independent variables. The mean absolute SHAP value plot indicates that e_0 exerts the greatest average impact on the model's output, followed by w, with PI and LL having comparatively lesser influence. These results underscore the critical role of the initial void ratio and natural water content in predicting the compression index of clays, thereby emphasizing the significance of these geotechnical properties in the predictive modeling of clay behavior.

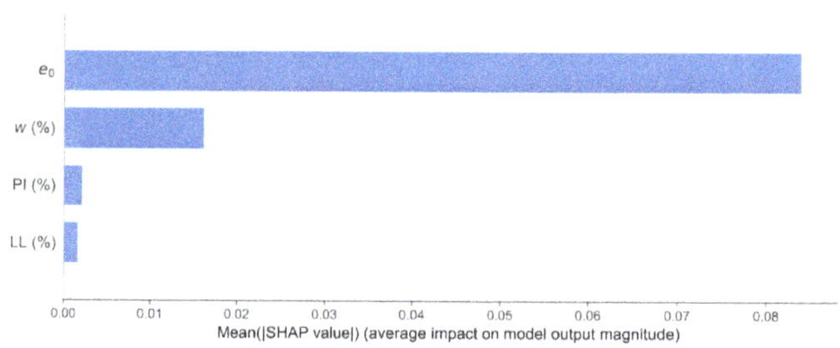

Figure 6. SHAP feature importances.

The SHAP summary plot presented in Figure 7 offers a comprehensive overview of feature importance and their effects on model predictions. In this plot, the y-axis represents the features, while the x-axis quantifies the SHAP values. Each dot signifies a Shapley value for a specific feature and instance, with color indicating feature values (red for high, blue for low). The plot reveals that e_0 is the most influential feature; higher e_0 values (red points) correlate with higher SHAP values, suggesting that increased e_0 levels lead to higher predicted c_c values. In contrast, w exhibits a more complex impact, with both positive and negative values appearing across the SHAP value spectrum. This variability reflects its intricate relationship with the model's output. PI and LL show relatively minor effects on predictions, aligning with the patterns observed in the previous feature importance analysis.

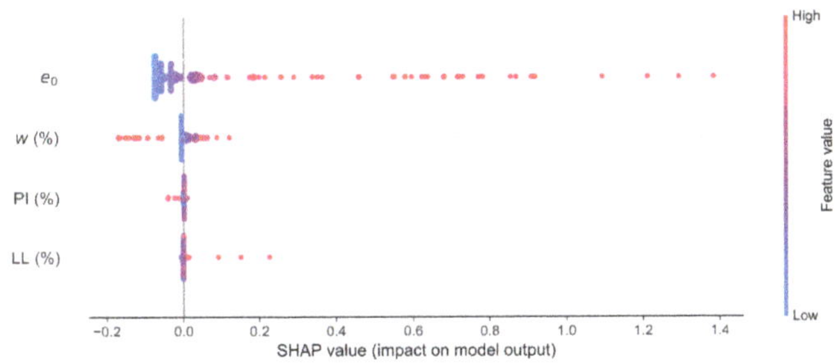

Figure 7. SHAP summary plot.

Finally, dependence plots for pairs of input features were created to show their joint influence on model predictions. Based on the correlation analysis in Figure 2, the four input features were grouped into two categories for detailed examination: e_0 and w, and LL and PI. In Figure 8, the horizontal axis shows input feature values and the vertical axis displays SHAP values. Additionally, the color bars on the right of each subplot indicate the values of the variable most dependent on the input feature shown on the horizontal axis.

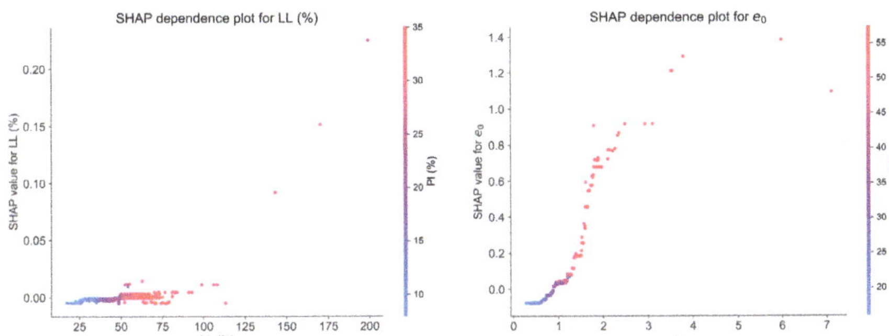

(**a**) LL and PI (**b**) e_0 and w

Figure 8. SHAP dependence plots.

As illustrated in Figure 8a, there is a notable positive correlation between LL and PI. Specifically, when the LL values are small (with data points on the left side of the x-axis), PI values are also low (indicated by blue data points), and vice versa. This aligns with the $r = 0.91$ correlation coefficient reported in Figure 2. However, LL has a minimal impact on the prediction model, as indicated by the relatively flat SHAP values despite rising LL. Although PI values increase (indicated by a gradual shift in color from blue to red) as LL increases, under the joint action of the two, the SHAP value for LL shows only a slight upward trend.

Likewise, Figure 8b demonstrates a strong positive correlation between e_0 and w. Unlike the LL and PI groups, the SHAP value rises sharply with increasing e_0. However, when e_0 surpasses 2, the rate of SHAP value growth slows. Despite this slowdown, there is no significant color change where growth decelerates (without substantial changes in w), suggesting that e_0 is the primary factor affecting the SHAP value, rather than the combined effect of e_0 and w.

5. Discussions

The results show that interpolating missing data, constructing ensemble models, and integrating additive interpretation algorithms lead to accurate and interpretable c_c predictions. Although we have compared various ensemble models and analyzed the interpretability of the optimal model, the varying degrees of missing data from different sources in practice highlight the need for effective data imputation methods. Therefore, a thorough evaluation of KNN interpolation for handling missing data in our study is necessary.

The dataset utilized in this study comprises 1080 groups of observations, with 169 instances of missing data for the input feature PI. In addition to employing the KNN interpolation method previously used, we also applied two traditional interpolation techniques—polynomial interpolation and spline interpolation—to address the missing PI data. Figure 9 illustrates the distribution patterns and characteristics of the PI raw data alongside the results after using the three interpolation methods. By analyzing the shape, peak values, symmetry, and other characteristics in histograms and density maps, we can intuitively discern the similarities and differences among different datasets.

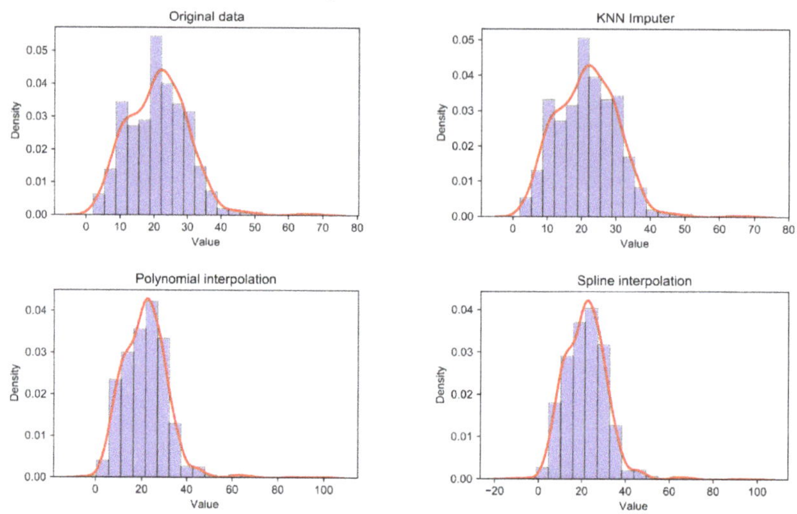

Figure 9. Histogram and density plot of PI data after different methods of interpolation.

The figure indicates that the PI dataset interpolated using KNN closely mirrors the distribution characteristics of the original dataset. This observation is corroborated by the quantitative statistics presented in Table 3, which show that the dataset handled with KNN exhibits high consistency with the original data in key statistical measures such as mean, maximum and minimum values, median, and variance. Although polynomial and spline interpolations preserve mean and median values relatively well, they exhibit notable discrepancies in maximum and minimum values and data variance, thereby failing to replicate the original data's distribution effectively.

Table 3. Comparison of different interpolation methods.

PI Values	Mean	Maximum	Minimum	Median	Variance
Original data with missing values	21.3	69.0	2.0	21.5	81.4
After KNN interpolation	21.6	69.0	2.0	22.0	79.6
After polynomial interpolation	21.8	100.8	−4.8	22.0	102.6
After spline interpolation	21.8	99.8	−12.2	22.0	107.1

Consequently, KNN interpolationoutperforms traditional methods by generating imputed data that accurately preserve the original dataset's distribution, thereby providing a robust foundation for the subsequent development of high-performance machine learning models.While the K-Nearest Neighbors (KNN) imputation method was effective, it introduces some uncertainty, especially in highly variable datasets like global soil samples. Future research could explore alternative imputation methods or more complete datasets to address this issue.

6. Conclusions

This study introduces an innovative and interpretable ensemble learning framework for predicting the compression index of clays, a key parameter for geotechnical infrastructure design in clay-rich regions. This research underscores the effectiveness of integrating a comprehensive global dataset with ensemble learning models such as RF, GBDT, XGBoost, and Stacking. Among these models, the Stacking approach emerged as the most accurate, demonstrating minimal prediction errors and robust generalization capabilities.

The use of SHAP for model interpretation identified the initial void ratio as the most influential predictor of c_c, providing valuable insights into the geotechnical behavior of clays. This interpretable framework not only enhances prediction reliability but also offers geotechnical engineers a transparent and practical tool for real-world applications.

Moreover, this study highlights the effectiveness of the KNN imputation method for generating missing data, which preserves the original dataset's distribution characteristics and contributes to the development of high-performance predictive models.

Although our model demonstrates strong predictive performance, future research could explore additional geotechnical parameters that affect clay compressibility. Addressing limitations such as missing data by investigating alternative imputation methods or incorporating diverse data sources could further improve model accuracy.

Author Contributions: Conceptualization, Q.G. and J.L.; methodology, Q.G., Y.X. and J.L.; software, Y.X. and J.L.; validation, Y.X. and J.S.; formal analysis, Y.X. and J.L.; investigation, J.S.; resources, H.S.; data curation, Q.G.; writing—original draft preparation, Q.G.; writing—review and editing, J.L. and H.S.; visualization, Y.X. and J.S.; supervision, H.S.; project administration, Q.G., J.L. and H.S.; funding acquisition, Q.G. and J.L. All authors have read and agreed to the published version of the manuscript.

Funding: This work was supported by the National Natural Science Foundation of China (Grant Nos. 42407222 and No. 82302352) and the Natural Science Foundation of Jiangsu Province (Grant No. BK20220421).

Data Availability Statement: The data used in this study can be found at: https://github.com/jli0117/Ensemble-Learning-Clay-Characteristics (accessed on 26 August 2024).

Acknowledgments: We thank the authors and institutions of the publicly available data used in this study for their valuable contributions. We thank the anonymous reviewers who helped to improve the paper.

Conflicts of Interest: All authors declare that this research was conducted in the absence of any commercial or financial relationships that could be construed as a potential conflict of interest.

References

1. Bugnot, A.; Mayer-Pinto, M.; Airoldi, L.; Heery, E.; Johnston, E.; Critchley, L.; Strain, E.; Morris, R.; Loke, L.; Bishop, M.; et al. Current and projected global extent of marine built structures. *Nat. Sustain.* **2021**, *4*, 33–41. [CrossRef]
2. Li, J.; Chen, H.; Yuan, X.; Shan, W. Analysis of the effectiveness of the step vacuum preloading method: A case study on high clay content dredger fill in Tianjin, China. *J. Mar. Sci. Eng.* **2020**, *8*, 38. [CrossRef]
3. Guo, X.; Fan, N.; Zheng, D.; Fu, C.; Wu, H.; Zhang, Y.; Song, X.; Nian, T. Predicting impact forces on pipelines from deep-sea fluidized slides: A comprehensive review of key factors. *Int. J. Min. Sci. Technol.* **2024**, *34*, 211–225. [CrossRef]
4. Chen, X.; Yu, Y.; Wang, L. Assessing scour prediction models for monopiles in sand from the perspective of design robustness. *Mar. Struct.* **2024**, *93*, 103532. [CrossRef]
5. Wang, J.; Dai, M.; Cai, Y.; Guo, L.; Du, Y.; Wang, C.; Li, M. Influences of initial static shear stress on the cyclic behaviour of over consolidated soft marine clay. *Ocean Eng.* **2021**, *224*, 108747. [CrossRef]
6. Zheng, J.; Hu, X.; Gao, S.; Wu, L.; Yao, S.; Dai, M.; Liu, Z.; Wang, J. Undrained cyclic behavior of under-consolidated soft marine clay with different degrees of consolidation. *Mar. Georesources Geotechnol.* **2024**, *42*, 176–183. [CrossRef]
7. Jiang, N.; Wang, C.; Wu, Q.; Li, S. Influence of structure and liquid limit on the secondary compressibility of soft soils. *J. Mar. Sci. Eng.* **2020**, *8*, 627. [CrossRef]
8. Guo, X.; Nian, T.; Zhao, W.; Gu, Z.; Liu, C.; Liu, X.; Jia, Y. Centrifuge experiment on the penetration test for evaluating undrained strength of deep-sea surface soils. *Int. J. Min. Sci. Technol.* **2022**, *32*, 363–373. [CrossRef]
9. Shimobe, S.; Spagnoli, G. A general overview on the correlation of compression index of clays with some geotechnical index properties. *Geotech. Geol. Eng.* **2022**, *40*, 311–324. [CrossRef]

10. Alzabeebee, S.; Alshkane, Y.M.; Rashed, K.A. Evolutionary computing of the compression index of fine-grained soils. *Arab. J. Geosci.* **2021**, *14*, 2040. [CrossRef]
11. Mawlood, Y.; Mohammed, A.; Hummadi, R.; Hasan, A.; Ibrahim, H. Modeling and statistical evaluations of unconfined compressive strength and compression index of the clay soils at various ranges of liquid limit. *J. Test. Eval.* **2022**, *50*, 551–569. [CrossRef]
12. Sridharan, A.; Nagaraj, H. Compressibility behaviour of remoulded, fine-grained soils and correlation with index properties. *Can. Geotech. J.* **2000**, *37*, 712–722. [CrossRef]
13. Spagnoli, G.; Shimobe, S. Statistical analysis of some correlations between compression index and Atterberg limits. *Environ. Earth Sci.* **2020**, *79*, 532. [CrossRef]
14. Heo, Y.; Hwang, I.; Kang, C.; Bae, W. Correlations Between the Physical Properties and Consolidation Parameter of West Shore Clay. *J. Korean GEO-Environ. Soc.* **2015**, *16*, 33–40. [CrossRef]
15. Park, H.I.; Lee, S.R. Evaluation of the compression index of soils using an artificial neural network. *Comput. Geotech.* **2011**, *38*, 472–481. [CrossRef]
16. Bardhan, A.; Kardani, N.; Alzo'ubi, A.K.; Samui, P.; Gandomi, A.H.; Gokceoglu, C. A comparative analysis of hybrid computational models constructed with swarm intelligence algorithms for estimating soil compression index. *Arch. Comput. Methods Eng.* **2022**, *29*, 4735–4773. [CrossRef]
17. Saisubramanian, R.; Murugaiyan, V. Prediction of compression index of marine clay using artificial neural network and multilinear regression models. *J. Soft Comput. Civ. Eng.* **2021**, *5*, 114–124.
18. Benbouras, M.A.; Kettab Mitiche, R.; Zedira, H.; Petrisor, A.I.; Mezouar, N.; Debiche, F. A new approach to predict the compression index using artificial intelligence methods. *Mar. Georesources Geotechnol.* **2019**, *37*, 704–720. [CrossRef]
19. Zhang, P.; Yin, Z.Y.; Jin, Y.F.; Chan, T.H.; Gao, F.P. Intelligent modelling of clay compressibility using hybrid meta-heuristic and machine learning algorithms. *Geosci. Front.* **2021**, *12*, 441–452. [CrossRef]
20. Asteris, P.G.; Mamou, A.; Ferentinou, M.; Tran, T.T.; Zhou, J. Predicting clay compressibility using a novel Manta ray foraging optimization-based extreme learning machine model. *Transp. Geotech.* **2022**, *37*, 100861. [CrossRef]
21. Long, T.; He, B.; Ghorbani, A.; Khatami, S.M.H. Tree-based techniques for predicting the compression index of clayey soils. *J. Soft Comput. Civ. Eng.* **2023**, *7*, 52–67.
22. Lee, S.; Kang, J.; Kim, J.; Baek, W.; Yoon, H. A Study on Developing a Model for Predicting the Compression Index of the South Coast Clay of Korea Using Statistical Analysis and Machine Learning Techniques. *Appl. Sci.* **2024**, *14*, 952. [CrossRef]
23. Díaz, E.; Spagnoli, G. A super-learner machine learning model for a global prediction of compression index in clays. *Appl. Clay Sci.* **2024**, *249*, 107239. [CrossRef]
24. Castelvecchi, D. Can we open the black box of AI? *Nat. News* **2016**, *538*, 20. [CrossRef] [PubMed]
25. Guidotti, R.; Monreale, A.; Ruggieri, S.; Turini, F.; Giannotti, F.; Pedreschi, D. A survey of methods for explaining black box models. *ACM Comput. Surv. (CSUR)* **2018**, *51*, 1–42. [CrossRef]
26. Rios, A.J.; Seghier, M.E.A.B.; Plevris, V.; Dai, J. Explainable ensemble learning framework for estimating corrosion rate in suspension bridge main cables. *Results Eng.* **2024**, *23*, 102723. [CrossRef]
27. Zaman, M.W.; Hossain, M.R.; Shahin, H.; Alam, A. A study on correlation between consolidation properties of soil with liquid limit, in situ water content, void ratio and plasticity index. *Geotech. Sustain. Infrastruct. Dev.* **2016**, *5*, 899–902.
28. McCabe, B.A.; Sheil, B.B.; Long, M.M.; Buggy, F.J.; Farrell, E.R. Empirical correlations for the compression index of Irish soft soils. *Proc. Inst. Civ.-Eng. Eng.* **2014**, *167*, 510–517. [CrossRef]
29. Saha, A.; Nath, A.; Dey, A.K. Multivariate geophysical index-based prediction of the compression index of fine-grained soil through nonlinear regression. *J. Appl. Geophys.* **2022**, *204*, 104706. [CrossRef]
30. Widodo, S.; Ibrahim, A. Estimation of primary compression index (Cc) using physical properties of Pontianak soft clay. *Int. J. Eng. Res. Appl.* **2012**, *2*, 2231–2235.
31. Kalantary, F.; Kordnaeij, A. Prediction of compression index using artificial neural network. *Sci. Res. Essays* **2012**, *7*, 2835–2848. [CrossRef]
32. Alhaji, M.M.; Alhassan, M.; Tsado, T.Y.; Mohammed, Y.A. Compression Index Prediction Models for Fine-grained Soil Deposits in Nigeria. In Proceedings of the 2nd International Engineering Conference, Charleston, SC, USA, 1–3 April 2017; Federal University of Technology: Minna, Nigeria, 2017.
33. Amagu, A.; Eze, S.; Jun-Ichi, K.; Nweke, M. Geological and geotechnical evaluation of gully erosion at Nguzu Edda, Afikpo Sub-basin, southeastern Nigeria. *J. Environ. Earth Sci.* **2018**, *8*, 148–158.
34. Amagu, C.A.; Enya, B.O.; Kodama, J.i.; Sharifzadeh, M. Impacts of Addition of Palm Kernel Shells Content on Mechanical Properties of Compacted Shale Used as an Alternative Landfill Liners. *Adv. Civ. Eng.* **2022**, *2022*, 9772816. [CrossRef]
35. Murti, D.M.P.; Pujianto, U.; Wibawa, A.P.; Akbar, M.I. K-nearest neighbor (k-NN) based missing data imputation. In Proceedings of the 2019 5th International Conference on Science in Information Technology (ICSITech), Yogyakarta, Indonesia, 23–24 October 2019; IEEE: Piscataway, NJ, USA, 2019; pp. 83–88.
36. Jadhav, A.; Pramod, D.; Ramanathan, K. Comparison of performance of data imputation methods for numeric dataset. *Appl. Artif. Intell.* **2019**, *33*, 913–933. [CrossRef]

37. Bhattacharya, G.; Ghosh, K.; Chowdhury, A.S. Test point specific k estimation for kNN classifier. In Proceedings of the 2014 22nd International Conference on Pattern Recognition, Stockholm, Sweden, 24–28 August 2014; IEEE: Piscataway, NJ, USA, 2014; pp. 1478–1483.
38. Breiman, L. Random forests. *Mach. Learn.* **2001**, *45*, 5–32. [CrossRef]
39. Ge, Q.; Liu, Z.q.; Sun, H.y.; Lang, D.; Shuai, F.x.; Shang, Y.q.; Zhang, Y.q. Robust design of self-starting drains using Random Forest. *J. Mt. Sci.* **2021**, *18*, 973–989. [CrossRef]
40. Schapire, R.E. The boosting approach to machine learning: An overview. In *Nonlinear Estimation and Classification*; Springer: New York, NY, USA, 2003; pp. 149–171.
41. Dong, X.; Guo, M.; Wang, S. GBDT-based multivariate structural stress data analysis for predicting the sinking speed of an open caisson foundation. *Georisk Assess. Manag. Risk Eng. Syst. Geohazards* **2024**, *18*, 333–345. [CrossRef]
42. Chen, T.; He, T.; Benesty, M.; Khotilovich, V.; Tang, Y.; Cho, H.; Chen, K.; Mitchell, R.; Cano, I.; Zhou, T.; et al. Xgboost: Extreme Gradient Boosting. R Package Version 0.4-2. 2015. Available online: https://rdocumentation.org/packages/xgboost/versions/0.4-2 (accessed on 8 October 2015).
43. Kavzoglu, T.; Teke, A. Advanced hyperparameter optimization for improved spatial prediction of shallow landslides using extreme gradient boosting (XGBoost). *Bull. Eng. Geol. Environ.* **2022**, *81*, 201. [CrossRef]
44. Naimi, A.I.; Balzer, L.B. Stacked generalization: An introduction to super learning. *Eur. J. Epidemiol.* **2018**, *33*, 459–464. [CrossRef]
45. Lundberg, S.M.; Lee, S.I. A unified approach to interpreting model predictions. *Adv. Neural Inf. Process. Syst.* **2017**, *30*, 4765–4774.
46. Ma, Y.; Zhao, Y.; Yu, J.; Zhou, J.; Kuang, H. An interpretable gray box model for ship fuel consumption prediction based on the SHAP framework. *J. Mar. Sci. Eng.* **2023**, *11*, 1059. [CrossRef]
47. Baptista, M.L.; Goebel, K.; Henriques, E.M. Relation between prognostics predictor evaluation metrics and local interpretability SHAP values. *Artif. Intell.* **2022**, *306*, 103667. [CrossRef]
48. Ben Seghier, M.E.A.; Knudsen, O.Ø.; Skilbred, A.W.B.; Höche, D. An intelligent framework for forecasting and investigating corrosion in marine conditions using time sensor data. *Npj Mater. Degrad.* **2023**, *7*, 91. [CrossRef]

Journal of
Marine Science and Engineering

MDPI

Article

An Experimental Investigation of the Effect of Interfacial Waves on the Evolution of Sliding Zones in a Liquefied Seabed

Xiaolei Liu [1,2], Xingyu Li [1], Hong Zhang [3,*], Yueying Wang [1], Qiang Zhang [4], Haoqiang Wei [3] and Xingsen Guo [1,3,5]

[1] Shandong Provincial Key Laboratory of Marine Environment and Geological Engineering, Ocean University of China, Qingdao 266100, China; xiaolei@ouc.edu.cn (X.L.); 21201031155@stu.ouc.edu.cn (X.L.); wyy98@stu.ouc.edu.cn (Y.W.); xingsen.guo@ucl.ac.uk (X.G.)
[2] Laboratory for Marine Geology, Qingdao Marine Science and Technology Center, Qingdao 266237, China
[3] College of Engineering, Ocean University of China, Qingdao 266404, China; 21230911130@stu.ouc.edu.cn
[4] Shandong Provincial Bureau of Geology & Mineral Resources, Jinan 250013, China; qddkzq@163.com
[5] Department of Civil, Environmental, and Geomatic Engineering, University College London, London WC1E 6BT, UK
* Correspondence: zhanghong9645@ouc.edu.cn

Abstract: The sliding process of liquefied submarine landslides is generally regarded as being induced by the coupling of excess pore pressure accumulation and shear stress under surface wave action. However, the significant role of interfacial waves formed over the seabed surface upon liquefaction has been largely ignored. The characteristics of interfacial waves and their effect on the development of a seabed sliding zone are poorly understood. Wave flume experiments were conducted to observe the occurrence and evolution of the interfacial wave and sliding zone, combined with image analysis to extract interfacial wave parameters. The results show that the shear action of interfacial waves can cause progressive liquefaction sliding of the seabed and the formation of a sliding zone. The specific location and thickness of the sliding zone are always dynamically changing during the liquefaction development process and are consistent with the liquefaction depth. The wave height of liquefaction interfacial waves increases with liquefaction depth, and the maximum ratio of interfacial wave height to surface wave height can reach 0.175, corresponding to a maximum longitudinal width ratio of the sliding zone of 0.25. The continuously developing interfacial waves transfer the energy of surface waves to deeper areas, expanding the limit depth of sliding zone evolution. This study can provide theoretical guidance for the prevention and control of seabed instability and sliding disasters under extreme storm conditions.

Keywords: liquefaction; interfacial wave; sliding zone; submarine landslides; seabed stability

1. Introduction

Seabed stability poses a significant importance to the safety of marine engineering facilities such as offshore drilling platforms and submarine pipelines [1]. However, multiple external factors (waves, earthquakes, hydrate decomposition, etc.) cause seabed instability such as liquefaction, scour, and shear slide to occur frequently [2–4]. The sliding zone, generated through folding and shearing on the sliding surface during seafloor instability, is an essential part of landslides [5]. The physical and mechanical characteristics of sediments in sliding zones, including sediment type, particle composition, and microscopic structure, are closely related to the evolutionary development and landslide stability evaluation [6,7]. Therefore, the study of the sliding zone holds immense importance in identifying the deformation and failure mechanism of submarine landslides, mitigating marine geological disasters, and ensuring the sustainable development of the marine economy.

Wave-induced seabed instability is the most common sliding geohazard in delta and shallow water regions [8,9]. A multitude of systematic studies of the seafloor instabilities in this region have been conducted [10–12]. Taking the Yellow River Subaqueous Delta as an illustration, earlier research frequently attributed seafloor instability in the area to wave-induced shear failure [10]. The shear stress exerted by waves on the seabed is proved to be a key factor for triggering submarine landslides in the Yellow River Subaqueous Delta through laboratory tests and in situ observation [1,10]. Subsequently, scholars gradually recognized that liquefaction induced by waves is also an important factor leading to the deformation and sliding of the seabed [13,14]. Model experiments about the liquefaction of saturated non-viscous or silty sediments found that the liquefaction is related to the gradual development of pore pressure [15–19]. The reason for wave-induced seabed liquefaction is identified to be the redistribution of excess pore water pressure caused by damping and phase lag [17,18]. The threshold of pore water pressure for the liquefaction of soft sediment through laboratory flume experiments was established [19]. The pore water pressure in seabed sediments showed different degrees of lag accumulation characteristics at different depths during wave action through tidal flat tests, and the strength of sediments was not completely lost when the excess pore water pressure exceeded the overlying pressure through a wave flume test [20,21]. It was found that the existing methods may underestimate the potential for seabed liquefaction and that the nonlinear effect of waves on promoting seabed liquefaction is more significant in progressive waves than in standing waves. The coupling effect between pore water pressure development and seabed stress evolution was investigated and used to improve the existing decoupling methods [17,18]. A model for evaluating the seabed instability (e.g., liquefaction, erosion, seepage, and shear sliding) and geological hazards (e.g., shallow surface landslides, pockmarks, mudflows, disturbance layers, and hard shells) of sandy seabed caused by storm surges was established based on the coupling effect in the Yellow River Subaqueous Delta [22]. To sum up, previous studies generally analyzed seafloor sliding from the perspective of shear caused by surface waves and liquefaction, with little attention paid to the interfacial waves formed on the seafloor by liquefied sediment.

Wave-induced liquefied sediment is assumed to be a kind of fluid with mass density [23,24]. The two-layer fluid theory was introduced first to consider the liquefied sediment as a fluid based on the theory proposed by Lamb (1932) [25]. A two-layer fluid model for the liquefaction problem was established when analyzing the propagation law of the sand liquefaction zone under the action of waves and currents [26]. The model was modified by considering parameters such as viscosity [26–31]. The movement trajectory of liquefied sediment particles under shallow-water small-amplitude waves was observed to be elliptical based on the new two-layer fluid model and wave flume experiments, and analytical solutions were given for the movement and displacement of sandy liquefied sediment particles under wave action using small-amplitude wave theory and fluid wave depth reference surface setting methods [27–29]. However, the change law of the interfacial wave and the internal shear stress caused by it, as well as its influence on the sliding zone during the development of seafloor liquefaction, are still unclear. In this paper, the wave flume experiment simulated the liquefied sliding phenomenon of the seabed was conducted to obtain the evolution characteristics of the sliding zone, and then the parameters of the interfacial wave were obtained through the image processing of liquefied seabed video images through the Matlab program platform. Finally, the influence of the interfacial wave on the evolution of the sliding zone was analyzed according to the changes in the interfacial wave during the liquefied sliding process. The results of this study could provide theoretical support for disaster prevention and control under extreme storm conditions.

2. Materials and Methods

2.1. Wave Flume Experiment

The experiment was carried out within a wave flume with dimensions of 5 m in length, 0.4 m in width, and 1 m in depth. Figure 1 illustrates the layout of the experimental flume. A piston-type wave generator was used to produce progressive surface waves. At the far end of the flume, a wedge-shaped wave absorber was installed to dissipate the surface wave energy. In the middle of the flume, an iron bed measuring 1.2 m long and 0.3 m deep was placed to simulate the seabed. To ensure a smooth transition in the overall water depth, a false-bottom platform was set up between the seabed and the wave generator. Pore pressure transducers were embedded in the sediment at depths of 5 cm and 15 cm below the seabed surface to monitor changes in pore pressure under dynamic wave loading, which is critical for analyzing seabed stability. The transducers used in this study were model 86 from Joint Sensor Instruments, piezoresistive sensors developed by the Nanjing Hydraulic Research Institute, China, with a diameter of 80 mm and a height of 20 mm. A turbidity meter was set up on the seabed surface to measure the change in seawater turbidity during the liquefaction process. The turbidity was measured in NTU (Nephelometric Turbidity Units) using RBR T. D designed by the Canada RBR Company. Equipped with conductivity sensors, temperature sensors, and pressure sensors, the instrument can be used to correct turbidity data. Considering the linear correlation between the undrained shear strength and the penetration resistance, the shear strength of the sediment was represented by the penetration resistance measured by one portable cone penetration test probe with a diameter of approximately 15 mm.

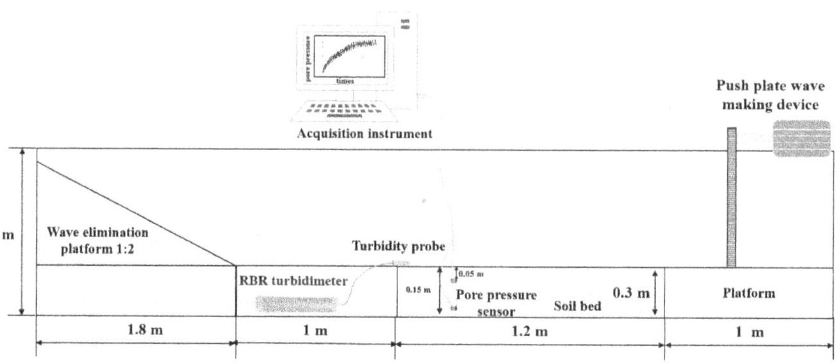

Figure 1. Sketch of the wave flume.

The sediment samples collected from the Yellow River estuary abandoned lobe were used to prepare the seabed in the experiment. The sediment samples were dried, ground, and sieved to remove impurities like shells before the experiment. Then, the weighed sediment samples were mixed with water in a blender with a desired water content of 33%, which is the average natural water content of the Yellow River Delta sediment. The mixture was stirred into a homogeneous slurry and was then slowly poured along the flume wall into the sediment tank up to about 30 cm thick. Water, 55 cm in depth, was then gradually added into the wave flume after the seabed was laid. The physical parameters of the initial seabed, such as bulk density, water content, liquid limit, and particle size composition, were measured before the experiment (Table 1). The seabed sediments were left to consolidate under gravity for 48 h before the flume experiment started. A wave loading of approximately 12 cm high was imposed using the wave generator mounted at the end of the flume to let the seabed reach the liquefaction state. The wave loading was stopped when the seabed reached different liquefaction depths (11 cm, 14 cm, 17 cm, and 20 cm; retracted to 16 cm was selected in this experiment), and the penetration resistance was tested. High-concentration fluid was extracted at the corresponding stages and subjected to density measurement in a 100 mL density bottle. In the liquefaction retraction phase, the columnar samples of a stable seabed, sliding seabed, and re-consolidation seabed were collected to investigate the physical and mechanical characteristics of the sediment. The difference between sliding zone sediment and sediment in the upper and lower regions of the sliding zone was compared. To avoid taking samples too close to the edge wall and to ensure that the liquefaction depth of the seabed in the center of the sediment tank was close to the bottom of the tank, the sampling position was selected at one-third of the distance from the left side of the sediment tank.

Table 1. Basic physical properties of the sediment sample used in the experiment.

Density (kg/m^3)	Water Content (%)	Liquid Limit	Plastic Index	Grain Composition		
				Clay (%)	Silt (%)	Sand (%)
1870	33	27	11	15	54.2	30.8

The liquefaction criterion proposed by Zen et al. (1998) was adopted in this study [32], which states that when the excess pore pressure P_e is greater than or equal to the initial mean normal effective stress σ_0' of the overlying sediment, liquefaction occurs at that location. The expression is as follows:

$$\sigma_0' \leq P_e \tag{1}$$

$$\sigma_0' = \gamma' z \frac{1+2K_0}{3} \tag{2}$$

$$P_e = P - P_b \tag{3}$$

where γ' is the effective unit weight of the sediment, and the value of 790 kg/m^3 was adopted; z is the thickness of the overlaying sediments; K_0 is lateral pressure modulus given by $K_0 = \sigma_h'/\sigma_v'$, and was taken as 0.5 as the average value of the tested sediment; P is the pore water pressure induced by a wave; and P_b is the wave pressure at the seabed.

2.2. Measurement of the Interfacial Waves

The interfacial waves that appeared in the experiment were recorded with a mobile terminal (Canon R5) with a high-definition video recording function at a speed of 30 fps. The image processing technology was used to realize the quantitative measurement of the interfacial wave under the experimental conditions. Based on the Matlab program platform, morphological processing methods were used to process the test process recorded by video, and the wave height, period, amplitude, and other interfacial wave experimental data were obtained. The recorded video was converted into a still image with a pixel of 1980 × 1080. The image was extracted at a speed of 3 Hz, and binary conversion, noise reduction, filtering, edge detection, and other operations were performed to obtain the fluctuation boundary of each frame image. According to the scale set in the sediment tank, coordinate transformation was performed to obtain the fluctuation parameters in the data acquisition area of the image. According to the frame rate of the video, the corresponding relationship between the frame position and time was established, and the time series of a fixed position fluctuation in the image acquisition area was obtained.

The liquefied sediment in this study is viewed as a kind of fluid with mass density, and interfacial wave research was conducted based on the two-layer fluid model (Figure 2). The wave-induced shear stress on the seabed is mainly caused by changes in the direction of internal pressure in the seabed [19]. According to the linear wave theory of progressive waves, the pressure gradient along the direction of wave propagation on the seabed surface is given by the following equation:

$$\frac{\partial p}{\partial x} = \frac{\rho g a k}{coshkd} \cos(kx - \omega t) \tag{4}$$

where ρ is fluid density; g is taken as 9.81 m/s^2; a is amplitude; k is the wave number equal to $2\pi/L$; d represents fluid depth, which consists of water depth d_w and the liquefied seabed depth d_{ls}; and ω represents wave frequency. The maximum shear stress generated by interfacial waves can be calculated by the following equation [33]:

$$\tau_{max} = \hat{p} k z e^{-kz} \tag{5}$$

$$\hat{p} = \frac{\rho g a}{coshkd}\left(\frac{F'}{F' - tanhkd}\right) \quad (6)$$

$$F' = \frac{Gk^2}{\rho\omega^2(1 - \varnothing)} \quad (7)$$

where $\hat{p}$ is the amplitude of the pressure at the sediment–water interface; z is the depth, which is positive downward at the water–sediment interface; G is the shear modulus; and $\varnothing$ is Poisson's ratio.

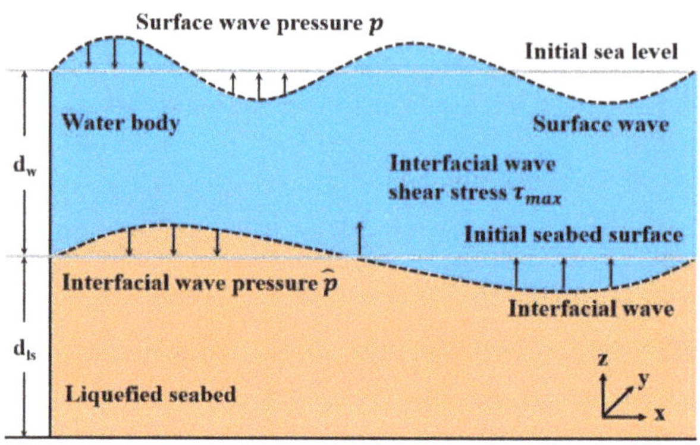

Figure 2. Configuration of the two-layer fluid system (after Sassa et al., 2001 [26]).

3. Results and Analysis

3.1. Development of the Sliding Zone during the Wave Flume Experiment

Figure 3 shows the alteration in the pore pressure at diverse depths and the turbidity of the seabed surface throughout the entire experiment process. With the advancement of the 12 cm high wave action, the pore pressures at the depths of 5 cm and 15 cm rose steeply at the beginning and declined slightly afterward. The excess pore water pressure peaked at 0.75 kPa and 1.4 kPa approximately at 6 min and then decreased gradually. According to Equation (2), the estimated initial mean normal effective stress of the sediment was approximately 0.26 kPa and 0.79 kPa for overlaying sediment thicknesses of 5 cm and 15 cm, respectively. Figure 3 reveals that the excess pore pressure at depths of 5 cm and 15 cm were both larger than the initial mean normal effective stress of the sediment, indicating that the sediment layer had undergone liquefaction. The turbidity data demonstrate a corresponding tendency. The turbidity rose rapidly along with the sudden increase in pore pressure, reaching a maximum of 100 NTU at 32 min, followed by a gradual decrease.

The array of features can be observed in the flume experiments (Figure 4). When excess pore pressure initiated its accumulation, minoroscillatory mud ripples emerged on the seabed surface (Figure 4A). Under the continued application of wave loading, the mud ripples gradually extended across the seabed surface, resulting in the liquefaction of the seabed (Figure 4B). Figure 4C shows the evolution of the liquefaction boundary expanding with time. The maximum liquefaction depth in this experiment is 0.2 m (Figure 4D). The sliding zone formed when the liquefied sediment slid left and right along the concave interface (Figure 4C,D). The liquefaction bottom boundary began to rise, and the stratum that passed through was re-stratified (Figure 4E). The process of liquefaction stopped, and the seabed sank by approximately 5 cm (Figure 4F).

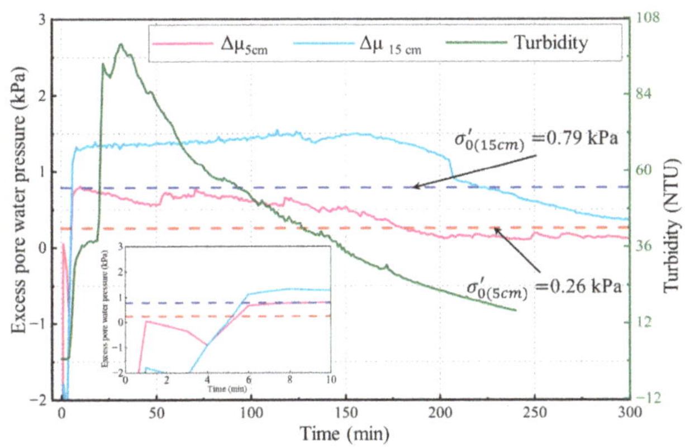

Figure 3. Development of pore pressure at different depths of seabed and turbidity at the seabed surface.

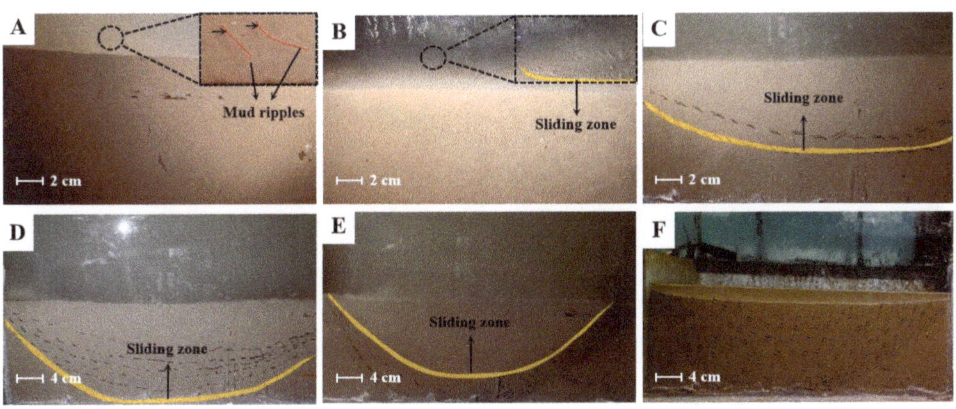

Figure 4. Records of phenomena at various stages of the flume test. (**A**) The mud ripples appeared at the initial stage of the flume test. (**B**) The seabed began to liquefy, and the sliding zone came to being. (**C**) The sliding zone expand downward as the evolution of the liquefaction boundary expanded with time. (**D**) The largest sliding zone formed when the maximum liquefaction depth was reached. (**E**) The sliding zone moved upward when the liquefaction depth decreased with time. (**F**) Both the liquefaction and sliding process were over.

Figure 5 presents the vertical profile of the sliding zone along with the grain size distribution and the penetration strength of the sediment. Specific locations of the sliding seabed, sliding zone, reconsolidated sediment layer, and the original sediment layer are also marked. The trend in the penetration strength of the sediment shows an initial increase followed by a decrease with depth. The CT image and grain size distribution both indicate that the sediment was re-stratified. The stable sediment layer at the bottom, which did not slide, consists of uniform sandy silt particles. The sliding seabed at the top is dominated by clayey sediment particles of relatively uniform size. The reconsolidated stable layer above the non-sliding layer has relatively fewer clay particles compared to the sliding zone, and between these two layers is the sliding zone where clay content gradually decreases with depth. The penetration strength of the original sediment layer is less than 3 N, and it is close to 0 N. The sliding zone is a transition zone with a sudden increase in strength from around 3 N to 8.3 N. The strength value of the sediment at the reconsolidated sediment layer increases to 10 N, whereas the strength of the original soil layer is approximately 3.5 N.

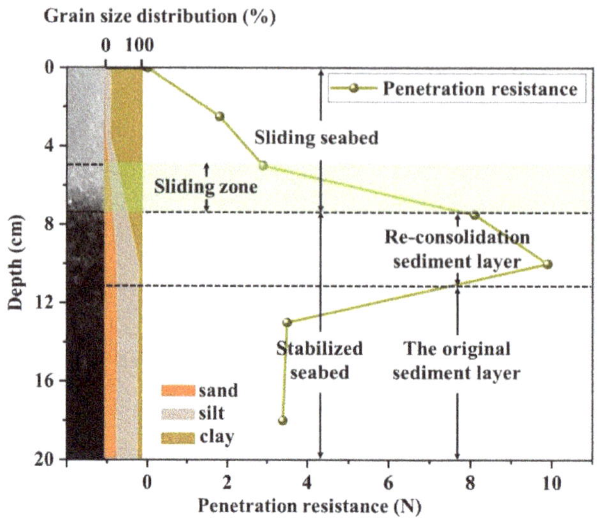

Figure 5. Sediment particle composition and penetration resistance in the upper and lower regions of the sliding zone.

3.2. Characteristics of the Liquified Interfacial Wave

The typical wave time series of the liquefied interfacial waves were extracted when the liquefaction depth was 11 cm, 14 cm, 17 cm, and 20 cm and during the onset of liquefaction contraction through the image method (as shown in Figure 6). The wave parameters for each stage are recorded in Table 2.

Table 2. Surface and interfacial wave parameters.

	Wave Height/h (m)	Wavelength/λ (m)	Period/T (s)	Density/ρ (kg/m^3)	Wave Number/k	Fluid Depth/d (m)	Shear Stress/τ (Pa)
Surface wave	0.120	2.28	1.27	1025	2.76	0.55	697
Stage 1	0.004	1.32	1.33	2104	4.76	0.11	172
Stage 2	0.005	1.48	1.33	2093	4.25	0.14	184
Stage 3	0.007	1.61	1.33	2091	3.90	0.17	228
Stage 4	0.020	1.72	1.33	2081	3.65	0.20	578
Stage 5	0.010	1.69	1.33	2067	3.72	0.19	299

In this experiment, the position of the sliding zone was located at the interface of the liquefied seabed and initial seabed, which corresponded to the liquefaction depth. When the liquefaction depth was 11 cm, the interfacial wave was relatively unstable with a small amplitude of around 4 mm (Figure 6A). As the sliding zone continued to move downward, the waveform became more stable. When the liquefaction depth was 14 cm, the amplitude of the interface wave increased slightly to about 5 mm (Figure 6B). When the liquefaction depth reached 17 cm, the amplitude of the interfacial wave continued to increase slowly, reaching approximately 7 mm (Figure 6C). When the liquefaction depth reached the peak value of 20 cm, the amplitude of the interface wave increased rapidly to around 20 mm (Figure 6D). After the liquefaction started to retract, the amplitude of the interfacial wave also decreased. Under the surface wave with an amplitude of 12 cm and a period of 1.27 s, the period of the interfacial wave was generally around 1.33 s throughout the evolution of the sliding zone, which was slightly higher than the surface wave and generally consistent it (Figure 6E). Overall, the liquefaction interfacial wave presented a relatively stable waveform in the time sequence. The frequency of the interfacial wave was nearly consistent with the surface wave, slightly lower than it, while the wavelength and amplitude were much smaller than the surface wave. The interfacial wave was in essence a subharmonic resonance caused by the surface wave resonating on the seabed.

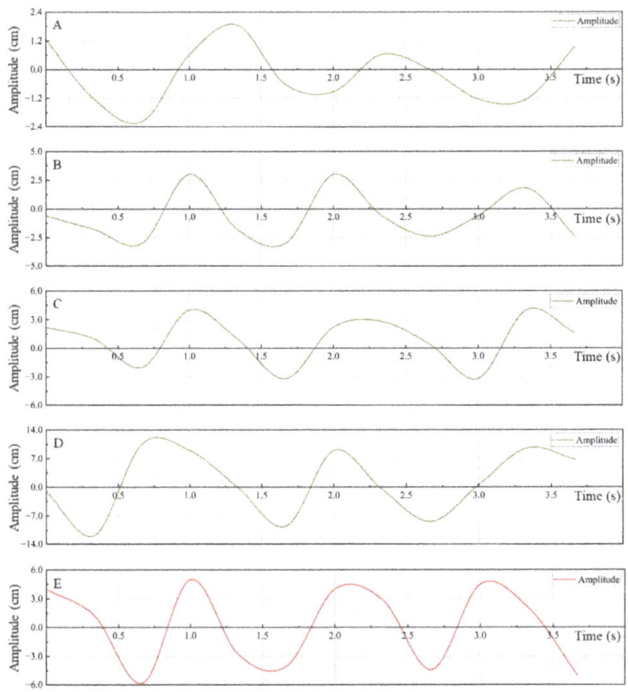

Figure 6. Time series of interfacial waves at a fixed position. Time series of an interfacial wave when the depth of liquefaction is (**A**) 11 cm, (**B**) 14 cm, (**C**) 17 cm, and (**D**) 20 cm. (**E**) Time series of an interfacial wave when liquefaction began to retract.

4. Discussion

4.1. The Evolution of Sliding Zone

The sliding zone is generated because of the liquefaction-induced sliding of sediment, and its thickness progressively augments during the downward progression of the sliding zone. As shown in Figure 5, the thickness of the sliding zone can increase to about 2 cm under the experimental conditions. Figure 7A illustrates the evolution process of the interfacial wave and the sliding zone. The wave height of the interfacial wave increased first and then decreased, and the sliding zone also has a corresponding evolution trend. A series of features observed in the flume experiments can be incorporated into a new model of interfacial wave-induced evolution of the sliding zone. As mentioned above, liquefaction sliding is caused by the coupling effect of wave-induced liquefaction and shearing. This model integrates the process of interfacial wave-induced sediment sliding—including the evolution of the sliding zone and the development of interfacial waves. Under the wave action, mud ripples first appear on the seabed surface, resulting in local interfacial waves. As the liquefied sediment begins to fluctuate, sliding zones initially form beneath it. The shear stress on the seabed sediment is directly applied by the surface wave in this stage. A continuous fluctuating stable interfacial wave with a period of about 1.33 s was generated due to the sliding of the liquefied sediment after the mud ripples spread over the entire surface (stage 1 in Figure 7A). The local sliding zone evolved into a continuous intact sliding zone due to the relative sliding between the fluctuating liquefied sediment and the stable seabed. The bottom boundary of liquefied sediment continued to decline under the action of waves and reached a maximum depth (0.2 m), and the thickness of the sliding zone increased gradually from 0 to the highest level (about 2 cm) (stage 2 in Figure 7A). The depth and width of the sliding zone began to retract after reaching the maximum depth (stage 3 in Figure 7A). In a practical seabed, due to the non-uniform distribution of interfacial waves, the liquefied sediment will slide in the form of flow slides in the evolution process under the action of many factors such as slope and waves. The sliding zone undergoes secondary evolution in the process (stage 4 in Figure 7A).

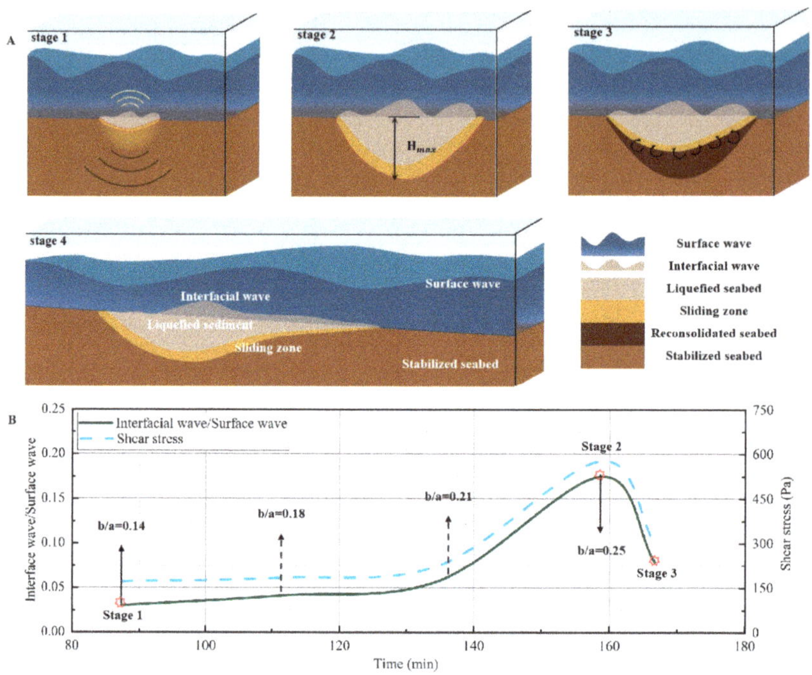

Figure 7. (**A**) Schematic of the evolution of the sliding zone induced by interfacial waves during the development of an interfacial wave. In stage 1, the interfacial wave and the sliding zone were generated due to the sliding of the liquefied sediment. The bottom boundary of liquefied sediment reached a maximum depth, and the thickness of the sliding zone increased gradually from 0 to the highest level in stage 2. Then the depth and width of the sliding zone began to retract in stage 3. The sliding zone undergoes secondary evolution under the action of many factors such as slope and waves. (**B**) Curves of the ratio of the interfacial wave height to the surface wave height and the corresponding shear stress.

4.2. The Effect of Interfacial Waves

The interactions between surface waves and interfacial waves have been addressed by many studies [27–29]. The energy of surface waves can only affect shallow sediment layers due to the dissipation of wave energy in the seabed [34–37]. The action of surface wave can only cause liquefaction in a certain range of the seabed surface when the interfacial wave can continuously shear the underlying sediment layers. According to Figure 3, the sustained action of interfacial waves leads to a slow dissipation of pore pressure [38]. During the period in which interfacial waves exist, high pore pressure values can still be maintained, leading to the downward movement of the sliding zone along with the continuous development of the range of liquefaction slide. Therefore, interfacial waves, to some extent, delayed the attenuation of energy transfer to deeper sediment layers [39,40].

According to Equations (4) and (5), the maximum shear stress generated by surface waves and liquefaction interfacial waves was estimated (Table 2). The shear stress caused by the surface waves within the sediment was about 697 Pa, while the shear stress induced by interfacial waves was similar in magnitude to that induced by surface waves and increased with wave height within the measurement range, ranging from a minimum of 172 Pa to a maximum of 578 Pa. Figure 7B shows the variation in the interfacial wave height and corresponding shear stress with the evolution of the sliding zone. As the sliding zone evolved deeper into the seabed, the wave height gradually increased and then decreased during the retraction of liquefaction. Previous simulation experiments based on traditional two-layer fluid models suggested that the propagation of interfacial waves would be limited when the phase angle between the surface and interface increased when liquefaction reached deeper [41]. This study suggests that the growth of interfacial waves is also influenced by sediment strength and other properties. During the initial development of the sliding zone, the growth of interfacial wave height was slow, as the original seabed strength was low, and the penetration resistance was only about 3 N. The interfacial wave continuously sheared the seabed, resulting in a continued increase in the depth of the liquefaction slide and an increase in the interfacial wave height. The rate of increase in the interfacial wave height accelerated when the depth of the sliding zone reached 17 mm until the sliding zone evolved to its deepest (20 mm), and then the interfacial wave height decreased. The stratum was re-stratified due to the differentiation and re-deposition of sediment particles [40–43]. The sediment strength is significantly enhanced, and a high-strength reconsolidated sediment gradually forms, which is not easily broken by the shear stress induced by the interfacial wave. The growth of the interfacial wave was limited by the reconsolidated sediment.

The depth b of the sliding zone and width a of the trench are defined as the longitudinal width ratio of the sliding zone, and the interfacial wave height is non-dimensionalized through the utilization of the ratio of the interfacial wave height to the surface wave height. As presented in Figure 7B, within the first 130 min before the onset of liquefaction, the ratio of the interfacial wave height to the surface wave height was less than 0.05, signifying the initial liquefaction stage. The liquefied area was small, and during this stage, the interfacial wave gradually stabilized, and it also constituted a fast-evolving stage of the sliding zone. As liquefaction continued to evolve, the particle coarsening effect continued, and the influence of interfacial wave height growth on the sliding zone evolution weakened until the maximum value of the ratio of interface wave height to surface wave height reached 0.175. Subsequently, a high-strength reconsolidated soil was produced, and the shear force caused by the interfacial wave was insufficient to break the soil, resulting in the contraction of the sliding zone.

Under the conditions of this experiment, the maximum ratio of interfacial wave height to surface wave height was found to be 0.175, corresponding to a maximum sliding zone aspect ratio of 0.25. In actual observation surveys, sliding zones on the seafloor are difficult to identify. Therefore, the maximum ratio of interface wave height to surface wave height determined through experiments can be used in conjunction with geophysical data to indirectly determine the width, depth, and extent of sliding zones in the survey area.

5. Conclusions

This paper studied the influence of liquefaction–induced interface waves on sliding zones through a series of flume experiments. The conclusions are as follows:

(1) Liquefaction-induced interfacial waves are formed on the seabed surface due to wave-induced liquefaction, with a slightly smaller period than surface waves. The size of the interfacial waves is mainly controlled by the amplitude.

(2) The growth of interfacial waves is significantly affected by the strength of the seabed. In low-strength original seabed, interfacial waves can continue to grow until they are suppressed and recede after the formation of high-strength reconsolidated sediment.
(3) The development of sliding zones under the influence of interfacial waves shows a pattern from non-existence to existence, and the influence of interfacial wave height on the evolution rate of sliding zones gradually decreases with the development of liquefaction.
(4) In a homogeneous silty seabed, the maximum ratio of the interfacial wave height to surface wave height is 0.175, corresponding to the maximum longitudinal width ratio of sliding zones of 0.25. Combined with geophysical data in the study area, the depth and range of sliding zones can be determined.

Author Contributions: Conceptualization, H.Z. and X.L. (Xiaolei Liu); methodology, X.L. (Xingyu Li) and Y.W.; formal analysis, X.L. (Xingyu Li), Y.W. and H.W.; investigation, X.L. (Xingyu Li), Y.W., Q.Z. and H.Z.; data curation, X.L. (Xingyu Li) and X.G.; writing—original draft preparation, X.L., X.L. (Xingyu Li) and H.Z.; writing—review and editing, H.Z., X.G. and X.L. (Xiaolei Liu); visualization, Y.W. and H.W.; supervision, X.L. (Xiaolei Liu) and X.G.; project administration, X.L. (Xiaolei Liu) and H.Z. All authors have read and agreed to the published version of the manuscript.

Funding: This research was funded by the National Natural Science Foundation of China (42207181), the Shandong Postdoctoral Science Foundation (SDCX-ZG-202303021), the Fundamental Research Funds for the Central Universities (202441003), and the Shandong Province National-Level Leading Talent Supporting Project (2022GJJLJRC-15).

Institutional Review Board Statement: Not applicable.

Informed Consent Statement: Not applicable.

Data Availability Statement: All data from this paper are available upon request. Please contact Hong Zhang at zhanghong9645@ouc.edu.cn.

Acknowledgments: The authors would like to thank Weijia Li, Heyu Yu, and Xiaotian Xie for their help in the process of the flume experiment.

Conflicts of Interest: The authors declare no conflicts of interest.

References

1. Guo, X.S.; Liu, X.L.; Zheng, T.Y.; Zhang, H.; Lu, Y.; Li, T.T. A mass transfer-based LES modelling methodology for analyzing the movement of submarine sediment flows with extensive shear behavior. *Coast. Eng.* **2024**, *191*, 104531. [CrossRef]
2. Fan, N.; Jiang, J.X.; Nian, T.K.; Dong, Y.K.; Guo, L.; Fu, C.W.; Tian, Z.C.; Guo, X.S. Impact action of submarine slides on pipelines: A review of the state-of-the-art since 2008. *Ocean Eng.* **2023**, *286*, 115532. [CrossRef]
3. Liu, X.L.; Wang, Y.Y.; Zhang, H.; Guo, X.S. Susceptibility of typical marine geological disasters: An overview. *Geoenviron. Disasters* **2023**, *10*, 10. [CrossRef]
4. Jamil, M.; Siddiqui, N.A.; Umar, M.; Usman, M.; Ahmed, N.; Rahman, A.H.; Zaidi, F.K. Aseismic and seismic impact on development of soft-sediment deformation structures in deep-marine sand-shaly Crocker fan in Sabah, NW Borneo. *J. King Saud Univ. Sci.* **2021**, *33*, 101522. [CrossRef]
5. Chen, Q.; Cui, D.S.; Chen, Y.; Tao, X.Y.; Xiang, W. Effect of prior cyclic loading on triaxial compression strength of sliding zone soil of the Huangtupo landslide. *Adv. Civ. Eng.* **2021**, *2021*, 9924995. [CrossRef]
6. Elger, J.; Berndt, C.; Rüpke, L.; Krastel, S.; Gross, F.; Geissler, W.H. Submarine slope failures due to pipe structure formation. *Nat. Commun.* **2018**, *9*, 715. [CrossRef] [PubMed]
7. Jamil, M.; Siddiqui, N.A.; Umar, M.; Usman, M.; Ahmed, N.; Rahman, A.H.; Zaidi, F.K. Facies analysis and distribution of Late Palaeogene deep-water massive sandstones in submarine-fan lobes, NW Borneo. *Geol. J.* **2022**, *57*, 4489–4507. [CrossRef]
8. Yin, M.; Rui, Y. Laboratory study on submarine debris flow. *Mar. Georesour. Geotechnol.* **2018**, *36*, 950–958. [CrossRef]
9. Zhang, W.C.; Randolph, M.F.; Puzrin, A.M.; Wang, D. Transition from shear band propagation to global slab failure in submarine landslides. *Can. Geotech. J.* **2019**, *56*, 554–569. [CrossRef]
10. Prior, D.B.; Suhayda, J.N.; Lu, N.Z.; Bornhold, B.D.; Keller, G.H.; Wiseman, W.J. Storm wave reactivation of a submarine landslide. *Nature* **1989**, *341*, 47–50. [CrossRef]
11. Quartau, R.; Ramalho, R.S.; Madeira, J.; Santos, R.; Rodrigues, A.; Roque, C. Gravitational, erosional and depositional processes on volcanic ocean islands: Insights from the submarine morphology of Madeira Archipelago. *Earth Planet. Sci. Lett.* **2018**, *482*, 288–299. [CrossRef]

12. Zhang, H.; Lu, Y.; Liu, X.; Li, X.; Wang, Z.; Ji, C. Morphology and origin of liquefaction-related sediment failures on the Yellow River subaqueous delta. *Mar. Petrol. Geol.* **2023**, *153*, 106262. [CrossRef]
13. Zen, K.K.; Yamazaki, H. Mechanism of wave-induced liquefaction and densification in seabed. *Soils Found.* **1990**, *30*, 90–104. [CrossRef] [PubMed]
14. Rahman, M.S. Wave-induced instability of seabed: Mechanism and conditions. *Mar. Georesour. Geotechnol.* **1991**, *10*, 277–299. [CrossRef]
15. Ulker, M.B.C.; Rahman, M.S.; Jeng, D.-S. Wave-induced response of seabed: Various formulations and their applicability. *Appl. Ocean Res.* **2009**, *31*, 12–24. [CrossRef]
16. Qi, W.; Li, C.; Jeng, D.; Gao, F.; Liang, Z. Combined wave-current induced excess pore-pressure in a sandy seabed: Flume observations and comparisons with theoretical models. *Coast. Eng.* **2019**, *147*, 89–98. [CrossRef]
17. Tong, L.; Zhang, J.; Zhao, J.; Zheng, J.; Guo, Y. Modelling study of wave damping over a sandy and a silty bed. *Coast. Eng.* **2020**, *61*, 103756. [CrossRef]
18. Ren, Y.P.; Xu, G.H.; Xu, X.B.; Zhao, T.L.; Wang, X.Z. The initial wave induced failure of silty seabed: Liquefaction or shear failure. *Ocean Eng.* **2020**, *200*, 106990. [CrossRef]
19. De Wit, P.J.; Kranenburg, C. The wave-induced liquefaction of cohesive sediment beds. *Estuar. Coast. Shelf Sci.* **1997**, *45*, 261–271. [CrossRef]
20. Zhang, H.; Liu, X.L.; Jia, Y.G.; Du, Q.Z.; Sun, Y.F.; Yin, P.; Shan, H.X. Rapid consolidation characteristics of Yellow River-derived sediment: Geotechnical characterization and its implications for the deltaic geomorphic evolution. *Eng. Geol.* **2020**, *270*, 105578. [CrossRef]
21. Dimitrova, R.S.; Yanful, E.K. Factors affecting the shear strength of mine tailings/clay mixtures with varying clay content and clay mineralogy. *Eng. Geol.* **2012**, *125*, 11–25. [CrossRef]
22. Wang, H.; Liu, H.J. Evaluation of storm wave-induced silty seabed instability and geo-hazards: A case study in the Yellow River delta. *Appl. Ocean Res.* **2016**, *58*, 135–145. [CrossRef]
23. Mathew, J.; Baba, M.; Kurian, N.P. Mudbanks of the southwest coast of India. I: Wave characteristics. *J. Coast. Res.* **1995**, *11*, 168–178.
24. Yu, H.Y.; Liu, X.L.; Lu, Y.; Li, W.J.; Gao, H.; Wu, R.Y.; Li, X.Y. Characteristics of the sediment gravity flow triggered by wave-induced liquefaction on a sloping silty seabed: An experimental investigation. *Front Earth Sci.* **2022**, *10*, 909605. [CrossRef]
25. Lamb, H. *Hydrodynamics*; Cambridge University Press: Cambridge, UK, 1932.
26. Sassa, S.; Sekiguchi, H.; Miyamoto, J. Analysis of progressive liquefaction as a moving-boundary problem. *Geotechnique* **2001**, *51*, 847–857. [CrossRef]
27. Eric, M.; Robert, A.D. Experimental observation of standing interfacial waves induced by surface waves in muddy water. *Phys. Fluids* **2011**, *23*, 096603. [CrossRef]
28. Guo, X.S.; Fan, N.; Zheng, D.F.; Fu, C.W.; Wu, H.; Zhang, Y.J.; Song, X.L.; Nian, T.K. Predicting impact forces on pipelines from deep-sea fluidized slides: A comprehensive review of key factors. *Int. J. Min. Sci. Technol.* **2024**, *34*, 211–225. [CrossRef]
29. Hsu, W.Y.; Hwung, H.H.; Yang, R.Y.; Liu, C.M. Interfacial wave motion caused by wave-mud interaction. *J. Vis.* **2012**, *15*, 215–224. [CrossRef]
30. Miyamoto, J.; Sassa, S.; Sekiguchi, H. Progressive solidification of a liquefied sand layer during continued wave loading. *Geotechnique* **2004**, *54*, 617–629. [CrossRef]
31. Ren, Y.P.; Xu, X.B.; Xu, G.H.; Liu, Z.Q. Measurement and calculation of particle trajectory of liquefied soil under wave action. *Appl. Ocean Res.* **2020**, *101*, 102202. [CrossRef]
32. Zen, K.K.; Jeng, D.S.; Hsu, J.R.C.; Ohyama, T. Wave-induced seabed instability: Difference between liquefaction and shear failure. *Soils Found.* **1998**, *38*, 37–47. [CrossRef]
33. Egan, G.; Cowherd, M.; Fringer, O.; Monismith, S. Observations of Near-Bed Shear Stress in a Shallow, Wave- and Current-Driven Flow. *J Geophys. Res. Oceans* **2019**, *124*, 6323–6344. [CrossRef]
34. Michallet, H.; Mory, M.; Piedra-Cueva, I. Wave-induced pore pressure measurements near a coastal structure. *J. Geophys. Res.* **2009**, *114*. [CrossRef]
35. Phillips, O.M. Wave interactions- the evolution of an idea. *J. Fluid Mech.* **1981**, *106*, 215. [CrossRef]
36. Jamali, M.; Lawrence, G. Viscous wave interaction due to motion of a surface wave over a sediment bed. *J. Offshore Mech. Arct. Eng.* **2006**, *128*, 276. [CrossRef]
37. Elgar, S.; Raubenheimer, B. Wave dissipation by muddy seafloors. *Geophys. Res. Lett.* **2008**, *35*. [CrossRef]
38. Wang, B.; Zen, K.; Chen, G.Q.; Zhang, Y.B.; Kasama, K. Excess pore pressure dissipation and solidification after liquefaction of saturated sand deposits. *Soil Dyn. Earthq. Eng.* **2013**, *49*, 157–164. [CrossRef]
39. Tzang, S.Y. Unfluidized soil responses of a silty seabed to monochromatic waves. *Coast. Eng.* **1998**, *35*, 283–301. [CrossRef]
40. Cui, L.; Jeng, D.S. Seabed liquefaction around breakwater heads at a river mouth: An integrated 3D model. *Ocean Eng.* **2021**, *242*, 110036. [CrossRef]
41. Aleebrahim, M.A.; Jamali, M. Experimental investigation of instability of fluid mud layer under surface wave motion. *Phys. Fluids* **2022**, *34*, 036602. [CrossRef]

42. Xu, G.H.; Liu, Z.Q.; Sun, Y.F.; Wang, X.; Lin, L.; Ren, Y.P. Experimental characterization of storm liquefaction deposits sequences. *Mar. Geol.* **2016**, *382*, 191–199. [CrossRef]
43. Yang, Z.N.; Zhu, Y.M.; Liu, T.; Sun, Z.Q.; Ling, X.Z.; Yang, J.M. Pumping effect of wave-induced pore pressure on the development of fluid mud layer. *Ocean Eng.* **2019**, *189*, 106391. [CrossRef]

Journal of
Marine Science and Engineering

MDPI

Review

A State-of-the-Art Review of the Hydrodynamics of Offshore Pipelines Under Submarine Gravity Flows and Their Interactions

Cheng Zhang [1], Tao Tang [1,*], Fan Zhang [1], Chengjiao Ren [1], Hongcao Zhang [2] and Guochao Wu [3,4]

1 School of Marine Science and Engineering, South China University of Technology, Guangzhou 511442, China; zhangcheng@scut.edu.cn (C.Z.); msezhangfan@mail.scut.edu.cn (F.Z.); chengjiaoren@scut.edu.cn (C.R.)
2 School of Engineering, The University of Western Australia, 35 Stirling Highway, Crawley, WA 6009, Australia; hongcao.zhang@research.uwa.edu.au
3 Second Institute of Oceanography, Ministry of Natural Resources, Hangzhou 310012, China; wugc@sio.org.cn
4 Key Laboratory of Submarine Geosciences, Ministry of Natural Resources, Hangzhou 310012, China
* Correspondence: 20259213@scut.edu.cn

Abstract: Submarine gravity flows, e.g., debris flows and turbidity currents, pose a significant threat to offshore pipeline integrity. This risk primarily manifests through the imposition of substantial dynamic loads on pipelines or their large displacement when impacted by such flows. To enhance our understanding of these threats and facilitate the development of more robust pipeline design and protection strategies, this work reviewed the interactions between submarine gravity flows and offshore pipelines. For an individual pipeline, critical focus lies in characterizing the influence of key parameters—including Reynolds number, span height, impact angle, pipe geometry, ambient temperature, and surface roughness—on both the resultant impact forces and the fluid-structure interaction dynamics. Then, investigations into the interactions between gravity flows and multiple pipes are summarized, where the in-line spacing distance between two pipes is a key factor in reducing the impact force. Further, flow-induced vibration responses of a single pipeline and two tandem pipelines under gravity flows are presented. Building upon a thorough review, we conducted overall evaluations. There are few experimental studies and most investigations ideally treat the seabed to be horizontal, which does not always occur in practical engineering. Choosing empirical formulas to evaluate hydrodynamic loads should carefully consider the specific working conditions. An appropriate non-Newtonian fluid model is significantly important to avoid uncertainties. Some practical risk reduction measures such as streamlined structures and reduction in roughness are recommended. Finally, suggestions for future study and practice are proposed, including the requirement for three-dimensional numerical investigations, assessment of fatigue damage by flow-induced vibrations, consideration of flexible pipeline, and more attention to multiple pipelines.

Keywords: gravity flows; offshore pipeline; hydrodynamics; vibrations

1. Introduction

Submarine gravity flows represent underwater currents carrying suspended sediments. Gravity is the primary energy source of submarine gravity flows, and it converts the potential energy of unstable slope sediments into kinetic energy, which facilitates the transport of eroded materials from continental slopes to abyssal plains. Gravity flows can not only generate impact forces on pipelines but also scour several meters deep, threatening even buried pipelines. Gravity flows form when sediment accumulation exceeds slope stability [1–5], often triggered by earthquakes, volcanic activity, and other internal or external mechanisms [6–8]. As shown in Figure 1, under the effect of gravity, the water content increases while density and strength decrease [9–13], resulting in various sliding movement forms, typically including plastic deformation, non-Newtonian fluid, transition fluid, and Newtonian fluid. Among the most significant types are debris flows and turbidity currents. Debris flows are cohesive, high-density mixtures of sediment (clay to boulders) and water, where particles are supported primarily by matrix strength rather than fluid turbulence. They exhibit poor sorting, chaotic deposition, and unstratified masses, and typically move as slow, viscous slurries confined to steep slopes or proximal basin settings. In contrast, turbidity currents are turbulent, dilute flows in which sediment is suspended by fluid turbulence, enabling long-distance transport. Their depositions are well-structured to form graded beds (Bouma sequences) with sharp basal contacts and sedimentary structures reflecting waning flow energy [14–16]. Key distinctions between debris flows and turbidity currents lie in their support mechanisms (matrix strength vs. turbulence), depositional fabrics (disorganized vs. stratified), and run-out distances (short vs. extensive). Debris flows may evolve into turbidity currents via dilution and turbulence incorporation, producing hybrid deposits with basal chaotic units overlain by graded turbidites [17]. Both processes are critical in shaping submarine fans, distributing organic-rich sediments, and triggering geohazards.

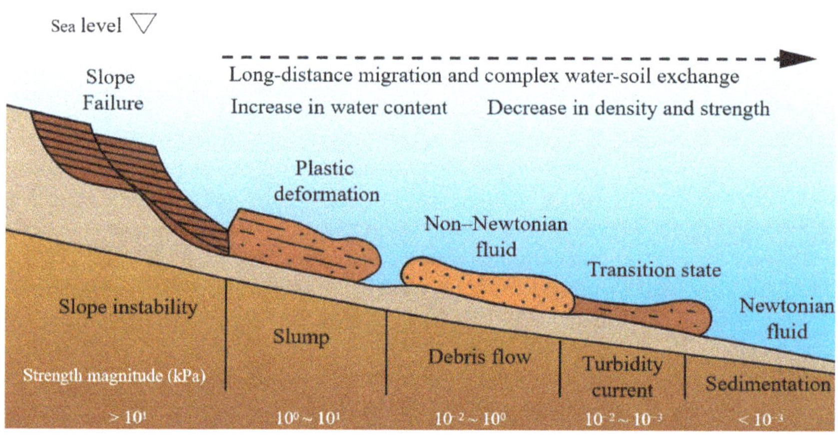

Figure 1. Typical process of gravity flows. (The panel was modified from Guo et al. [18]).

Gravity currents can transport large amounts of submarine sediment from the continental shelf to the deep ocean floor due to their general characteristics of massive volumes, large influence area, high velocity, and large run-out distance [19–21]. For example, the velocity of high-density turbidity flows can surprisingly reach up to 30 m/s [22], and move extensive run-out distance covering hundreds of kilometers [23], both of which have a significant influence on sediment behavior and a substantial geohazard to submarine installations. The unique phenomenon, hydroplaning, is the key mechanism that allows gravity currents to travel long distances, a phenomenon that is attributed to the reduction in frictional resistance at the interface between the flow and the seafloor [24–26]. Hydroplaning occurs when the hydrodynamic pressure at the flow head reaches sufficient magnitude to balance or exceed the dynamic pressure exerted by the sediment load, resulting in fluidization and subsequent water intrusion beneath the gravity flow. The hydroplaning phenomenon induces frontal acceleration in gravity flows, leading to progressive separation from the main flow body and ultimately surging behavior [27,28]. Hydroplaning can happen easily at a very small velocity of around 1 m/s for small-scale flows in laboratory experiments [27].

Clear classifications of submarine gravity currents are significantly important for analyses in flow dynamics and associated interactions with deep-sea installations. Dott [29] identified the elastic, plastic, and viscous flow behavior with the development of submarine landslide. This classification provides a simple framework to understand the sediment transportation and deposition processes by gravity currents. Based on the particle support methods, Middleton [30] classified gravity flows into four types including particle flow, debris flow, turbidity flow, and pure fluid flow. Further, researchers [31–41] found that gravity currents cannot be treated as one single fluid or adopt only one deposition mechanism in practical sediment and transportation process. Later, Lowe [31,32] proposed a new classification framework by considering both the rheological characteristics of fluids and the supporting mechanisms of sediments. Firstly, based on rheological features, gravity currents are divided into plastic debris flow and pure liquid flow. Then, according to different transport mechanisms, sediment gravity currents are classified into five types including turbidity flow, fluidized flow, liquefied flow, grain flow and debris flow [31,32]. This classification scheme, which not only analyzes rheological characteristics but also takes into consideration the influence of deposition mechanisms, has been widely used for a long time. Now, the most commonly adopted classification method is from the work of Shanmugam [33], as depicted in Figure 2. The concept of sandy debris flow encompasses the types of gravity flow that are unclear or controversial in traditional classifications, such as the transitional type between high-density and low-density turbidity current, turbidity current types with a higher sand content, sediment types with a mixture of mud and sand components, and types with the characteristics of a mixture of turbidity current and debris flow. This method has greatly improved the accuracy and practicality of classification and has been widely applied in the research of deep-water gravity flows. After complex water–sediment exchange and long-distance transportation, the submarine landslide can gradually evolve into debris flows with high density and large velocity, and, ultimately, develop into fluidized turbidity flows. Compared with fluidized turbidity flows (Newtonian fluid), the debris flows—which show higher shear stress and density—are generally considered non-Newtonian fluids [13,34–36], possibly causing significant damage to marine facilities [37]. Non-Newtonian fluid models such as Bingham plastic [37,38] and Herschel–Bulkley models [12,35,39–41] have been commonly adopted to model the yield stress behavior and viscous properties of debris flows.

The progressive development of offshore operations in deep water over 1000 m across continental slope regions increases submarine pipeline systems' vulnerability to geohazard impacts from submarine gravity flows [37,42]. Historical evidence demonstrates their destructive potential. The most catastrophic recorded incident remains the 1929 Grand Banks turbidity current, which destroyed twelve transatlantic cables sequentially across 600 km, with estimated speeds of 60 km/h [43]. The Vancouver Island M7.3 earthquake in 1946 caused significant gravity flows, destroying coastal facilities and shearing submarine pipelines [44]. In 2006, submarine turbidity current triggered by the Pingtung earthquake caused damage to nine submarine optical cables in the nearby deep-water area [45]. Similarly, in Taiwan's Gaoping Submarine Canyon, turbidity current velocities reaching 20 m/s have severed multiple international communication cables during typhoon events [46]. These cases underscore the urgent need to understand interactions between gravity flows and pipelines/cables to protect offshore infrastructure and marine environments.

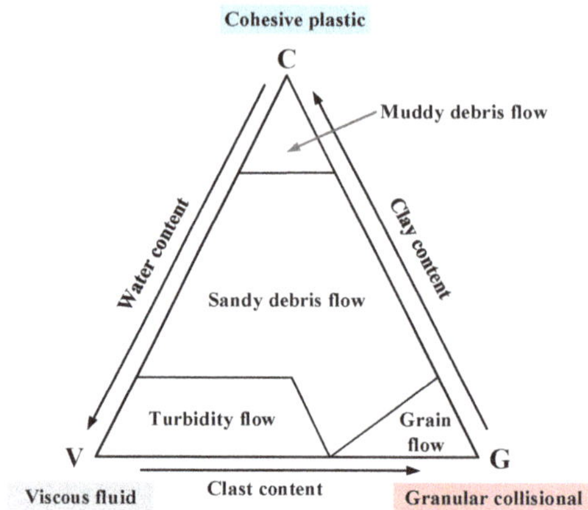

Figure 2. Classification of sediment gravity currents, showing properties of sandy debris flows. The panel was modified from Shanmugam [33].

In this work, we focus on the hydrodynamics of offshore pipelines under submarine gravity flows. By reviewing the published literature, key factors influencing interactions between gravity flows and a single pipe are first clarified, then multiple pipelines are considered, and, finally, the flow-induced vibrations of pipes are discussed. The main objectives of this work are to (1) provide evidence on how to choose the working conditions and design the pipelines under the impact of gravity currents, and (2) identify directions for future work that researchers can pursue.

2. Factors Influencing the Interactions Between Gravity Flows and a Single Pipe

2.1. *Effect of Reynolds Number*

Reynolds number is a significantly important non-dimensional parameter in fluid mechanics, particularly affecting the vortex shedding behavior and associated hydrodynamic forces. The definition of the Reynolds number for Newtonian fluids, e.g., flows around an airfoil where gravitational effects are ignored, is presented as follows:

$$\mathrm{Re}_{\mathrm{Newtonian}} = \frac{\rho U_\infty D}{\mu_{\mathrm{abs}}}. \quad (1)$$

where μ_{abs} is the absolute fluid viscosity, D is the characteristic length scale of the structure, ρ is fluid density, and U_∞ is the free stream velocity. However, for non-Newtonian fluids flowing around a typical circular-sectional body, Equation (1) is not applicable. The hydrodynamic characteristics of shear-thinning, non-Newtonian fluids was advanced by [34] through their proposal of a specialized Reynolds number formulation. This dimensionless parameter was specifically developed for fluids exhibiting power-law or Herschel–Bulkley rheological behavior. The Reynolds number for non-Newtonian fluids ($\mathrm{Re}_{\text{non-Newtonian}}$) is defined as follows:

$$\mathrm{Re}_{\text{non-Newtonian}} = \frac{\rho {U_\infty}^2}{\mu_{\mathrm{app}} \dot{\gamma}} = \frac{\rho {U_\infty}^2}{\tau} \quad (2)$$

where μ_{app} is the apparent viscosity, defined as the ratio of shear stress to the rate of shear of a non-Newtonian fluid, and it is closely related to the changing rates of shear and ambient temperature; $\dot{\gamma} = U_\infty / D$ is the shear rate outside the boundary layer; τ is the fluid shear stress at the strain rate. To calculate $\mathrm{Re}_{\text{non-Newtonian}}$, an appropriate μ_{app} should first be considered, defined as follows:

For power-law fluids:

$$\begin{cases} \tau = K \cdot \dot{\gamma}^n \\ \mu_{app} = K \cdot \dot{\gamma}^{n-1} \end{cases} \tag{3}$$

For Herschel–Bulkley fluids:

$$\begin{cases} \tau = \tau_C + K \cdot \dot{\gamma}^n \\ \mu_{app} = \frac{\tau_C}{\dot{\gamma}} + K \cdot \dot{\gamma}^{n-1} \end{cases} \tag{4}$$

where K is the consistency index, n is the flow index, and τ_C is the yield stress. The rheological behavior for most clay-rich debris flows can be described by the Herschel–Bulkley model [47].

The Reynolds number of Newtonian fluids can vary from 10^1 to 10^6. As Reynolds number increases, the fluid flow exhibits the subcritical, critical, and supercritical regimes [48,49]. In contrast, the Reynolds number of non-Newtonian fluids is much smaller. According to [34], in engineering practice, the submarine pipeline diameter is 0.1–1 m, a submarine landslide velocity is 1–10 m/s, and, consequently, the $Re_{non\text{-}Newtonian}$ ranges from 0.6 to 240, as evidenced in Table 1. For flow past a circular pipe, the Reynolds number determines the vortex shedding mode behind the pipe. Figure 3 compares the typical vortex shedding behavior at low Reynolds numbers for Newtonian [50] and non-Newtonian fluids [51]. For both fluids, three distinct stages including symmetrical flow, vortex formation, and vortex shedding are clearly identified. At very low Reynolds numbers, the flow exhibits a symmetrical wake with fully attached streamlines, indicating a stable and predictable laminar flow pattern. Increasing Reynolds number leads to flow separation phenomenon, where two identical recirculation regions are observed while no vortex shedding occurs. When the Reynolds number exceeds the critical threshold, the flow undergoes a transition characterized by the emergence of periodic vortex shedding in the wake region downstream of the pipe. As evidenced in Figure 3, it is clearly seen that the boundaries of Reynolds numbers for non-Newtonian fluids in three stages are larger than those for Newtonian fluids, indicating the significant difference in fluid properties. For example, overcoming the yield stress of Herschel–Bulkley fluid requires additional energy, resulting in a delay in vortex formation and shedding.

Table 1. Reynolds numbers in previously published studies on the effect of gravity flows on pipelines.

Fluids	Authors	Fluid Density (kg/m^3)	Reynolds Number
Debris flows Non-Newtonian fluids	Zakeri et al. [34]	1681~1694	0.6~240
	Zakeri et al. [28]	1681~1694	2~320
	Perez-gruszkiewicz [52]	670	1.5~34
	Haza et al. [53]	2630	0.18~92.64
	Sahdi et al. [54]	/	0.00001~110
	Liu et al. [38]	1681~1694	2.25~353.47
	Dong et al. [12]	1500	10~100
	Fan et al. [55]	1681~1694	10~10,000
	Nian et al. [39]	1312~1468	0.24~410.12
	Sahdi et al. [50]	/	0.5~40
	Zhang et al. [56]	1681~1694	1.06~345.68
	Guo et al. [51]	1312~1468	0.2~130
	Guo et al. [57–61]	1312~1468	0.45, 3.4, 86.7, 346.47
	Qian et al. [62]	1681~1694	0.3~847.8
	Zhao et al. [63]	1687.7~1694	1000~15,000
	Fan et al. [64]	/	0.001~1241
	Fan et al. [65]	/	0.00001~100
	Fan et al. [41]	/	0.36~287
	Guo et al. [66]	1312~1468	1.81~723.42
	Guo et al. [67]	1312~1468	4~130

Table 1. *Cont.*

Fluids	Authors	Fluid Density (kg/m^3)	Reynolds Number
Turbidity flows Newtonian fluids	Ermanyuk and Gavrilov [68]	1008.081~1038.024	1000
	Gonznlez-Juze et al. [69]	/	2000, 6000
	Gonznlez-Juze et al. [70]	/	2000~45,000
	Jung and Yoon [71]	/	6000
	Xie et al. [72]	1008.081~1038.024	840, 23,500
	Guo et al. [73]	1466.3	95, 500
	Guo et al. [74]	1312	1112~333,559

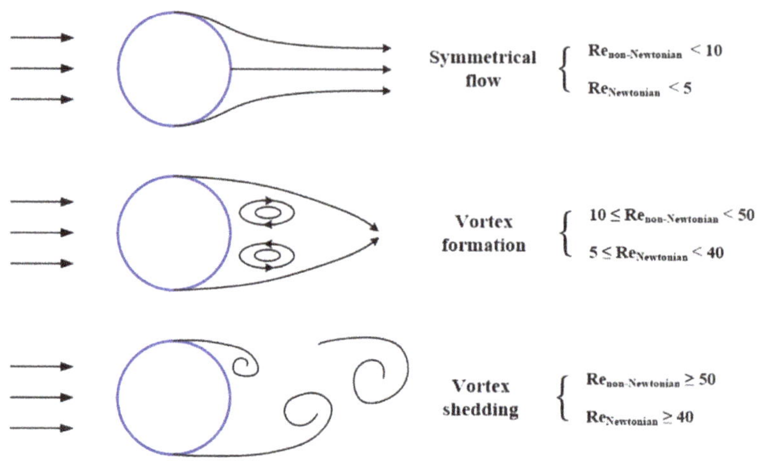

Figure 3. Process of vortex formation and shedding behavior for non-Newtonian and Newtonian fluids. The panel was modified from Guo et al. [51] and Sahdi et al. [50].

For non-Newtonian fluids past a circular pipeline, the relationship between the drag and Reynolds number is attracting more and more attention, enabling engineers to estimate drag forces based on flow conditions and pipeline positioning. From the perspective of fluid framework, the drag force (F_D) can be expressed as follows:

$$F_D = \frac{1}{2}\rho U_\infty{}^2 A \cdot C_D \tag{5}$$

where A is the projected area and C_D is the drag coefficient. From Equation (5), we can find that the calculation of drag coefficient C_D is the key problem. In the experiments of Zakeri et al. [34], slurries were supplied by a head tank and flowed into a water compartment through a chute, and the fully developed debris flows finally interacted with the pipeline. According to the experimental results, Zakeri et al. [34] proposed the first C_D-$Re_{\text{non-Newtonian}}$ relationship for the interactions between the submarine debris flows and pipelines.

For a suspended pipeline, the following is true:

$$C_D = 1.4 + \frac{17.5}{Re_{\text{non-Newtonian}}^{1.25}} \tag{6}$$

For a laid-on-seafloor pipeline, the following is true:

$$C_D = 1.25 + \frac{11}{Re_{\text{non-Newtonian}}^{1.15}} \tag{7}$$

This finding provides significant insights into interactions between a submarine pipeline and debris flows, highlighting the importance of drag force estimation and the necessity of integrating non-Newtonian rheological models into offshore pipeline designs to account for the complex shear-dependent behavior of debris flows. By using a lock-exchange system, Haza et al. [53] proposed a new equation to estimate the drag coefficient for a suspended pipeline:

$$C_D = 0.7 + \frac{29.1}{Re_{\text{non-Newtonian}}^{1.3}} \tag{8}$$

The distinct difference between the results of Zakeri et al. [34] and Haza et al. [53] indicates the significant effect of experimental conditions. According to the Morison equation, the total drag force applied on a suspended pipeline can be divided into two components, including velocity-related drag force and acceleration-induced inertia force. Qian et al. [62] suggested that the inertia force should be taken into consideration to provide accurate estimations of the total drag forces, which is calculated as follows:

$$F_D = \frac{1}{2}\rho U_\infty{}^2 A \cdot C_D + \frac{\pi}{4}\rho D A \ddot{U}_\infty \cdot C_M \tag{9}$$

where the following is true:

$$C_D = 0.58 + \frac{18}{Re_{\text{non-Newtonian}}^{0.9}} \tag{10}$$

$$C_M = \begin{cases} 0 & , \quad Re_{\text{non-Newtonian}} < 17.6 \\ 0.2975 \ln(Re_{\text{non-Newtonian}}) - 0.8533, & Re_{\text{non-Newtonian}} \geq 17.6 \end{cases} \tag{11}$$

Apart from the abovementioned factors, the span height ratio [41], ambient temperature [39], impact angle [28,38,75], and surface roughness [60] also play crucial roles in estimating the drag forces, and they will be discussed later.

2.2. Effect of Span Height Ratio

Submarine free-spanning pipelines are important sections of offshore pipelines that are unsupported by the seabed over certain lengths due to uneven topography, scour erosion, or anthropogenic activities [76]. These suspended spans are subjected to complex hydrodynamic forces, including submarine landslides, vortex-induced vibrations (VIV), wave-current interactions, and potential fatigue damage, all of which may compromise structural integrity over time. The key parameter, span height ratio $H^* = H/D$, is defined as the vertical distance between the seabed line and the lowest point of the pipeline. Investigations of the single-phase Newtonian fluids like pure water flowing past a submarine suspended pipeline have been extensively reported [77–83]. However, considerations of multi-phase Newtonian/non-Newtonian fluids, such as debris and turbidity flows past free-spanning pipelines, are insufficient, and the latest studies are summarized in Table 2.

Table 2. Studies of debris and turbidity flows past a submarine suspended pipeline.

Flow Type	Authors	Slope Angle of the Seabed (°)	H^*
Debris flow	Zakeri et al. [34]	3, 6	1
	Sahdi et al. [54]	/	6.2
	Zhang et al. [56]	0	0~3
	Guo et al. [51]	0	0~3
	Fan et al. [41,64,65]	0	0.08~10
	Guo et al. [59]	0	0~2.5
	Guo et al. [61]	0	0~1

Table 2. *Cont.*

Flow Type	Authors	Slope Angle of the Seabed (°)	H^*
Turbidity flow	Ermanyuk and Gavrilov [68]	0	0.03~4.5
	Gonznlez-Juze et al. [69]	0	0.067~1.33
	Gonznlez-Juze et al. [70]	0	0.15~1.5
	Jung and Yoon [71]	0	0.2~1.4
	Xie et al. [72]	0	0.05~1
	Ding et al. [84]	0	0.9~1.7
	Guo et al. [73]	0	1

For different span heights, the flow around the pipeline varies. Figure 4 plots the velocity field and the shedding vortices when gravity flows past a pipeline under three typical span heights. For a very low span height case, as shown in Figure 4c, the space between the pipe and seabed is partially or fully occupied by the submarine pipe. In this case, the pipeline creates a complete flow obstruction, resulting in zero underflow penetration, and the whole mass flux of the gravity-driven flow must be diverted over the pipeline structure. Consequently, two velocity deceleration regions on the lower front area and an acceleration region on the upper front area of the pipe are observed. This velocity distribution will lead to great pressure differences between the two surfaces of the pipe and consequently large lift forces [59], as well as an asymmetrical vortex shedding mode. When the span height increases to a moderate level, as sketched in Figure 4b, the flow field changes significantly. The approaching flow experiences significant deceleration upstream of the pipeline due to the obstruction effect, creating a stagnation zone. Simultaneously, the pressure potential energy of the gravity flow on the lower front surface of the pipeline undergoes conversion to kinetic energy, leading to a strong acceleration. The flow on the upper front surface forms a relatively weak acceleration. Vortices are observed to be alternately shed from both sides of the pipeline, while they are not identical due to the different flow channels. For a very large span height, the seabed effect can be ignored, and the flow field is considered to be infinite. In this case, as depicted in Figure 4a, two equal vortices with opposite signs are shed alternately, forming the classical Karman vortex street.

Since the span height has a significant effect on the flow field, the associated hydrodynamic loads should be carefully checked. For debris flows, Guo et al. [51] numerically studied the impact force coefficients of a suspended pipeline under a series of span heights, and the results are plotted in Figure 5. Three distinct modes are identified. At low Reynolds numbers less than 1 (mode I), the maximum drag and lift coefficients usually appear at $H^* = 2.5{\sim}3$, and their minimum values often occur at $H^* = 0$. The span height ratio has a significant effect on the hydrodynamic loads when $Re_{non\text{-}Newtonian} < 1$, especially for the lift. When $1 \leq Re_{non\text{-}Newtonian} < 10$ (mode II), the effect of span height is gradually weakening as the differences between the maximum and minimum hydrodynamic coefficients decrease as the Reynolds number increases. In mode III where Reynolds number is greater than 10, the span heights corresponding to the maximum and minimum forces have changed. The maximum and minimum drag appear at $H^* = 0{\sim}0.5$ and $H^* = 2.5{\sim}3$, respectively. The lift coefficient exhibits similar trends, with a critical transition threshold emerging at $Re_{non\text{-}Newtonian} = 10$. The finding that the lift coefficient decreases with increasing span height in mode III can be further confirmed by the results of Fan et al. [64].

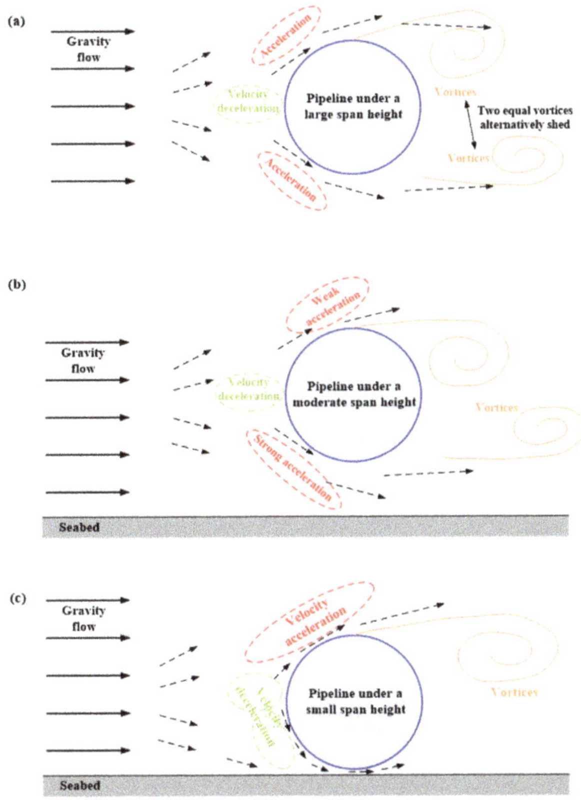

Figure 4. Conceptual diagram of the velocity field and the shedding vortices for a gravity flow past a pipeline under at (**a**) a very large span height, (**b**) a moderate span height, and (**c**) a very low span height. The panel was modified from Guo et al. [59].

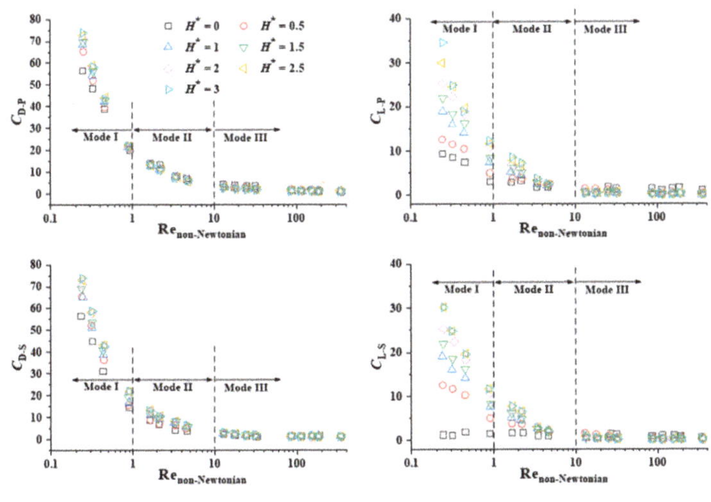

Figure 5. Variations of peak drag ($C_{D\text{-}P}$), stable drag ($C_{D\text{-}S}$), peak lift ($C_{L\text{-}P}$), and stable lift ($C_{L\text{-}S}$) values against Reynolds number under different span heights, where mode I, II, and III represent the significant, weakening, and insignificant effects of span height ratios. The data were obtained from Guo et al. [51].

For turbidity flows, Ermanyuk and Gavrilov [68] conducted flume experiments to study the pipeline loads with various span heights. The results revealed that hydrodynamic loads reach their maximum magnitude when the pipeline is in direct contact with the channel bed, while they decrease rapidly with increasing span height. Specifically, the rapid decrease in hydrodynamic loads occurs at $H^* = 1.5$. Gonznlez-Juze et al. [69], Xie et al. [72], and Ding et al. [84] employed numerical methods to investigate the variations of peak drag and lift coefficients against span height ratio, and their results are compared in Figure 6. With increasing span height ratio, the peak drag coefficients generally decrease. However, clear fluctuations in $C_{D\text{-}P}$ vs. H^* can be found at a very high Reynolds number of 23,500, which belongs to the subcritical regime. These fluctuations reflect the significant Reynolds number effect. The results of Ding et al. [84] show that the peak lift coefficient of the pipeline at Re = 840 is very large when the pipeline is close to the seabed, while it rapidly decreases to a low level and does not show significant fluctuations with further increases in the span height ratio. In contrast, at a large Reynolds number of 2000, totally different variations are observed, where the lift coefficients show a nearly linear increasing trend with the span height ratio. The different trends of $C_{L\text{-}P}$ vs. H^* also reflect the Reynolds number effect.

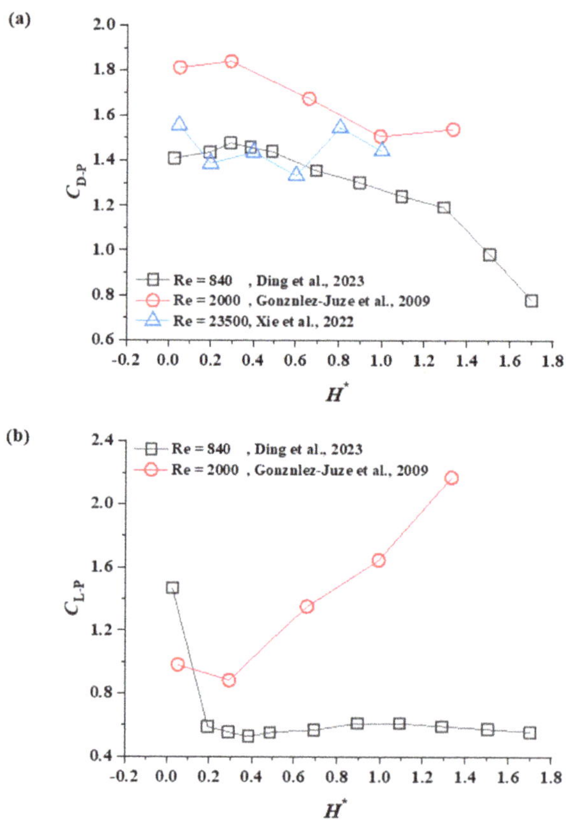

Figure 6. Hydrodynamic coefficients at various span heights under the impact of turbidity flows, (**a**) peak drag, and (**b**) peak lift [69,72,84].

2.3. Effect of Impact Angle

In practical engineering, submarine gravity flows may not always strike the pipeline vertically. Instead, there exists an impact angle (θ, defined as the angle between the oncoming flow and the pipe axis), varying from 0° to 90°. In that case, the normal and axial drag forces on the pipeline under gravity flows can be expressed as follows (Cohen and Kundu [85]):

$$F_{D\text{-}90} = C_{D\text{-}90} \cdot \frac{1}{2}\rho U_\infty{}^2 (A \cdot \sin\theta) \tag{12}$$

$$F_{D\text{-}0} = C_{D\text{-}0} \cdot \frac{1}{2}\rho U_\infty{}^2 (A \cdot \cos\theta) \tag{13}$$

Comparisons of the drag estimation methods under the effect of impact angle and span height ratio are listed in Table 3. Scientific researchers and engineers should note that Equations (14)–(21) are empirically obtained based on specific conditions, so it is important to choose an appropriate one according to practical applications.

Table 3. Conditions for drag estimation methods under the effects of impact angle as well as span height ratio.

Authors	$Re_{non\text{-}Newtonian}$	Oncoming Flow Velocity (m/s)	θ (°)	H^*
Zakeri [28]	1.5~317.1	0.35~3.4	0, 30, 45, 60, 90	1
Liu et al. [38]	2.25~353.47	0.35~3.4	0, 30, 45, 60, 90	1
Wang et al. [86]	1~350	Below 0.2 for the smallest velocity	0, 15, 30, 45, 60, 75, 90	1
Zhang et al. [56]	1.23~345.68	0.2~5	0, 15, 30, 45, 60, 75, 90	0~3

Zakeri [28] employed computational fluid dynamics analysis to numerically simulate the impact of clay-rich submarine debris flows on a suspended pipeline at various impact angles. A primary method is presented for estimating the normal and axial drag forces of a pipeline, expressed as follows:

$$F_{D\text{-}90} = \left(1.4 + \frac{17.5}{Re_{non\text{-}Newtonian}^{1.25}}\right) \cdot \frac{1}{2}\rho U_\infty{}^2 (A \cdot \sin\theta) \tag{14}$$

$$F_{D\text{-}0} = \left(0.08 + \frac{9.2}{Re_{non\text{-}Newtonian}^{1.1}}\right) \cdot \frac{1}{2}\rho U_\infty{}^2 (A \cdot \cos\theta) \tag{15}$$

Based on the results of Zakeri [28], Liu et al. [38] examined more cases by expanding the flow velocity range and increasing impact angle conditions. The experimental findings demonstrated that while the debris flow impact angle significantly influences the normal drag coefficient, it exhibits a negligible effect on the axial drag coefficient. Therefore, the drag forces are written as follows:

$$F_{D\text{-}90} = \left(\frac{0.46 + (23.55/Re_{non\text{-}Newtonian})}{(\sin\theta)^{0.6}}\right) \cdot \frac{1}{2}\rho U_\infty{}^2 (A \cdot \sin\theta) \tag{16}$$

$$F_{D\text{-}0} = \left(\frac{9.13}{Re_{non\text{-}Newtonian}}\right) \cdot \frac{1}{2}\rho U_\infty{}^2 (A \cdot \cos\theta) \tag{17}$$

Later, Wang et al. [86] considered lower flow velocities, and further improved the estimation method of drag forces as follows:

$$F_{D\text{-}90} = \left(\frac{0.65 + (19.16/Re_{non\text{-}Newtonian})}{(\sin\theta)^{0.55}}\right) \cdot \frac{1}{2}\rho U_\infty{}^2 (A \cdot \sin\theta) \tag{18}$$

$$F_{D\text{-}0} = \left(\frac{8.361}{Re_{non\text{-}Newtonian}} \cdot (\cos\theta)^{0.22}\right) \cdot \frac{1}{2}\rho U_\infty{}^2 (A \cdot \cos\theta) \tag{19}$$

By considering the effect of both impact angle and span height ratio (H^*), Zhang et al. [56] proposed an improved method to estimate the normal and axial drag force:

$$F_{\text{D-90}} = \left(\frac{23.56 - \frac{25}{4.6+e^{3H*}}}{\text{Re}_{\text{non-Newtonian}}} \right) \cdot \frac{1}{2} \rho U_\infty{}^2 A (\sin\theta)^{0.45} \tag{20}$$

$$F_{\text{D-0}} = \left(\frac{8.96 - 0.5e^{-2H*}}{\text{Re}_{\text{non-Newtonian}}} \right) \cdot \frac{1}{2} \rho U_\infty{}^2 A (\cos\theta)^{1.22} \tag{21}$$

2.4. Effect of Pipe Geometry

To reduce the impact loads on offshore pipelines under submarine gravity flows, researchers have proposed several novel methods, including using streamlined profiles such as wedge, airfoil, arc-angle hexagon, and double ellipse (Perez-Gruszkiewicz [52]; Fan et al. [55]) and changing the pipe surface by arranging uniform honeycomb holes (Guo et al. [57,59]), as shown in Figure 7. Both experimental results by Perez-Gruszkiewicz [52] and numerical simulations by Fan et al. [55] demonstrated that the drag coefficients of all four streamlined pipelines are effectively reduced compared with a regular pipe. As evidenced in Figure 8, the drag reduction is also greatly affected by the Reynolds number of non-Newtonian fluids, where significant reductions are successfully achieved in the low $\text{Re}_{\text{non-Newtonian}}$ range. Additionally, streamlined profiles are helpful for stabilizing lift force responses, which is attributed to the delay in the separation of boundary layers (Figure 9). Specifically, the wedge and double-ellipse pipelines exhibit the greatest reduction among the four streamlined pipelines. According to the C_D-$\text{Re}_{\text{non-Newtonian}}$ relationship, Fan et al. [55] proposed estimation equations for the drag coefficients of different streamlined devices.

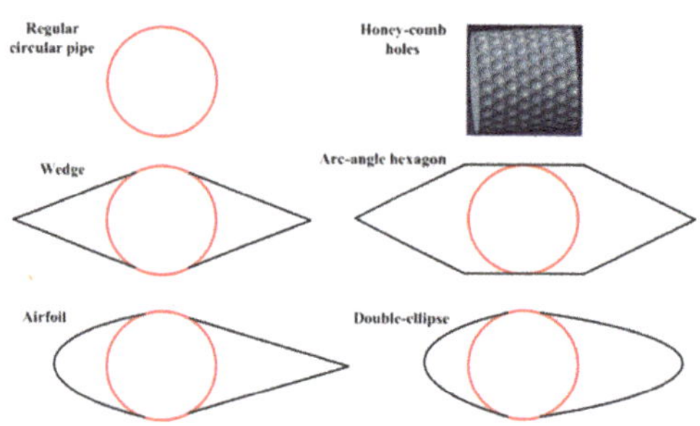

Figure 7. Streamlined pipelines and altered pipe surface.

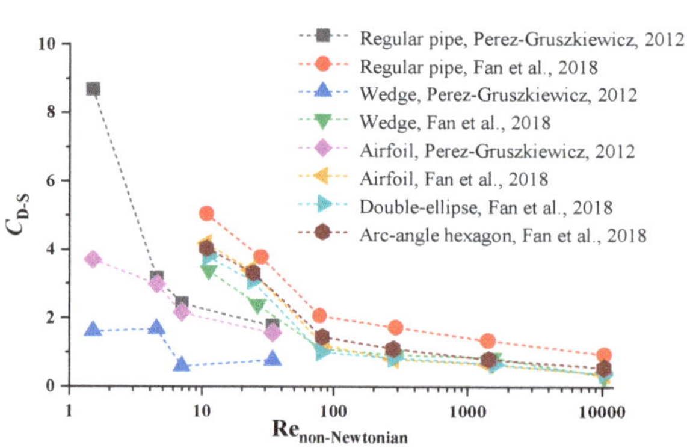

Figure 8. Variations in drag coefficients against Reynolds number of non-Newtonian fluids for different streamlined pipelines [52,55].

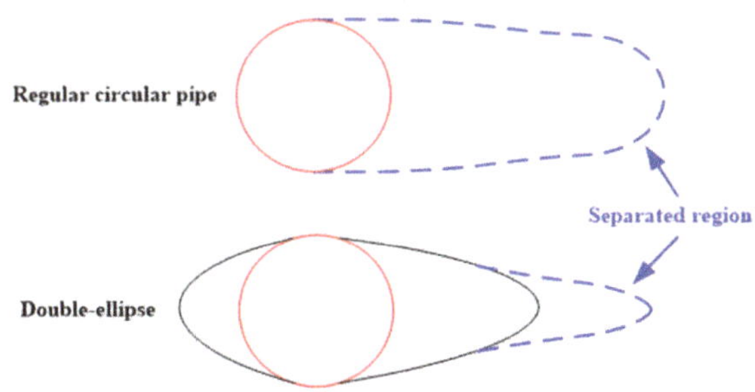

Figure 9. Comparison of the separation regions for circular and double-ellipse pipelines. The panel was modified from Fan et al. [55].

For wedge pipelines:

$$C_D = 0.75 + \frac{31.28}{Re_{non\text{-}Newtonian}} \tag{22}$$

For double-ellipse pipelines:

$$C_D = 0.7 + \frac{38.14}{Re_{non\text{-}Newtonian}} \tag{23}$$

For airfoil pipelines:

$$C_D = 0.71 + \frac{42.35}{Re_{non\text{-}Newtonian}} \tag{24}$$

For arc-angle hexagon pipelines:

$$C_D = 0.93 + \frac{38.7}{Re_{non\text{-}Newtonian}} \tag{25}$$

Focusing on the innovate pipelines, Guo and his team proposed pipelines with honeycomb holes [57] and dimples [59]. Both devices can achieve a significant reduction of about 20% in the peak drag forces of the pipeline under the impact of submarine debris flows. The drag reduction mechanism is mainly attributed to the reduced maximum pressure difference around the pipeline. In addition, the honeycomb holes and dimples are helpful for suppressing lift force fluctuations and effectively reducing the force amplitude, preventing potential pipeline damage from flow-induced vibrations.

2.5. Effect of Ambient Temperature

Wang et al. [86] conducted low-temperature rheology tests of debris flows, and they found the environmental temperature plays an important role in rheological properties of submarine debris flows. Based on this finding, Nian et al. [39] established a low-temperature rheological model to study the impact force on suspended pipelines under debris flows. Five ambient temperatures—0.5 °C, 4.5 °C, 8.5 °C, 12 °C, and 22 °C—were considered in their work. The results indicated that the drag and lift forces on the pipeline at 0.5 °C increase by 26.0% and 70.3%, respectively, compared with those at 22 °C. Moreover, calculations of the peak ($C_{D\text{-}P}$) and stable ($C_{D\text{-}S}$) drag coefficients were established as follows:

$$C_{D\text{-}P} = 1.52 + \frac{17.21}{Re_{non\text{-}Newtonian}} \tag{26}$$

$$C_{D\text{-}S} = 1.2 + \frac{16.06}{Re_{non\text{-}Newtonian}} \tag{27}$$

Later, Guo et al. [51] compared the impact of debris flows past a pipeline under low (0.5 °C) and normal (22 °C) ambient temperature. A similar conclusion was obtained: the loads on pipelines at 0.5 °C are significantly larger than those at 22 °C. Interestingly, the maximum rate of change of $C_{D\text{-}P}$ and $C_{L\text{-}P}$ (the peak lift coefficient) can reach 30% and 49%, respectively. For a laid-on-seafloor pipeline, a peak enhancement of 123% in the stable lift coefficient ($C_{L\text{-}S}$) is observed. Therefore, low temperature has a vital effect on the forces acting on the pipeline and must be carefully considered.

2.6. Effect of Roughness

In practice, the outer materials of submarine pipelines can be concrete, steel, rubber, asphalt, polyethylene, etc. These surfaces, with varying roughness, have a significant effect on the flow field around submarine pipelines. In addition, the attachment of microorganisms can alter the pipeline's surface roughness, further affecting the flow and associated dynamic loads on the pipeline. Guo et al. [60] numerically investigated the characteristics of debris flow around circular pipelines with surface roughness using an equivalent sand-grain roughness model. The results demonstrated that increasing surface roughness leads to a significant increase in peak lift force, while peak drag force decreases. Specifically, for a suspended pipeline with surface roughness of 0.15 mm, the peak lift force increases by 62% and the peak drag force decreases by 17% compared to smooth pipelines. Based on these findings, the researchers established a chart methodology considering pipeline roughness effects for estimating impact forces, expressed as:

$$F_D = \frac{1}{2}\rho U_\infty{}^2 (D + k_s) \cdot C_D \tag{28}$$

where k_s is the roughness height.

3. Interactions Between Gravity Flows and Multiple Pipes

With the significant increase in the demand for submarine pipelines and cables, multiple pipelines in close proximity are increasingly common in order to meet the demands of the growth of offshore activities. For instance, the "Nord Stream 2" pipeline was constructed in parallel to the "Nord Stream 1" pipeline through the Baltic Sea to transport natural gas. Single-medium fluid flow with uniform incoming velocity past two circular pipes has been systematically reviewed, identifying several distinct flow regimes and interference modes [87]. In contrast, the hydrodynamic characteristics of multiple pipes under gravity flows involving at least two different fluids have been less studied. Guo et al. [73] employed the large-eddy simulation (LES) framework coupled with density transport equations to investigate the gravity-driven high-density turbidity current dynamics and their interactions with parallel suspended pipelines. The effect of the streamwise spacing distance between two pipes on the flow field and associated hydrodynamic loads is examined. The results indicated that the turbulent wake generated by high-density fluid flowing past the front pipe promotes velocity fluctuations, leading to increased impact forces on the rear pipe with increasing streamwise spacing distance up to eight times the pipe diameter. The results also suggested that the streamwise spacing distance should be less than two times the pipe diameter to minimize hydrodynamic loads on the rear pipe. Ding et al. [84] conducted LES simulations to study the interactions between gravity currents and two tandem pipes. Results showed that the distribution of the maximum dynamic pressure and vortices around two pipes is closely related to the spacing distance. The drag force of the front pipe is considerably smaller than that of an individual pipe when the spacing distance is small. Unfortunately, although three-dimensional simulations using the LES method were used in the studies of both Guo et al. [73] and Ding et al. [84], only two-dimensional results were analyzed, and the vorticity distribution and process in the spanwise direction were not discussed.

4. Flow-Induced Vibrations of Pipes Under Gravity Flows

Previous investigations mainly focused on gravity flows past a stationary pipeline. However, the pipeline may vibrate in both the in-line and transverse directions, commonly known as flow-induced vibrations (FIV). The studies about FIV responses of pipes under gravity flows are scarce. Zhao et al. [63] adopted an air-water-debris model coupled with the one-degree flow-induced vibrating pipe module to study the pipe dynamics under the debris flow. Specifically, the effects of debris flow (volume, density, and location) and pipe characteristics (stiffness coefficient, location, and number) on the hydrodynamic loads and vibrations are systematically investigated. As depicted in Figure 10, before the debris flow interacts with the pipe at t = 20.4 s, some vortices appear due to the turbulence of the two fluids, while flow fields are much similar for cases with and without a pipe. At t = 28 s when the debris flow passes the pipe, the pipe is pushed upward because of the large density difference between the debris flow and water. The head shape of the debris flow is similar for both cases, but the pipe-flow interactions lead to more vortices around the pipe, thereby enhancing turbulence intensity. At t = 28.8 s, the debris flow is located above the pipe, pushing the pipe downward. After the debris flow passes through the pipe, many pairs of fish-eye-like swirls are generated, strengthening the turbulence of both the water and the debris flow. At the same time, the pipe starts to rise again, exhibiting a reciprocating vibration behavior. As compared with perfectly periodic oscillations of the constant-density flow past a circular pipe [88], the displacements under debris flows show strong time-dependent features [63] (Zhao et al., 2021).

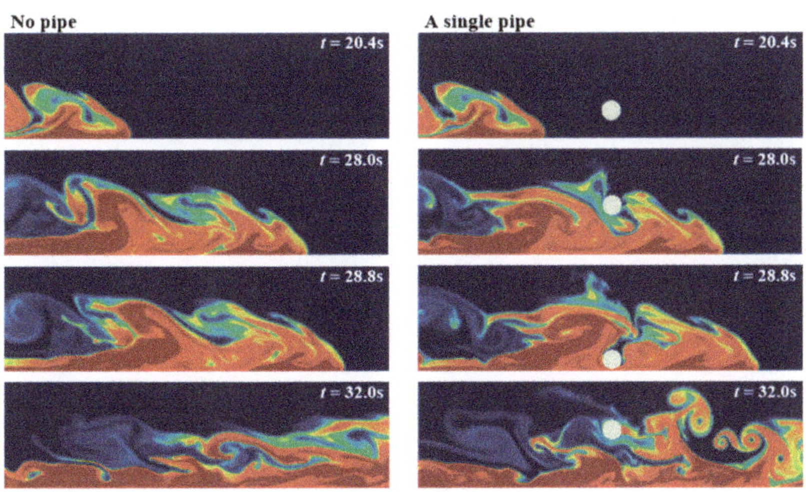

Figure 10. Volume fraction contour of debris flows with and without a single pipeline, where the range of volume fraction is from 0.05 (blue) to 0.95 (red). The panel was modified from Zhao et al. [63].

Compared with a single pipe, the mutual interference between two tandem pipes and their interactions with debris flows create a more complex flow field with strong turbulence, as shown in Figure 11. The displacement variations in Figure 11 show that (1) the vibration trajectory of the rear pipe closely mirrors that of the front pipe for a period of time; (2) the front pipe maintains its elevated position longer than the rear pipe, owing to the combined effects of direct debris flow impact and the shielding effect provided by the front pipe to its downstream counterpart; and (3) the vibration frequency of the rear pipe is significantly higher than that of the front pipe, a phenomenon that is mainly attributed to the wake interference.

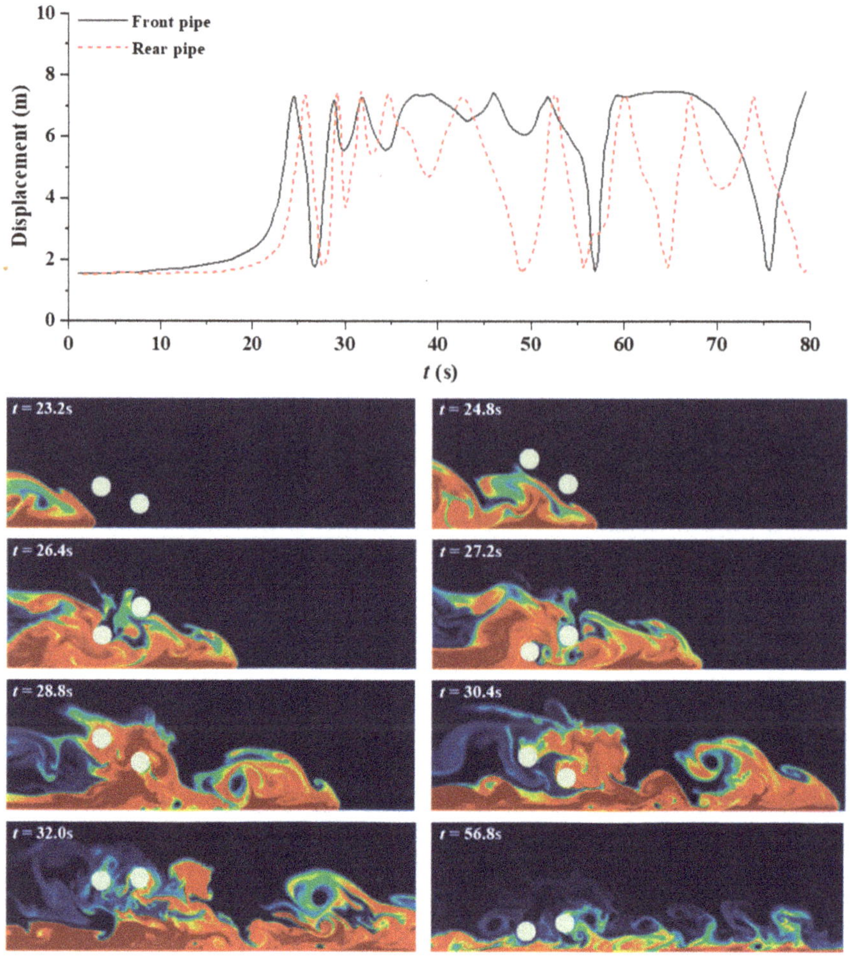

Figure 11. Vertical displacements and interactions between debris flows and two tandem pipelines, where the range of volume fraction is from 0.05 (blue) to 0.95 (red). The panel was modified from Zhao et al. [63].

5. Overall Evaluations

This work presents a review of interactions between submarine gravity flows and offshore pipelines. By synthesizing relevant experimental and computational findings, key factors (Reynolds number, span height, impact angle, pipe geometry, ambient temperature, and surface roughness) influencing the dynamics of a single pipeline and the resulting impact forces under gravity flow impact are first discussed. Further, flow fields and hydrodynamic characteristics of multiple pipelines under the impact of submarine gravity flows are presented. Finally, flow-induced vibrations of suspended pipelines are depicted. Some substantial advancements have been achieved in comprehending and mitigating submarine gravity flow effects on offshore pipelines, while there are still shortcomings in the studies referenced.

Numerous numerical studies have been conducted, while only a few flume experiments have been carried out, leading to a lack of physical experimental results for numerical validation. Moreover, in many studies, the seabed slope is commonly idealized and assumed to be horizontal (0°) for simplicity in modelling and analysis. However, in reality, natural seafloors often exhibit complex and varying degrees of slope, which can significantly influence hydrodynamic and geotechnical processes.

Based on experimental and numerical results, empirical formulas evaluating the gravity flow loads on the offshore pipelines have been proposed. These formulas are closely related to key parameters, such as Reynolds number, span height ratio, impact angle, pipe geometry, ambient temperature, and surface roughness. The establishment of empirical formulas provides important guidance for engineers to design offshore pipelines. Nevertheless, the prerequisite for using these formulas is to carefully consider the specific working conditions. For example, in Section 2.3, the varied Reynolds numbers, impact angles, and height ratios determine distinct methods to estimate the forces acting on the pipeline. Therefore, multi-parameter uncertainty evaluation and the sensitivity of empirical constants deserve more attention.

In most previous reports, clay-rich debris flows are described by the Herschel–Bulkley model. However, choosing an appropriate non-Newtonian fluid model is critically important. Modeling non-Newtonian flows is fraught with significant uncertainties, primarily stemming from the challenge of selecting an appropriate constitutive model that can accurately capture complex rheological behaviors like shear-thinning, yield stress, and viscoelasticity. This uncertainty is compounded by the inherent difficulty in precisely measuring the model's parameters, which are highly sensitive and prone to experimental artifacts like wall slip. To mitigate these issues, a comprehensive rheological characterization is essential to guide model selection, while advanced uncertainty quantification techniques should be employed to calibrate parameters and assess their confidence intervals. Furthermore, the critical uncertainty arising from the imperfect coupling between a fluid's evolving microstructure and its macro-scale flow behavior can be addressed by developing multi-scale modeling frameworks that dynamically link these scales. Finally, the numerical uncertainties from solving highly nonlinear equations are alleviated by using robust algorithms, adaptive mesh refinement, and validating predictions against detailed experimental data from advanced flow visualization techniques like Particle Image Velocimetry (PIV).

To mitigate the impact of gravity flows on offshore pipelines, some practical risk reduction measures can be directly used according to the results of previous reports. Choosing an appropriate span height ratio is helpful for reducing hydrodynamic loads on pipelines. Installing streamlined structures like double-ellipse and wedge around the pipelines can effectively recover base pressures and stabilize wake flows, resulting in reduced impact forces. Increasing ambient temperature and reducing roughness are good choices to weaken gravity flow effects.

6. Suggestions for Future Study and Practice

6.1. Three-Dimensional Numerical Studies

Transitioning from two-dimensional to three-dimensional numerical modeling approaches is essential for enhancing the understanding of submarine gravity flow dynamics and their complex interactions with offshore pipelines. Most existing numerical studies on gravity flow–pipeline interactions rely on simplified two-dimensional models, which suffer from inherent limitations. In two-dimensional simulations, out-of-plane effects, such as non-uniform sediment stresses, three-dimensional fluid flow patterns, and asymmetric pipeline deformations, are neglected or oversimplified. This can lead to inaccurate predictions of key behaviors, including localized scour around pipelines, uneven load distribution, and the true failure mechanisms of the pipeline-sediment system. Moreover, two-dimensional models often fail to account for the influence of seabed slope, transverse slope variations, and complex boundary conditions, which are critical in real-world scenarios. In contrast, three-dimensional numerical models can overcome these limitations by fully resolving spatial interactions. A three-dimensional framework enables the explicit representation of turbulent flow dynamics around pipelines, multi-directional sediment displacement, and the true geometry of the pipeline (e.g., joints and bends). It also allows for a more realistic simulation of phenomena such as vortex shedding, sediment transport heterogeneity, and the progressive development of failure zones in the surrounding sediment. Future research should prioritize the development of validated three-dimensional models, incorporating multi-physics couplings (e.g., fluid-structure-sediment interactions) and experimental data for calibration. Such efforts would bridge the gap between theoretical studies and practical engineering applications, ultimately improving the design and risk assessment of pipeline systems in gravity-driven flow environments.

6.2. Fatigue Damage by Flow-Vibration Responses

While previous studies on gravity flow-pipeline interactions have primarily focused on estimating impact forces under static pipeline conditions, the critical aspect of flow-induced vibration (FIV) remains largely unexplored. The assumption of fixed pipelines neglects important dynamic effects, such as vortex-induced vibration (VIV), periodic loading from turbulent fluctuations, and potential resonance phenomena that may lead to fatigue damage or structural failure over time. Future research should systematically investigate FIV mechanisms in gravity flow environments through coupled fluid-structure interaction (FSI) modeling, which can enable safer pipeline designs in gravity flow systems, particularly in scenarios involving unsteady flows.

6.3. Flexible Pipelines

Existing studies on interactions between the gravity flow and suspended pipeline have predominantly modeled pipelines as rigid, non-deformable structures. While this simplification facilitates computational analysis, it overlooks critical aspects of flexible pipeline behavior that are increasingly relevant in modern engineering applications. Future research should systematically investigate the dynamic response of flexible suspended pipelines under gravity flow conditions, with particular focus on (1) the coupled hydroelastic vibration mechanisms induced by unsteady flow separation and vortex shedding; (2) nonlinear deformation characteristics, including sagging effects and geometric nonlinearity, under combined hydrodynamic and self-weight loading; and (3) the fatigue performance under cyclic loading conditions. Advanced computational frameworks combining large-eddy simulation with nonlinear structural dynamics should be developed, incorporating both material nonlinearity (e.g., hyperelastic or viscoelastic behavior) and geometric nonlinearity (e.g., large deflection theory). These investigations will provide fundamental insights for optimizing the design and operation of suspended pipeline systems in applications such as deep-sea mining tailings disposal, subaqueous sediment transport, and other scenarios where pipeline flexibility significantly influences system performance and safety.

6.4. Multiple Pipelines

Real-world engineering applications often involve multiple pipelines arranged in complex configurations (e.g., parallel, staggered, or bundled systems). The hydrodynamic interference between adjacent pipelines and their collective impact on flow patterns, sediment transport, and local scour development remain poorly understood. Future studies should pay more attention to multiple pipeline systems under gravity flow conditions: (1) wake interference effects and their influence on hydrodynamic loading characteristics; (2) the amplification or sheltering effects on local scour depth and erosion patterns; and (3) coupled vibration mechanisms that may lead to collision risks or fatigue damage. Such research will provide essential insights for the optimal spacing design, risk assessment, and maintenance strategy of pipeline networks in gravity flow environments, particularly for offshore mining operations and subsea cable–pipeline crossings where multiple pipelines coexist.

Author Contributions: Conceptualization, C.Z., T.T., H.Z. and G.W.; funding acquisition, C.Z.; resources, C.Z.; writing—original draft preparation, T.T. and C.Z.; writing—review and editing, F.Z., H.Z., C.Z. and C.R.; supervision, C.Z. and C.R. All authors have read and agreed to the published version of the manuscript.

Funding: The work presented here was supported by the National Natural Science Foundation of China [grant number 52201310], the GJYC Program of Guangzhou [grant numbers 2024D01J0076, 2024D03J0022], and the International Cooperation Program in SCUT [grant number K525018008].

Data Availability Statement: Data can be obtained from the first author upon request.

Acknowledgments: The authors gratefully acknowledge the support from the funding bodies listed above. We also appreciate the anonymous reviewers, whose comments helped improve the paper.

Conflicts of Interest: The authors declare no conflicts of interest.

References

1. Biscara, L.; Hanquiez, V.; Leynaud, D.; Marieu, V.; Mulder, T.; Gallissaires, J.M.; Crespin, J.P.; Braccini, E.; Garlan, T. Submarine slide initiation and evolution offshore pointe odden, Gabon—Analysis from annual bathymetric data (2004–2009). *Mar. Geol.* **2012**, *299–302*, 43–50. [CrossRef]
2. Locat, J.; Lee, H.J. Submarine landslides: Advances and challenges. *Can. Geotech. J.* **2002**, *39*, 193–212. [CrossRef]
3. Wang, F.; Dai, Z.; Zhang, S. Experimental study on the motion behavior and mechanism of submarine landslides. *Bull. Eng. Geol. Environ.* **2018**, *77*, 1117–1126. [CrossRef]
4. Zhang, Y.; Xie, J.; Gu, K.; Zhao, H.; Li, C.; Hao, X. Sedimentary characteristics and model of lacustrine deep water gravity flow in the third member of paleogene shahejie formation in niuzhuang sag, bohai bay basin, China. *J. Mar. Sci. Eng.* **2023**, *11*, 1598. [CrossRef]
5. Zhu, C.; Li, S.; Chen, J.; Wang, D.; Song, X.; Li, Z.; Chen, B.; Shan, H.; Jia, Y. Nepheloid layer generation by gas eruption: Unexpected experimental results. *J. Oceanol. Limnol.* **2023**, *41*, 769–777. [CrossRef]
6. He, Z.; Okon, S.U.; Hu, P.; Zhang, H.; Ewa-Oboho, I.; Li, Q. Submarine gravity flows and their interaction with offshore pipelines: A review of recent advances. *Eng. Geol.* **2025**, *347*, 107914. [CrossRef]
7. Huang, J.; Yuan, R.; Zhu, J. Numerical simulation and analysis of water and suspended sediment transport in Hangzhou bay, China. *J. Mar. Sci. Eng.* **2022**, *10*, 1248. [CrossRef]
8. Mojtahedi, F.F.; Yousefpour, N.; Chow, S.H.; Cassidy, M. Offshore turbidity currents forecasting (part I): Integrating deep learning and computational fluid dynamics. *Ocean Eng.* **2025**, *331*, 121360. [CrossRef]
9. Breien, H.; Pagliardi, M.; De Blasio, F.V.; Issler, D.; Elverhøi, A. Experimental studies of subaqueous vs. subaerial debris flows—Velocity characteristics as A function of the ambient fluid. In *Submarine Mass Movements and Their Consequences: 3 International Symposium*; Lykousis, V., Sakellariou, D., Locat, J., Eds.; Springer: Dordrecht, The Netherlands, 2007; pp. 101–110, ISBN 978-1-4020-6512-5.
10. Sassa, S.; Sekiguchi, H. Dynamics of submarine liquefied sediment flows: Theory, experiments and analysis of field behavior. In *Submarine Mass Movements and Their Consequences*; Springer: Dordrecht, The Netherlands, 2012; pp. 405–416.
11. Sassa, S.; Sekiguchi, H. Liqsedflow: Role of two-phase physics in subaqueous sediment gravity flows. *Soils Found.* **2010**, *50*, 495–504. [CrossRef]
12. Dong, Y.; Wang, D.; Randolph, M.F. Investigation of impact forces on pipeline by submarine landslide using material point method. *Ocean Eng.* **2017**, *146*, 21–28. [CrossRef]
13. Guo, X.; Nian, T.; Gu, Z.; Li, D.; Fan, N.; Zheng, D. Evaluation methodology of laminar-turbulent flow state for fluidized material with special reference to submarine landslide. *J. Waterw. Port Coast. Ocean Eng.* **2021**, *147*, 4020048. [CrossRef]
14. Alhaddad, S.; Labeur, R.J.; Uijttewaal, W. Breaching flow slides and the associated turbidity current. *J. Mar. Sci. Eng.* **2020**, *8*, 67. [CrossRef]
15. Alhaddad, S.; de Wit, L.; Labeur, R.J.; Uijttewaal, W. Modeling of breaching-generated turbidity currents using large eddy simulation. *J. Mar. Sci. Eng.* **2020**, *8*, 728. [CrossRef]
16. Wells, M.G.; Dorrell, R.M. Turbulence processes within turbidity currents. *Annu. Rev. Fluid Mech.* **2021**, *53*, 59–83. [CrossRef]

17. Kneller, B.; Buckee, C. The structure and fluid mechanics of turbidity currents: A review of some recent studies and their geological implications. *Sedimentology* **2000**, *47*, 62–94. [CrossRef]
18. Guo, X.; Liu, X.; Zhang, H.; Li, M.; Luo, Q. Evaluation of instantaneous impact forces on fixed pipelines from submarine slumps. *Landslides* **2022**, *19*, 2889–2903. [CrossRef]
19. Hance, J.J. Development of a Database and Assessment of Seafloor Slope Stability Based on Published Literature. Master's Thesis, University of Texas at Austin, Austin, TX, USA, 2003.
20. Clare, M.; Lintern, D.G.; Rosenberger, K.; Hughes, C.J.E.; Paull, C.; Gwiazda, R.; Cartigny, M.J.B.; Talling, P.J.; Perara, D.; Xu, J.; et al. Lessons learned from the monitoring of turbidity currents and guidance for future platform designs. *Geol. Soc. Spec. Publ.* **2020**, *500*, 605–634. [CrossRef]
21. Fan, N.; Jiang, J.; Nian, T.; Dong, Y.; Guo, L.; Fu, C.; Tian, Z.; Guo, X. Impact action of submarine slides on pipelines: A review of the state-of-the-art since 2008. *Ocean Eng.* **2023**, *286*, 115532. [CrossRef]
22. De Blasio, F.V.; Elverhøi, A.; Issler, D.; Harbitz, C.B.; Bryn, P.; Lien, R. Flow models of natural debris flows originating from overconsolidated clay materials. *Mar. Geol.* **2004**, *213*, 439–455. [CrossRef]
23. Xiu, Z.; Liu, L.; Xie, Q.; Li, J.; Hu, G.; Yang, J. Runout prediction and dynamic characteristic analysis of a potential submarine landslide in Liwan 3-1 gas field. *Acta Oceanol. Sin.* **2015**, *34*, 116–122. [CrossRef]
24. Elverhi, A.; Harbitz, C.B.; Dimakis, P.; Mohrig, D.; Marr, J.; Parker, G. On the dynamics of subaqueous debris flows. *Oceanography* **2000**, *13*, 109–117. [CrossRef]
25. Du, J.; Choi, C.E.; Yu, J.; Thakur, V. Mechanisms of submarine debris flow growth. *JGR Earth Surf.* **2022**, *127*, e2021JF006470. [CrossRef]
26. Nourmohammadi, Z.; Afshin, H.; Firoozabadi, B. Experimental observation of the flow structure of turbidity currents. *J. Hydraul. Res.* **2011**, *49*, 168–177. [CrossRef]
27. Mohrig, D.; Ellis, C.; Parker, G.; Whipple, K.X.; Hondzo, M. Hydroplaning of subaqueous debris flows. *GSA Bull.* **1998**, *110*, 387–394. [CrossRef]
28. Zakeri, A.; Høeg, K.; Nadim, F. Submarine debris flow impact on pipelines—Part II: Numerical analysis. *Coast. Eng.* **2009**, *56*, 1–10. [CrossRef]
29. Dott, R.H. Dynamics of subaqueous gravity depositional processes. *AAPG Bull.* **1963**, *47*, 104–128. [CrossRef]
30. Middleton, G.V. Subaqueous sediment transport and deposition by sediment gravity flows. In *Marine Sediment Transport and Environmental Management*; Wiley: New York, NY, USA, 1976; pp. 197–218.
31. Lowe, D.R. Sediment gravity flows: Their classification and some problems of application to natural flows and deposits. In *Geology of Continental Slopes*; Doyle, L.J., Pilkey, O.H., Eds.; SEPM Society for Sedimentary Geology: Tulsa, OK, USA, 1979; Volume 27, ISBN 9781565761575.
32. Lowe, D.R. Sediment gravity flows; II, Depositional models with special reference to the deposits of high-density turbidity currents. *J. Sediment. Res.* **1982**, *52*, 279–297. [CrossRef]
33. Shanmugam, G. High-density turbidity currents; are they sandy debris flows? *J. Sediment. Res.* **1996**, *66*, 2–10. [CrossRef]
34. Zakeri, A.; Høeg, K.; Nadim, F. Submarine debris flow impact on pipelines—Part I: Experimental investigation. *Coast. Eng.* **2008**, *55*, 1209–1218. [CrossRef]
35. Guo, X.; Nian, T.; Wang, Z.; Zhao, W.; Fan, N.; Jiao, H. Low-temperature rheological behavior of submarine mudflows. *J. Waterw. Port Coast. Ocean Eng.* **2020**, *146*, 4019043. [CrossRef]
36. Boukpeti, N.; White, D.J.; Randolph, M.F.; Low, H.E. Strength of fine-grained soils at the solid–Fluid transition. *Géotechnique* **2012**, *62*, 213–226. [CrossRef]
37. Bruschi, R.; Bughi, S.; Spinazzè, M.; Torselletti, E.; Vitali, L. Impact of debris flows and turbidity currents on seafloor structures. *Nor. J. Geol.* **2006**, *88*, 317–336.
38. Liu, J.; Tian, J.; Yi, P. Impact forces of submarine landslides on offshore pipelines. *Ocean Eng.* **2015**, *95*, 116–127. [CrossRef]
39. Nian, T.; Guo, X.; Fan, N.; Jiao, H.; Li, D. Impact forces of submarine landslides on suspended pipelines considering the low-temperature environment. *Appl. Ocean Res.* **2018**, *81*, 116–125. [CrossRef]
40. Guo, X.; Zheng, D.; Nian, T.; Lv, L. Large-scale seafloor stability evaluation of the northern continental slope of South China Sea. *Mar. Geores. Geotechnol.* **2020**, *38*, 804–817. [CrossRef]
41. Fan, N.; Jiang, J.; Dong, Y.; Guo, L.; Song, L. Approach for evaluating instantaneous impact forces during submarine slide-pipeline interaction considering the inertial action. *Ocean Eng.* **2022**, *245*, 110466. [CrossRef]
42. Randolph, M.F.; White, D.J. Interaction forces between pipelines and submarine slides—A geotechnical viewpoint. *Ocean Eng.* **2012**, *48*, 32–37. [CrossRef]
43. Zhang, M.; Xie, A.; He, H.; Lu, R.; Tang, M. Mechanism of deep-water international submarine cables damage: Submarine earthquakes. *J. Mar. Sci.* **2024**, *42*, 100–113. [CrossRef]
44. Mosher, D.C.; Monahan, P.A.; Barrie, J.V.; Courtney, R.C. Coastal submarine failures in the strait of georgia, british columbia: Landslides of the 1946 vancouver island earthquake*. *J. Coast. Res.* **2004**, *20*, 277–291. [CrossRef]

45. Hsu, S.; Kuo, J.; Chung-Liang, L.; Ching-Hui, T.; Doo, W.; Ku, C.; Sibuet, J. Turbidity currents, submarine landslides and the 2006 Pingtung earthquake off SW Taiwan. *Terr. Atmos. Ocean. Sci.* **2008**, *19*, 767–772. [CrossRef]
46. Zhang, Y.; Liu, Z.; Zhao, Y.; Colin, C.; Zhang, X.; Wang, M.; Zhao, S.; Kneller, B. Long-term in situ observations on typhoon-triggered turbidity currents in the deep sea. *Geology* **2018**, *46*, 675–678. [CrossRef]
47. Locat, J. Normalized rheological behavior of fine muds and their flow properties in a pseudoplastic regime. In *Debris-Flow Hazards Mitigation: Mechanics, Prediction, and Assessment*; Millpress: Bethlehem, PA, USA, 1997; pp. 260–269.
48. Lienhard, J.H. *Synopsis of Lift, Drag, and Vortex Frequency Data for Rigid Circular Cylinders*; Technical Extension Service, Washington State University: Pullman, WA, USA, 1966.
49. Coutanceau, M.; Defaye, J. Circular cylinder wake configurations: A flow visualization survey. *Appl. Mech. Rev.* **1991**, *44*, 255–305. [CrossRef]
50. Sahdi, F.; Gaudin, C.; Tom, J.G.; Tong, F. Mechanisms of soil flow during submarine slide-pipe impact. *Ocean Eng.* **2019**, *186*, 106079. [CrossRef]
51. Guo, X.; Zheng, D.; Nian, T.; Yin, P. Effect of different span heights on the pipeline impact forces induced by deep-sea landslides. *Appl. Ocean Res.* **2019**, *87*, 38–46. [CrossRef]
52. Perez-Gruszkiewicz, S.E. Reducing underwater-slide impact forces on pipelines by streamlining. *J. Waterw. Port Coast. Ocean Eng.* **2012**, *138*, 142–148. [CrossRef]
53. Haza, Z.F.; Harahap, I.S.H.; Dakssa, L.M. Experimental studies of the flow-front and drag forces exerted by subaqueous mudflow on inclined base. *Nat. Hazards* **2013**, *68*, 587–611. [CrossRef]
54. Sahdi, F.; Gaudin, C.; White, D.J.; Boylan, N.; Randolph, M.F. Centrifuge modelling of active slide–Pipeline loading in soft clay. *Géotechnique* **2014**, *64*, 16–27. [CrossRef]
55. Fan, N.; Nian, T.; Jiao, H.; Jia, Y. Interaction between submarine landslides and suspended pipelines with a streamlined contour. *Mar. Georesources. Geotechnol.* **2018**, *36*, 652–662. [CrossRef]
56. Zhang, Y.; Wang, Z.; Yang, Q.; Wang, H. Numerical analysis of the impact forces exerted by submarine landslides on pipelines. *Appl. Ocean Res.* **2019**, *92*, 101936. [CrossRef]
57. Guo, X.; Nian, T.; Wang, F.; Zheng, L. Landslides impact reduction effect by using honeycomb-hole submarine pipeline. *Ocean Eng.* **2019**, *187*, 106155. [CrossRef]
58. Guo, X.; Nian, T.; Fan, N.; Jia, Y. Optimization design of a honeycomb-hole submarine pipeline under a hydrodynamic landslide impact. *Mar. Georesources. Geotechnol.* **2021**, *39*, 1055–1070. [CrossRef]
59. Guo, X.; Nian, T.; Stoesser, T. Using dimpled-pipe surface to reduce submarine landslide impact forces on pipelines at different span heights. *Ocean Eng.* **2022**, *244*, 110343. [CrossRef]
60. Guo, X.; Stoesser, T.; Nian, T.; Jia, Y.; Liu, X. Effect of pipeline surface roughness on peak impact forces caused by hydrodynamic submarine mudflow. *Ocean Eng.* **2022**, *243*, 110184. [CrossRef]
61. Guo, X.; Nian, T.; Fu, C.; Zheng, D. Numerical investigation of the landslide cover thickness effect on the drag forces acting on submarine pipelines. *J. Waterw. Port Coast. Ocean Eng.* **2023**, *149*, 4022032. [CrossRef]
62. Qian, X.; Xu, J.; Bai, Y.; Das, H.S. Formation and estimation of peak impact force on suspended pipelines due to submarine debris flow. *Ocean Eng.* **2020**, *195*, 106695. [CrossRef]
63. Zhao, E.; Dong, Y.; Tang, Y.; Cui, L. Numerical study on hydrodynamic load and vibration of pipeline exerted by submarine debris flow. *Ocean Eng.* **2021**, *239*, 109754. [CrossRef]
64. Fan, N.; Sahdi, F.; Zhang, W.; Nian, T.; Randolph, M.F. Effect of pipeline-seabed gaps on the vertical forces of a pipeline induced by submarine slide impact. *Ocean Eng.* **2021**, *221*, 108506. [CrossRef]
65. Fan, N.; Zhang, W.; Sahdi, F.; Nian, T. Evaluation of horizontal submarine slide impact force on pipeline via a modified hybrid geotechnical—Fluid dynamics framework. *Can. Geotech. J.* **2021**, *59*, 827–836. [CrossRef]
66. Guo, X.; Liu, X.; Luo, Q.; Chen, B.; Zhang, C. Dimensional effect of CFD analysis for submarine landslides interactions with infinite suspension pipelines. *Ocean Eng.* **2022**, *266*, 113094. [CrossRef]
67. Guo, X.; Stoesser, T.; Zhang, C.; Fu, C.; Nian, T. Effect of opening and wall boundaries on CFD modeling for submarine landslide-ambient water-pipeline interaction. *Appl. Ocean Res.* **2022**, *126*, 103266. [CrossRef]
68. Ermanyuk, E.V.; Gavrilov, N.V. Interaction of an internal gravity current with a submerged circular cylinder. *J. Appl. Mech. Tech. Phys.* **2005**, *46*, 216–223. [CrossRef]
69. Gonzalez-Juez, E.; Meiburg, E.; Constantinescu, G. The interaction of a gravity current with a circular cylinder mounted above a wall: Effect of the gap size. *J. Fluids Struct.* **2009**, *25*, 629–640. [CrossRef]
70. Gonzalez-Juez, E.; Meiburg, E.; Tokyay, T.; Constantinescu, G. Gravity current flow past a circular cylinder: Forces, wall shear stresses and implications for scour. *J. Fluid Mech.* **2010**, *649*, 69–102. [CrossRef]
71. Jung, J.H.; Yoon, H.S. Effect of scour depth on flow around circular cylinder in gravity current. *Ocean Eng.* **2016**, *117*, 78–87. [CrossRef]

72. Xie, P.; Du, Y.; Chu, V.H. Impact forces by a gravity current on a circular cylinder above the seabed. *Appl. Ocean Res.* **2022**, *125*, 103243. [CrossRef]
73. Guo, X.; Luo, Q.; Stoesser, T.; Hajaali, A.; Liu, X. Evolution of high-density submarine turbidity current and its interaction with a pair of parallel suspended pipes. *Phys. Fluids* **2023**, *35*, 86608. [CrossRef]
74. Guo, X.; Liu, X.; Zhang, C.; Jing, S.; Hou, F. Impact of high-speed turbidity currents on offshore spanning pipelines. *Ocean Eng.* **2023**, *287*, 115797. [CrossRef]
75. Wang, Z.; Zhang, Y.; Yu, L.; Yang, Q. Centrifuge modelling of active pipeline-soil loading under different impact angle in soft clay. *Appl. Ocean Res.* **2020**, *98*, 102129. [CrossRef]
76. Schramm, K.D.; Marnane, M.J.; Elsdon, T.S.; Jones, C.M.; Saunders, B.J.; Newman, S.J.; Harvey, E.S. Fish associations with shallow water subsea pipelines compared to surrounding reef and soft sediment habitats. *Sci. Rep.* **2021**, *11*, 6238. [CrossRef]
77. Zdravkovich, M.M. Forces on a circular cylinder near a plane wall. *Appl. Ocean Res.* **1985**, *7*, 197–201. [CrossRef]
78. Price, S.J.; Sumner, D.; Smith, J.G.; Leong, K.; Païdoussis, M.P. Flow visualization around a circular cylinder near to a plane wall. *J. Fluids Struct.* **2002**, *16*, 175–191. [CrossRef]
79. Wang, X.K.; Tan, S.K. Near-wake flow characteristics of a circular cylinder close to a wall. *J. Fluids Struct.* **2008**, *24*, 605–627. [CrossRef]
80. Sarkar, S.; Sarkar, S. Vortex dynamics of a cylinder wake in proximity to a wall. *J. Fluids Struct.* **2010**, *26*, 19–40. [CrossRef]
81. Zhou, J.; Qiu, X.; Li, J.; Liu, Y. The gap ratio effects on vortex evolution behind a circular cylinder placed near a wall. *Phys. Fluids* **2021**, *33*, 37112. [CrossRef]
82. Zhu, H.; Zhao, H. Solid-structure impact pattern of a vortex-induced oscillating flexible pipeline in the vicinity of a bottom wall boundary. *Phys. Fluids* **2021**, *33*, 61704. [CrossRef]
83. Zhu, H.; Zhang, X.; Zhao, H.; Xie, Y.; Tang, T.; Zhou, T. Experimental investigation on the vortex-induced vibration of an inclined flexible pipe and the evaluation of the independence principle. *Phys. Fluids.* **2023**, *35*, 37115. [CrossRef]
84. Ding, W.; Cai, T.; Sun, H.; Zhao, X.; Fan, Y. Numerical simulation study on the interaction between a gravity current and horizontal cylinders. *Ocean Eng.* **2023**, *289*, 116236. [CrossRef]
85. Cohen, I.M.; Kundu, P.K. *Fluid Mechanics*; Elsevier: Amsterdam, The Netherlands, 2004; ISBN 0080470238.
86. Zhongtao, W.; Hanyang, W.; Yu, Z. CFD numerical analysis of submarine landslides impact on laid-on-seafloor pipeline. *Haiyang Xuebao* **2016**, *38*, 110–117. (In Chinese)
87. Zdravkovich, M.M. Flow around circular cylinders. *Fundamentals* **2003**, *1*, 216. [CrossRef]
88. Williamson, C.H.K.; Govardhan, R. Vortex-induced vibrations. *Annu. Rev. Fluid Mech.* **2004**, *36*, 413–455. [CrossRef]

MDPI AG
Grosspeteranlage 5
4052 Basel
Switzerland
Tel.: +41 61 683 77 34

Journal of Marine Science and Engineering Editorial Office
E-mail: jmse@mdpi.com
www.mdpi.com/journal/jmse

www.ingramcontent.com/pod-product-compliance
Lightning Source LLC
Chambersburg PA
CBHW040742250926
51850CB00007BB/1386